Niebuhr in Egypt

Niebuhr in Egypt

European Science in a Biblical World

Roger H. Guichard Jr.

PICKWICK Publications • Eugene, Oregon

NIEBUHR IN EGYPT
European Science in a Biblical World

Copyright © 2013 Roger H. Guichard Jr. All rights reserved. Except for brief quotations in critical publications or reviews, no part of this book may be reproduced in any manner without prior written permission from the publisher. Write: Permissions, Wipf and Stock Publishers, 199 W. 8th Ave., Suite 3, Eugene, OR 97401.

Pickwick Publications
An Imprint of Wipf and Stock Publishers
199 W. 8th Ave., Suite 3
Eugene, OR 97401

www.wipfandstock.com

ISBN 13: 978-1-62032-505-6

Cataloguing-in-Publication data:

Guichard, Roger H., Jr.

Niebuhr in Egypt : European science in a biblical world / Roger H. Guichard Jr.

xviii + 336 pp. ; 23 cm. Includes bibliographical references.

ISBN 13: 978-1-62032-505-6

1. Niebuhr, Carsten (1733–1815). 2. Arabian Peninsula—Description and travel—Early works to 1800. 3. Arabian Peninsula—Description and travel. I. Title.

DS206 G8 2013

Manufactured in the U.S.A.

Reprinted by permission of the publishers and Trustees of the Loeb Classical Library from MANETHO , Loeb Classical Library Volume 350, translated by W. G. Waddell, pp. xvi, 107, Cambridge Mass.: Harvard University Press Copyright 1940 by the President and Fellows of Harvard College. The Loeb Classical Library is a registered trademark of the President and Fellows of Harvard College.

Contents

Illustrations | *vi*
Preface | *vii*
Acknowledgments | *xvii*

1. Introduction | 1
2. The Conceit | 18
3. To the Orient | 36
4. Alexandria | 56
5. To Cairo | 77
6. The Mother of the World | 93
7. Government | 121
8. Inhabitants | 139
9. Commerce | 156
10. The Delta | 170
11. Manners and Customs | 188
12. The Antiquities of Egypt | 201
13. To Suez and Sinai | 228
14. Afterward | 261
15. The Results | 293

Appendices
   A *Questions* | *319*
   B *Keys to the Map of Cairo* | *325*
Bibliography | *331*
Subject Index | *337*

Illustrations

No.	Plate	in *Travels* or *Description*	Page Number
1	III	"*Grundris der Städte Constantinopel* . . ."	48
2	IV	Hieroglyphs	50
3	V	Geometric studies	63
4	XXIX	"*Abbildung der Araber in Egÿpten*"	66
5	XII	"*Urbis* Káhira . . ."	94–5
6	XIII	"*Bâb el fitûch* . . ."	102
7	X	Map of the Delta	175
8	XIX	Oriental headgear	192
9	XXVI	Musical instruments	197
10	XXVII	"*Abbildung der* Täntzerinnen . . ."	199
11	XLI	"*Verschiedene Zeichen und Figuren* . . ."	218
12	XXIV	"*Plan* . . . *du Golfe arabique* . . ."	234
13	XXIII	"*Tabula Itinera* . . . ad . . . Montem SINAI"	238
14	XLIV	"*Grundris eines Todtenackers* . . ."	243
15	XLV	"*Hieroglyphen auf einem Leichen* . . ."	246
16	XLVIII	"*Prospect des Klosters am Berge Sinai* . . ."	251
17	L	"*Inschriften am Wege von Sués* . . ."	256
18	XLIX	"*Inschriften am Wege von Sués* . . ."	257
19	LIII	"*Prospect der Stadt Dsjidda* . . ."	262
20	LIV	"*Abbildung eines türkischen Pilgrims*"	265
21	LXIII	"*Prospect beÿ dem Dorfe Bulgose* . . ."	268
22	LXVIII	"*Prospect des Castels* . . . *der Stadt Ierîm*"	275
23	X	"*Abbildung der Figur* . . ."	283
24	XXII	From *Icones Rerum Naturalium*	302

Note: Most plates are from Volume I of *Travels in Arabia*. No. 23 is from Volume II. No.'s 12 and 13 are from the *Description of Arabia*. No. 24 is as marked.

Preface

IN THE SPRING OF 1984, the Royal Library in Copenhagen, in cooperation with the Danish Foreign Ministry, sponsored an exhibition in Riyadh called "The Arabian Journey 1761–1767." The purpose of the exhibition was to commemorate a little-understood and long-forgotten episode in the relations between Europe and the Arab World, the Royal Danish Expedition to the Yemen. It was true, there were records of the expedition, including a multi-volume account left by its sole survivor, Carsten Niebuhr. These had appeared in a series of releases beginning in 1772. By 1778, Niebuhr's work of publication was largely complete, although a final posthumous volume would not be published until 1837. In the years that followed publication, translations from Niebuhr's original German would appear in French, Dutch, Italian, and even Farsi. Excerpts in English would be included in the travel compendia for which eighteenth-century Europe had developed a nearly insatiable appetite. More recently, there was a book—in Swedish, later translated into English—whose title, *Arabia Felix*, or "Happy Arabia," captured the quixotic, and ultimately unhappy, quest that the expedition represented. However, in 1984 what was *not* known about Carsten Niebuhr and the Danish expedition was out all of proportion to what was.

To the serious student of the European exploration of Arabia, Carsten Niebuhr had always been a name to conjure with. He was cited by many of the explorers and travelers who followed in his footsteps as their great predecessor, although specific references were surprisingly brief. John Lewis Burckhardt, the Swiss traveler who in the early part of the nineteenth century had been the first European in centuries to see Petra and then Abu Simbel, clearly had read Niebuhr, although textual citations were few. But Burckhardt died in Cairo in 1817 and his accounts—largely written without access to scholarly resources—were published posthumously, so the absence of references is understandable. Richard Burton appears to have read him carefully, although Burton was not a man to readily credit others laboring in the same field. Burton gave his grudging approbation to the "accurate"

Preface

Niebuhr, although references in his *Personal Narrative of a Pilgrimage to El-Madinah and Meccah* focused more on Niebuhr's occasional lapses than on his celebrated accuracy. Others were more generous. William Gifford Palgrave, the half-Jewish English Jesuit who traveled to the heart of the peninsula under the sponsorship of the Emperor of France, dedicated his *Central and Eastern Arabia* to Niebuhr, "in honor of that intelligence and courage which first opened Arabia to Europe." H. St.J. B. Philby opens *The Heart of Arabia* with a quote from the French edition of the *Voyage en Arabie* and calls Niebuhr "the father of Arabian exploration." In *A Pilgrimage to Nejd* the Blunts quote Niebuhr extensively on a matter close to their hearts, that of horse-breeding. J. G. Lorimer in his monumental *Gazetteer of the Persian Gulf, Oman and Central Arabia* says that in 1908 Niebuhr was still the most valuable source of information about the Gulf of the middle of the eighteenth century. Of the English-speaking travelers and writers, however, only the American divine Edward Robinson appears to have been familiar with Niebuhr in the original German. Robinson spent several months in Germany prior to his journey to Egypt, the Sinai and Palestine in 1838–39 and refers extensively to Niebuhr in his text.

In his classic *The Penetration of Arabia* David Hogarth devoted a chapter to Niebuhr in the Yemen. Hogarth recognized that Niebuhr had left the most complete account to date of that remote corner of the Arabian peninsula. But there was more to Niebuhr than the Yemen, and Hogarth was ample, if not unstinting, in his praise:

> If he was not the most brilliant of the party, if any of his fellows surpassed him in energy, courage, and endurance, in intelligence or in his measure of that scientific temper which is equally free from prejudice and from laxity, then a more remarkable mission was never dispatched to any land.

If the compliment is a bit left-handed, we will become used to it. Niebuhr, by common consent, appears to have been "intelligent" if not "brilliant." But we should probably reserve judgment until we have seen the complete man.

Of as much interest as the citations are the omissions. Edward William Lane, writing his *Manners and Customs of the Modern Egyptians* seventy years later, ignores Niebuhr completely although Niebuhr devotes over 200 pages of his *Travels in Arabia* to Egypt, including descriptions of the inhabitants, their dress, religion, diversions, musical instruments and games. In the voluminous notes to his translation of the *Thousand Nights and a Night* Burton gives us his usual encyclopedic treatment of subjects and sources but makes no mention of Niebuhr's treatment of some of the same material. Charles Doughty does not mention Niebuhr at all. Where he is cited as an

Preface

historical source in nineteenth-century accounts of the Yemen there are few textual references and a curious lack of precision about the dates and details of Niebuhr's sojourn in that country.

It should come as no surprise that all the references cited above are in English. Surely among the reasons for the lack of knowledge must be that the complete Niebuhr has never been available in anything but the original German, and the editors and abridgers have not done him justice. In addition, most of the references we see are the works of travelers and not academicians, for whom the monuments of German oriental scholarship were probably inaccessible. Niebuhr was also a traveler, although he was a traveler of unusual perspicacity. He certainly had access to the literature of the subject when he prepared his accounts, and his bibliography would include over 120 sources, from Herodotus to the latest eighteenth-century publications. But, like Burckhardt, Burton, Palgrave, Lorimer, and Philby, Niebuhr was also a first-hand observer of what he reported. What makes his account especially valuable is the quality of his insights: he was a particularly shrewd observer and recorded only what he saw with his own eyes. Where he did not see, but only heard at second hand, he tells us, lest we give him more credit than, in his eyes, he deserved.

But there is more to the puzzle than the lack of familiarity of English writers with an obscure, eighteenth-century German traveler. Because in its conception, the Royal Danish Expedition aspired to an end that lay beyond individual languages or narrow national concerns. Its frankly ecumenical appeal at the outset makes the parochial nature of the response all the more puzzling. The "Arabian Journey 1761–1767," commemorated by representatives of Denmark and Saudi Arabia in Riyadh in 1984, was nothing less than a multifaceted, pan-European undertaking devoted to the highest moral purpose. The expedition may have been sponsored by the King of Denmark, but it was made up of Germans and Swedes in addition to Danes—and a German, Prof. Johann David Michaelis, had been its prime mover. Another German, Carsten Niebuhr, was the only survivor and the only one that anyone really remembered. Among its objectives had been an understanding of Arabia in general, but its specific purpose had been to assist in the explication of the Hebrew Bible, and scholars throughout the Continent had been consulted in the drafting of its terms of reference. The goal of the expedition may have been "Happy Arabia" but, by the time Niebuhr returned to Copenhagen in 1767, his peregrinations had taken him to the west coast of India, Persia as far inland as Persepolis, then to Iraq, the Levant, Cyprus, Anatolia, and Rumelia, as well as the Yemen. In fact, due to a series of circumstances that can only be described as fortuitous, the

Preface

longest and most concentrated period of time the members spent together, as an expedition, was not in Arabia at all. It was in Egypt.

It was also in Egypt, at about the same time as the exhibition in Riyadh, that I found the 1766–80 editions, in French, of Niebuhr's *Travels in Arabia*. For many years I had been interested in the European exploration of Arabia and had seen references to Niebuhr in other works, including those listed above. Now, I had access to his works at first or, at least, at second hand. The books were all and more than Burton, Palgrave, Lorimer, Philby, and Hogarth promised, and as my knowledge deepened, so did my appreciation of the value of Niebuhr's contributions. They were not just another dry account of one man's travels, but represented the record of a serious intellectual enterprise involving Enlightenment science, sacred philology, the Bible as history, "Orientalism," Egyptology, and discovery. At the same time, they had all the ingredients of a first-rate story. And no one, at least in the English–speaking world, seemed to know about them.

Until the second half of the twentieth century there were only the Niebuhr volumes themselves, but very little else to memorialize the expedition. Then, a series of books appeared in Swedish, German, and Danish. The first, in 1962, was Thorkild Hansen's *Arabia Felix*, a translation of which appeared in English in 1964. It chronicled the sometimes contentious relationships between the members of the expedition and gave a lively, not to say breezy, account of their progress towards the Yemen and their misadventures there. This was followed in 1968 by reprints in German of the three volumes of Niebuhr's *Reisebeschreibung*, or Travels. Then in 1986—in response to the Hansen book, which he believed did not do justice to the expedition's achievements—Stig Rasmussen of the Royal Library in Copenhagen published a small paperback review and catalogue of the expedition titled *Carsten Niebuhr und die Arabische Reise 1761–1767*. He followed this in 1990 with an impressive memorial entitled *Den Arabiske Rejse 1761–1767*. It was not a small paperback, but a large tome consisting of instructions to the members of the expedition, excerpts of the printed works, maps, reprints of original plates (some in color) and scholarly essays on the contributions of the members. However, it was published only in Danish.

Missing for the English reader was any serious discussion of Johann David Michaelis, the foremost Oriental philologist of the eighteenth century and the real author of the expedition, and of his belief that in the highlands of the Yemen the travelers would find a variant of Arabic, an "eastern" dialect of the language that was closest to Hebrew, and so a link with the original language of the Scriptures. Missing also was the link to Enlightenment science, and the boundless self-confidence of those who believed that anything—including the Bible—could be understood if subjected to rigorous scientific examination.

Preface

Finally, there was little discussion of the elaborate pains to which Michaelis had gone to prepare the members of the expedition for their work of biblical scholarship, and of the hundreds of specific queries he drafted to guide their investigations. The fact that Michaelis ultimately failed to put his stamp on the results of the expedition in no way detracts from the fact that it represented a kind of milestone in European intellectual history. As we will see, what Niebuhr and his companions produced was, at the same time, much less and much more than Michaelis had hoped.

The sojourn in Egypt was an unexpected boon, the country not even appearing on the original itinerary of the expedition. But what an opportunity it presented to an undertaking with an avowedly biblical purpose! When Niebuhr and his companions were detained for a year in Egypt in 1761-62 it was, after all, in a place which some have called the cradle of the Jewish people. But, although Egypt had existed for millennia, with or without the Jews, the notion that its history served as little more than stage setting for the great drama of mankind as played out in the Hebrew scriptures was pervasive in eighteenth-century Europe. The notion persists to this day in the Christian West, and the Bible as history remains nearly as vexed subject at the outset of the twenty-first century as it was in the eighteenth—or indeed, any other—century. To his credit, Niebuhr approached the subject of Egypt with an open mind, without the preconceptions or credulity that had characterized much of the traditional European approach to the country.

What Niebuhr also gave the West was a first critical look at the Egypt of the middle of the eighteenth century, as well as the first detailed maps of the city of Cairo and the Delta. In 1761–62, Ali Bey Bulut Kapan—the "cloud catcher"—was maneuvering to establish his unchecked rule, becoming in the process a worthy precursor to Mohammed Ali. As seen through Niebuhr's eyes, Ali Bey was only one of the caste of military slaves, or Mamluks, that had ruled Egypt since the arrival of the first Central Asians in the thirteenth century. But Ali Bey would soon replace the unbridled rapacity of the Mamluks with his own more modern and systematic plunder of the wealthy province that Egypt had been at least since the Ptolemies. At about this same time, the study of the hieroglyphs and the ancient history of the country—or Egyptology—was beginning to free itself from the shackles of several odd but persistent notions that stood in the way of an understanding of ancient Egypt. One of these was a belief in the arcane nature of the Egyptian hieroglyphs, understandable only to initiates, that seemed to render fruitless any rigorous textual analysis.

It was in 1761, the year Carsten Niebuhr and his companions arrived in Egypt, that the Abbé J. J. Barthélémy took the first tentative steps towards an understanding of the hieroglyphs by suggesting that they contained

Preface

elements of a phonetic system. Niebuhr made his own modest contribution to the process of decipherment, a process that would be continued by another learned European excursion into Egypt some forty years later, the French Expedition. But there were other influences as well, a result of the focus on the Bible and of the tendency to subject evidence, first, to the test of biblical conformity before it passed muster as history. That the pyramids of Giza were the original corn storehouses of Joseph, and that they had been built by the Hebrews (the Egyptians not having the requisite technical skills), were among the least absurd of these notions.

In the Sinai, the Danish expedition would look for the odd inscriptions at "Gebel el Mokatab," first reported by the Bishop of Clougher forty years before. They sparked intense interest in Europe since they were thought to be the precursors of the square Hebrew script, learned by the Israelites during their wanderings in the wilderness. They were not, but in the process Niebuhr and his colleagues discovered something almost as interesting, the pharaonic temple at Serabit al-Khadem, where later researchers would find traces of the so-called proto-Sinaitic script, which *was* a precursor to Hebrew. All these things were afterthoughts in the original plan of the expedition. But there was hardly a thing in the world of the Orient that didn't interest Niebuhr and, freed for the year from the painstaking directions of Michaelis, he made very good use of his time in Egypt.

The book that follows—*Niebuhr in Egypt: European Science in a Biblical World*—is only a part of the story of the expedition and of Niebuhr's part in it. Hogarth may have focused on "Niebuhr in the Yemen," but he might just as well have added chapters on "Niebuhr in the Hejaz," "Niebuhr in Oman," "Niebuhr in the Arabian Gulf," or, had he permitted himself to expand his brief, "Niebuhr in India," or "Niebuhr in Persia." Or, for that matter, "Niebuhr in Egypt." Because it was largely in the years 1761–62—particularly in Egypt, but also in the Yemen—that the biblical nature of the expedition played itself out. As we will see below, when the survivors set sail from Mocha for Bombay in August of 1763, the expedition to "Happy Arabia" was technically over. However, much remained to be seen in the Orient, and the next four years would yield as much published material as the previous three. But, however much Niebuhr accomplished in the years 1763–67, the later period lacked the drama of those first years of promise, enthusiasm, and disappointment, followed by the premature death of the other members of the expedition. Not surprisingly, Michaelis himself seemed to lose all interest in the progress of the expedition after it left the Yemen. These other Niebuhrs deserve their own chapters, but they will be saved for another work.

Preface

The reader might ask "Why a book instead of a translation? Why not let Niebuhr speak for himself?" The answer, a least from several publishers, was that those really interested in Niebuhr would consult the original in German, and a translation was not necessary. And there were already excerpts available in English, products of the eighteenth-century enthusiasm for travel and discovery. But these were perhaps too conscious of the attention span of the audience, and the comment in one was typical:

> It would be unfair to neglect advertising the reader that the whole of Mr. Niebuhr's account of his travels and observations in Arabia is not comprised in these volumes . . . Various things seemed to be addressed so exclusively to men of erudition that they could not be expected to win the attention of the public in general and have therefore been left out.

Unfortunately, in addition to all the Arabic texts and the mapmaking—arguably, matters too recondite for the general reader—the above excerpt makes no mention of the Bible, Egyptology, or indeed of Egypt at all. But it is these things that give life to what might otherwise may seem a dry recitation of facts, of latitudes and compass headings, etymologies, and obscure place names.

But a better answer is that, without the perspective of Michaelis and his part in establishing the intellectual framework of the expedition, the story is incomplete. Only with an understanding of this framework can the value of Niebuhr's insights can be appreciated. Part of this is the fault of Niebuhr himself. He was a man who would as soon embellish a fact as tell an untruth, and his reticence did not always serve him well. There were things that simply did not belong in print, including all mention of conflict with other members of the expedition. But they lent a human touch to the story, one not only of jealousy, frailty, and disappointment, but also of ambition and ultimate triumph.

The book that follows is an account of the expedition's year in Egypt, with lengthy excursions into the several subplots—Enlightenment science, the Bible as history, and Egyptology—mentioned above. It makes no claim to being scholarly, and is aimed at the general reader, although it resorts to no gimmicks in its appeal. The above subjects *are* difficult and no attempt will be made to make them appear easy. But if the Bible as history, and its baleful effect on serious scholarship about Egypt, is considered recondite, it is also topical and should be of interest to the general reader concerned with the region today. The book makes no claim of access to original sources, other than the Niebuhr works themselves. A word about method is in order. The original Niebuhr is in German, although my introduction to him was through

Preface

the French translation, from which I made my own English translation. All citations, however, have been carefully reconciled with the original German. This involved the process of working with three texts and, in effect, looking over the shoulder of the anonymous French translator of 1776. While he was an invaluable guide to some of the German archaisms, he didn't always get it right. As a matter of interest, the French reads like a modern language, while the syntax of Niebuhr's eighteenth century Low German presents difficulties closer to translating Arabic than a Romance language.

Along with Niebuhr, the book makes no concession to the notion of the Orient as a place of mystery and sensuality, of strange practices and arcane knowledge, as if Orientals were somehow fundamentally different from human beings in other parts of the planet. And we will not see Niebuhr as a representative of a Europe intent on domination of the East. That presumption would violate every principle that he stood for. If he dealt with a part of the Orient with which the West had—and still has—an historical difficulty, he set a standard of openness and fairness that shines through the text. In that text we will see Orientals and Europeans—Jews, Christians, and Muslims alike—displayed through their own words and actions in various flattering, and not-so-flattering, guises. It would be a mistake to try to conform Niebuhr's observations to a twenty-first century standard of correctness, and his occasional lapses—departures from the high standard he set for himself—will be permitted. They will make him only more human, and his story more believable.

In dealing with a subject as contentious as the relationships between the three "people of the book"—in the twenty-first no less than the eighteenth century—it would be difficult to avoid trespassing on the sensibilities of one party or another to the conflict. The prejudices and suspicions are plain for all to see: the pervasive animus directed against Islam and its founder in the Christian West; anti-Jewish sentiments among both eighteenth-century Christians and Muslims; anti-Papist sentiments by northern European Protestants; lingering suspicion and mistrust by Orthodox Christians of their Latin coreligionists; a perceived Ottoman and Jewish conspiracy to subject native Egyptians; strictures enacted against Copts and Jews by the Mamluk authorities in Egypt; the fear of a fifth column of covert Muslims and Jews in Europe; anti-Frankish sentiments directed by eighteenth-century Semites at this particular European traveler. When they are reported by Niebuhr, they are done so openly and directly, and an attempt will be made to deal with them equally openly and directly in the book that follows. Niebuhr doesn't preach, and we will resist the temptation to sermonize. But with some of the most difficult issues—the Bible as history, the place of the Children of Israel in Egypt, the history of Egypt itself—we will see how many of our attitudes today

Preface

are unchanged from those of the eighteenth century. Where earlier scholars were wrong-headed or mistaken, we will see their errors, not in the sense of being triumphalist or wise after-the-fact, but rather to learn from their mistakes. But there should be no mistaking that some of the errors persist.

In his chapter on Niebuhr in the Yemen, David Hogarth remarked that it would be to tedious to quote "a hundredth part of Niebuhr's judicious observations." I hope, with this book, to bring to the reader interested in Egypt a portion of that trove.

The manuscript has been fortunate in its readers: an anonymous reviewer and sometime editor of the *American Journal of Romance Philology*; Suzanne L. Marchand, a profesor of European Intellectual History at Louisiana State University; and Mr. Michel-Pierre Detalle a long-time student of and expert in Niebuhr. Each of them reviewed the manuscript carefully and made many suggestions of great value. I would like to think that their reviews made up in quality for their relative lack of quantity. Any errors or misapprehensions that remain in the text that follows are, of course, my own.

A word on the transliteration of Arabic is in order. The rigorous and consistent use of a system of transliteration is alone an infallible guide to the determination of the original triliteral root of the word, and I am a great believer in such systems. However, I believe that the systematic use of diacritics here would serve only as a headache for the typesetter without adding much to an understanding of the text. I have consequently adopted forms closer to popular rather than scholarly usage. Hence Omar, Taizz and Koran, not 'Omar, Ta'izz, and Qur'an. I have also sometimes been inconsistent in my use of "sun" and "moon" letters: thus, Salah ad-Din, not Salah-al-Din, but Burg al-Zafar rather than Burg az-Zafar. They simply sounded better. In any case, the Arabic of Niebuhr's map of Cairo (with occasional irregularities) and of Michaelis's questions is listed in the Appendices and is available to those interested in the original language.

I would like to thank in particular the art department of the Royal Library in Copenhagen for the copies of the original plates that appear throughout the text.

Acknowledgments

I am grateful to copyright holders and publishers for their kind permission to quote from the following listed works:

Doris Behrens Abuseif for *Azbakiyya and its Environs from Azbek to Isma'il, 1476-1879*

Margaret S. Drower for *Flinders Petrie, A Life in Archaeology*

Afaf Lutfi al-Sayyid Marsot for *Egypt in the Reign of Mohammed Ali*

Stig T. Rasmussen for *Den Arabiske Rejse 1761-1767*

Stig. T. Rasmussen and Anne Haslund Hansen for *Frederik Christian von Havens REISEJOURNAL fra Den Arabiske Rejse 1760-1763*

Donald B. Redford for *Egypt Canaan and Israel in Ancient Times Harvard Theological Studies* for a quote from Vo. 22 (1966) by W. F. Albright

The University of Oklahoma Press considered the citation from *On Interpretation and Criticism* by August Boekh, as translated by John Paul Pritchard, to be "fair use." Nonetheless they requested that the following copyright statement be included: "Copyright 1968 by the University of Oklahoma University Press Publishing Division of the University. Composed and printed at Norman Oklahoma, U.S.A., by the University of Oklahoma. First edition."

I

Introduction

> Niebuhr was an accurate and careful observer, had the instincts of a scholar, was animated by a high moral purpose, and was rigorously conscientious and anxiously truthful in recording the results of his observation. His works have long been classics in the geography, the people, the antiquities and the archaeology of much of the district of Arabia which he traversed. *The Encyclopedia Britannica* (11th Edition, 1910–1911)

THE LONE EUROPEAN SAT at a low table as he put the finishing touches on the text. He wrote in German, the language of his native Saxony, and he was compiling the notes to accompany his drawings of the hieroglyphs he had seen on a watering trough near the Qalaat al-Kabsch, or "Fortress of the Ram," in Cairo. Some Egyptians traced the name of the fortress to the sacrifice by Abraham of a ram in place of his son, Isaac (Gen 22:1–19, Koran, XXXVII: 99–111). But like so many tales from the Old Testament, as they were later modified and passed through the Koran, it was impossible to verify their truth. In any case, the fortress was long gone and in its place lay the *madrasa* of Qaytbey, the greatest of the Mamluk builders of Egypt. It was not far from the mosque of Ibn Tulun, which lay beneath the range of the Muqattam and the Citadel to the east of the Fatimid city. The walls of the latter were visible through the morning haze and the *mashrabiyya* of the window in the study. It was already warm, but the early morning hours were still the most productive time for the sustained work of drafting and composition that constituted much of his daily routine. The temperature in Cairo on this day, August 19th, 1762, was already eighty-three degrees and it was only 7:30 in the morning. We know this from the records of the careful, thrice-daily readings on his Fahrenheit thermometer. That was another part of Carsten Niebuhr's routine, and the records of temperature would continue for the almost seven years of his travels in the Orient. We also know that there had been a violent rainstorm earlier in the week, and that it had probably tempered the heat.

But Egyptians were famous for their ability to use the elements to their advantage, and most houses had a kind of vent that was pointed to the north and admitted cooling breezes into the living area. In addition, the water from the Nile had just been introduced into the *khalig*, the canal that ran along the western boundary of the Fatimid city. That was a double blessing for those who lived in its vicinity. In the first place, when it was dry the canal bed was full of refuse, much of it organic, and this house was in the Harat al-Ifrang, or the Frankish Quarter, and directly overlooked the canal. The odor could be oppressive in the summer heat. But the swift flow of the Nile water had carried away the collected garbage and in its place provided another source of cooling breezes, although the respite was brief before effluents from the nearby houses would turn the canal into an open sewer. And the day before, on the 18th of August, the water from the *khalig* had been introduced into the Birkat al-Azbakiyya, the large pond that lay to the west of the canal. Soon, the grandees of the city would begin their seasonal fetes on the water. Planks would be laid down on the shore of the pond, over which people could walk, and lights would be hung on the nearby houses, their reflections twinkling in the water at night. The heat of the day was to be avoided in Cairo, but the city came alive at night and it was a colorful scene, with the multitudes, musicians, madmen, magicians, and jugglers. Pickpockets and prostitutes worked the crowd. Cairenes, no matter how poor or oppressed they were, had always known how to enjoy themselves.

But in the middle of the eighteenth century they were still insular and they disliked outsiders, particularly if the outsiders were Christians and Franks. Europeans were permitted to live in the city, but only three European powers were allowed to maintain consulates in Cairo—France, Venice, and Holland. It was through the French that Niebuhr and his colleagues had rented the house overlooking the canal. Along with Copts, or native Christians, and Jews, Europeans were subject to the conditions of the Caliph Omar: they couldn't ride horses, they had to dismount from a donkey in the presence of a Turk, and they couldn't publicly lament their dead. For the local Christians and Jews, the strictures were even more severe. So Europeans were very careful residents of Cairo indeed, and they avoided as much as possible contact with the representatives of the slave caste, or Mamluks, who tilted for control of the country. The year before, the residents of Damietta on the Mediterranean had taken exception to French merchants' mixing with Muslim women, and they had risen up and killed them to a man. The French were now forbidden to enter Damietta and other Europeans did so only with trepidation.

Introduction

However, if a man was careful, he could learn much, and Cairo and its environs probably had more in the way of monuments of recent and remote antiquity than any place on earth. Niebuhr had already walked the length and breadth of the city, pacing off distances and noting its geographical features and the location of its major structures. In the process, he would produce the first detailed map of the city by anyone, European or otherwise. So we already see the Niebuhr we have come to expect, having read vol. I of his *Reisebeschreibung nach Arabien* or *Travels in Arabia* (hereafter called the *Travels*). There would be the familiar sociological studies, the careful astronomical observations, a short history and detailed maps of the city and the Delta, drawings and commentary on the nearby Pharaonic antiquities, a discussion of the commercial activity, a description of the ruling class, and a survey of the polyglot population. In the relatively congenial atmosphere of the house overlooking the *khalig*, he was able to polish his observations, take sightings with his quadrant, complete his sketches, and dispatch to Copenhagen the material already collected by the members of the expedition. But Niebuhr was only beginning to hit his stride, and it would be another five years before he himself returned to Denmark. The expedition to Happy Arabia had theoretically not even begun, although Niebuhr had already collected enough material to fill several hundred pages of his *Travels*. But what brought him to Cairo in the first place? And what reserves of curiosity, dedication and scholarship would urge him on? Who was this man Carsten Niebuhr, and what was he doing in Egypt?

The short answer is that he was the German mathematician, for want of a better title, on the ill-fated Royal Danish Expedition to Happy Arabia of 1761–67. Born on March 17th, 1733[1] in West Ludingworth (today Cuxhaven) on the estuary where the Elbe exits into the North Sea, Carsten Niebuhr was the son of a free peasant farmer, a Frisian and a Saxon. The Frisians were a stubborn, independent people of Teutonic stock who had historically resisted outside domination, whether Roman, Frank, Christian, Hollandish, or Burgundian. They were:

> ... a free peasantry ... each man occupying and cultivating his own little freehold; and possessed the industry, frugality, and

---

1. See *Lives of Eminent Persons*, published by the Society for the Diffusion of Useful Knowledge; *Journal des Savans*, 1818; *The Life and Letters of Barthold George Niebuhr*; the *Algemiene Deutsch Biographie*, vol. 23; and the *Biographie Universelle Moderne*, vol. 30. Niebuhr's son, Barthold Georg Niebuhr, wrote a short life of his father that is used in all the above references. The most extensive was in the *Lives of Eminent Persons*, where this simple peasant from Friesland appeared as one of thirteen intellectual and historical giants, among them Gallileo, Kepler, Newton, Mohammed, Adam Smith, Michaelangelo, and Sir Christopher Wren.

sturdy independence which usually characterize their order. The circumstance that his childhood and youth were passed among such a population probably contributed to the strong interest and sympathy which Niebuhr always regarded this class.[2]

His son says that he was born a peasant—"in stature . . . rather under the middle size, of a very robust and sturdy make"—and remained a peasant to the end of his days. This apparently was evidence not of any lack of culture but of stubborn independence and an unwillingness to put on airs. His early life was scarred by the death of both parents, his mother when he was six months old and his father when he was an adolescent. Being a younger son, Niebuhr's share of the inheritance was small and his careful husbanding of the available funds had a decisive influence in his choice of careers. While he showed an inquisitiveness and promise at school, the uncertainties of his inheritance and the financial difficulties of his guardian made it necessary for him to work for extended periods during his youth. He originally studied music and he played the organ, the flute and the violin,[3] hoping to secure a post as an organist. It is interesting that this ambition, if realized, would probably have brought employment at one of the Lutheran churches in the canton and exposure to the work of Johann Sebastian Bach, then employed as a cantor in Leipzig, 150 miles to the southeast. Having largely fulfilled the passion of his middle years, the development of "a well-regulated church music to the glory of God," Bach had returned in this final phase to his original focus as an organist and musical theoritician. Unfortunately, we have little indication of Niebuhr's theological attitudes at the time. But subsequent evidence would suggest that he combined a certain latitudinarianism with the same commonsense approach to religion that he showed with regard to other matters. This was to have mixed results in an expedition whose purpose was expressly religious, as we will see below.

During a period of work on the farm of an uncle, and hearing the call by the government for a cadastral survey of the area, Niebuhr's patriotism was engaged and he decided to become a surveyor. So, at the age of twenty-one, he set about preparing himself for higher studies. He went to Hamburg where he spent the next eight months studying Latin and mathematics. This only whetted his appetite and in 1757 he was accepted at the University of Göttingen, 150 miles south of Hamburg and, incidentally, a center of Oriental studies. There, he studied mathematics and astronomy, both necessary

---

2. Winkworth, vol. I, 3.

3. As a matter of fact, he took the violin with him on the journey, and he and other members of the party occasionally treated their hosts to a European concert. The Arabs found the sound unendurable.

to an understanding of position-finding using celestial bodies. However, his funds were nearly exhausted, and to help make ends meet he enrolled in the engineering corps. His life might have taken a decidely different turn if, in Göttingen, his quiet intelligence had not come to the attention of a certain Professor Kästner, and it was through Kästner that Niebuhr was introduced to another Professor, Johann David Michaelis. It was rather by accident that Michaelis recommended that Niebuhr be appointed mathematician and surveyor on an expedition to be sponsored by the King of Denmark. The goal of the expedition was Arabia.

Johann David Michaelis and Philology

The long answer to the question as to what Niebuhr was doing in Egypt on that August morning in 1762 is rather more complicated. It involves an understanding of the discipline of philology as it was conceived in the middle of the eighteenth century, of the Hebrew and Christian scriptures as they were increasingly subjected to critical examination, and of the European study of the "Orient," itself a subject nearly as difficult to define as it was to describe. It also involves an awareness of advances in science in a seminal age of progress in the understanding of the physical world. In the end, it was the marriage of science with the study of the Orient that made the Danish expedition so unusual. It was a marriage that, in the eyes of its sponsors, could hardly be other than successful. In a curious way, it both succeeded and failed. Its success was due largely to the determination of Carsten Niebuhr who, throughout the years he was in the Orient and in the equally difficult years after his return, studiously applied himself to the principles set out in the instructions to the members of the expedition. To understand its failure we must understand the perhaps extravagant expectations of the man who conceived it in the first place.

The epic journey for which Niebuhr is almost alone remembered began in the fertile mind of the same Johann David Michaelis, professor at the University of Göttingen and the foremost philologist and Oriental linguist in Europe. He had been born to the study of Oriental languages. His father, Christian Benedict Michaelis, preceded him as an Orientalist and everyone agreed that the father was his superior as a Hebrew grammarian. To Hebrew, Greek, Latin, and the more common Oriental languages the younger Michaelis added Chaldean and Syriac, which he believed were critical to an understanding of the Hebrew scriptures. But his real love was history and, in an age when the study of so broad a subject could be undertaken by a single scholar, he combined his interests into what might be called the discipline of sacred philology. Philology, the historical study of written texts and determination of their authenticity and meaning, differs from what we call today linguistics. Linguistics, or the study of language in all its aspects, has since resolved itself into the separate disciplines of phonetics, phonology, morphology, syntax, and semantics. But the word philology carried a far greater burden in the eighteenth century than linguistics does today. It rested, coequal with philosophy, at the top of the scholarly hierarchy.[4] If philosophy, taken from the Greek, meant the love of knowledge, philology, taken from the same root, meant the love of words. But there was more to the relationship than appears in this simple distinction. Philosophy represented what *was*, the sum of knowledge as a whole. Philology represented the knowledge of what was *known*, the sum of what had been produced by the human mind. Where philosophy strove to know more, philology strove to understand what had previously been known. In this effort the subject matter of philology was literature, not narrowly defined but, in the minds of eighteenth-century philologists, an all-encompassing literature that included art, government, science—in fact, the entire written record of mankind. It was in this formulation that it stood alone with philosophy atop a pyramid representing everything produced by the human spirit.

The primary focus of philology was the past. The techniques it used could be applied to the present as well. But the present, with its incessant claims on the attention of the scholar, was too near in time for the kind of dispassionate study such a subject required. The most fertile ground was the remote past, where the study could be taken more leisurely and objectively. This might seem to consign the work of philologists to a perpetual reshuffling of knowledge, with no net increase in what was known. But that judgment would be superficial. The goal of philology was nothing less than

> to relearn what has been known, to present it in a pure state, to remove falsifications of time, to make an apparent into a real whole . . . these are necessary to the very life of knowledge.[5]

Such an activity represented an *addition* to what was known. Moreover, he who ignored the past would do so at his own risk for, however refined and complete one's knowledge at a particular time, it was still dependent on the vast body of knowledge that had gone before.

The full range of philological activity could be applied to specific areas: thus, there was Greek philology, Roman philology, and Oriental philology.

---

4. This organization of knowledge, and especially the place of philology within it, was probably best codified by the German philologist August Boeckh (1785–1867) in his *On Interpretation and Criticism*.

5. Ibid., 13.

The last had heavily religious overtones. Oriental philology could, of course, exist without a specifically Christian focus. In the general development of Orientalism beginning in the sixteenth century, source documents in the original languages were examined, lexica and dictionaries developed and refined, and treatises written, all with an ostensibly secular purpose. But it is remarkable the degree to which this activity served a religious, even polemical, purpose. Most of the pioneers in Oriental studies had taken orders. Among the near contemporaries of Michaelis, only Sale (see below) was not a man of the cloth. A refusal to subscribe to the Confession of Augsburg, a summary of the teachings of Martin Luther required of all Lutheran clergymen, deprived Michaelis of the career in the church that, everyone agreed, was his natural vocation. But this in no way diminished the religious nature of his concerns. In fact, his biographer in the *Biographie Universelle Moderne* faults him for too great a propensity to see in Scripture a foreshadowing of modern, secular knowledge and for a tendency to

> see the authors of sacred texts too often as scholars, naturalists, doctors, astronomers etc., and to search in the poetic tablets of Job, and in the writings of Moses and the prophets discoveries of modern times and the observations of Linnaeus.

In the untutored hands of his students and followers, the tendency to eclecticism became extreme. To be fair, the same biographer also remarks that this tendency should not detract from Michaelis's immense contribution to the study of the Orient. Again, to quote the entry, Michaelis

> found the edifices of human knowledge composed of bricks and he left them changed into gold or, better yet, he put together the fragments and the building materials into solid structures, regular and spacious, capable after their initial arrangement of accommodating all the additions which new knowledge would make necessary.

That the Bible was incontrovertible truth was accepted at the outset, and evidence that called the document into question, or did not support it, was unwelcome. As interesting as the myriad subjects of concern to the philologist Michaelis are those things that were *not* important, beginning with the language of another, even more ancient Oriental people, the Egyptians. As we will see below, it would be left to a rank amateur, Carsten Niebuhr, to suggest that an understanding of the language of the ancient Egyptians would come only with an understanding of Coptic, its ultimate successor. We would probably not be too far wrong if we assumed that one of the reasons for this lack of interest was that the language was "profane." Although classical philology came increasingly to constitute a triumvirate of Greek, Latin, and Hebrew, there was still a useful distinction between the sacred and the profane. The first two were considered important, indeed vital, to an understanding of the classical past. The last was vital to an understanding of Europe's putative religious past. For it was not just any text but that interested the sacred philologist, but the Bible itself, the Jewish and Christian Scriptures, the word of God as passed down to mankind through His prophets and messengers. Here the exegesis—or critical analysis—was of a document that was simultaneously accepted as the product of the human spirit and divinely inspired.

The problems of sacred philology were, by comparison with classical, much greater. Sacred Scripture as a written document presented enormous difficulties. Not the least of these was understanding the language in which it was originally written and the several languages, filters, through which the text had passed in the centuries before it reached Europe. Christians had come relatively late to the study of Hebrew and, by the middle of the eighteenth century, the discipline of sacred philology was a comparatively recent one. Unlike the Greek and Roman works that constituted the classical canon, the extant examples of Hebrew were far fewer and far less a part of the European intellectual tradition. Study of the Old Testament suffered from this paucity of sources, or at least of a Christian understanding of them. But with the rise of Protestantism and the emphasis on the importance of the Bible[6], the need arose for Christian scholars competent in Hebrew, still considered by many to be the first and most perfect of all languages. In a pamphlet published in 1740 Gregory Sharpe was profuse in his claims for Hebrew.[7] While we shouldn't necessarily attribute to Michaelis all of Sharpe's opinions, the essay was nonetheless a general statement of the prevailing eighteenth-century attitudes about the importance of the language of the ancient Jews. According to Sharpe, Hebrew was an invaluable aid in learning *all* eastern languages, including Greek, Turkish, and Arabic. The author appends his method of learning Hebrew without points, almost a "Hebrew in three easy steps" of the eighteenth century, as well as a list of some 5,000

---

6. In one of the most revolutionary contributions to the new religious dispensation, Martin Luther made his landmark German translations of the Bible from Hebrew and Greek, the two sacred languages. "In proportion as we value the gospel, let us zealously hold to the languages," he said, "for the languages are the sheath in which the sword of the Spirit (namely the gospel) is contained." See *Luther's Works*, 359–60. It should be noted that this "philological" approach to Bible translation was accompanied by Luther's firm belief in divine inspiration as a necessary guide to the translator. Only by means of the two together, with inspiration in the lead, could truth be arrived at.

7. See Sharpe, *TWO DISSERTATIONS*.

Latin and English words which, he claims, are derived from Hebrew. But his claim for the importance of the language does not stop there:

> But to say that *Hebrew* is the key to all the Oriental languages, and the source of the *Greek*, is not so say enough in its favor. It is also so simple in itself, and so easy to learn, that one may be forgiven for calling it the language of nature, or the first language of the world. (p. xiv).

In spite of the alleged simplicity of Hebrew, textual controversies had occupied Jewish scholars for centuries, and Christian scholars now lept into the fray. The seventeenth century had witnessed the growth of a body of Christian Hebraists,[8] assisted by Portuguese and Spanish *Marraños* and Jews who had moved northward to the more congenial atmosphere of the Protestant Netherlands. Hugo Grotius (1583-1645), Thomas Erpenius (1584-1624) and Jacob Golius (1596-1667) were early examples of the type. As we have seen, Michaelis's own father transmitted the tradition to his son.

The parties to this early exercise in Jewish-Christian cooperation had their own polemical interests. The two religions, after all, shared a sacred text but remained bitter enemies. To some Jews, the acquisition by Christian scholars of a knowledge of the Hebrew sources was a sacrilege. To others, it was felt that as long as Christians were interested in the Hebrew scriptures, they should at least be provided with Jewish guidance. To some early Christian scholars, the knowledge of Hebrew was a guarantee against Jewish tampering. Just as the more reasoned view of Islam often served only to place the basic hostility of Europe on firmer scholarly ground, the same held true for the study of Judaism. Even the earliest Hebraists, men such as Grotius or the Englishman John Seldon (1584-1654), studied Hebrew both to better inform themselves about scripture, and to arm themselves against the errors of the Jews. The Jews had, after all, for centuries been keepers of a part of the sacred books of Christianity. It was generally, if grudgingly, acknowledged that they had exercised this trust faithfully and had not, as some alleged, purposely corrupted the Hebrew text.[9] But one couldn't be too careful, and it was unthinkable that Christians should not have their own scholars, capable of understanding so important a part of their heritage. While relations between representatives of the two religions were not always the most edifying, at least this long-ignored area of scholarship was now opening to Christian Europe.

---

8. See Katchen, *Christian Hebraists and Dutch Rabbis*.

9. This suspicion had to do particularly with language prefiguring the coming of Christ. For a contemporary refutation of the charge of corruption, see Gill's *A Dissertation Concerning the Antiquity of the Hebrew Language*.

The textual obsession is surely understandable when we consider the reverence with which the text of the Koran is held by Muslims, where a misplaced diacritic or mistake in pronunciation can lead to the most egregious of errors. But instead of a normative sacred text, codified for all time in grammatical and rhetorical if not doctrinal purity, Christians were dependent on an uncertain text, one part of which—the Old Testament—was originally in Hebrew, a dead and little understood language, and the other—the New Testament—was in a debased Greek with Hebrew, Chaldean, and Syriac influences.[10] The existence of an accepted text was *not* the end of controversy, as centuries of Jewish commentary on their own sacred texts had shown. Masoretes, or those who zealously guarded the correct spelling, reading and writing of the Hebrew Bible, had for centuries concerned themselves with preserving a normative text, even if it contained irregularities, which they treated with marginal notes. But we can understand why the delicate consciences[11] of men like Michaelis found congenial the most minute study of the sacred books of Christianity.

In spite of an early reputation as a *religionspötter*,[12] or "scoffer at religion," Michaelis was a believing Christian who saw in the study of the cognate languages of Hebrew—Arabic, Syriac, Chaldean, and Samaritan—the key to Biblical exegesis. Having wrestled with a question as to the basis of his Christian faith, he apparently drew back from the precipice of unbelief into a literalism, a focus on the text itself, that condemned him to the status of one of the near great. Of two theses he submitted on graduation in 1739–40, one was designed to prove the antiquity and divine inspiration of the vowel points of the Hebrew scripture. To a believing rationalist like Michaelis, there could be no conflict between the Bible and science. God was the author of both. Both were therefore "true," and the challenge was to reconcile these two aspects of the truth. Indeed, since there could be no disagreement, apparent conflict could only be due to our lack of understanding of the context in which the Bible appeared. The explicit purpose of the Danish expedition to Arabia was to assist those interested in sacred philology to better understand this context.

---

10. Even with the New Testament there seemed to be grounds for suspicion that the "schismatic Greeks" might have corrupted parts of the text.

11. His biographer says that it was the 12th and the 24th verses of the 19th chapter of the Gospel according to St. Matthew that were most troubling to Michaelis. The first had to do with eunuchs and the second contains the statement that "It is easier for a camel to pass through the eye of a needle, than for a rich man to enter into the kingdom of God." A later and deeper knowledge of the texts apparently laid these scruples to rest.

12. See Flaherty, *The Quarrel of Reason with Itself.*

Introduction

The Context

An understanding of the context in which Michaelis himself lived and worked is also necessary for us to understand the mainsprings of his thinking. Europe was in the midst of that extraordinary outpouring of scientific and philosophic thought that we know as the Enlightenment, where reason, not faith, was the guide to truth. Philosophy now represented not the love of wisdom or knowledge of the ancients, but that branch of knowledge or speculation dealing with the nature of the universe. The age led to important discoveries in astronomy, chemistry, mathematics, physics, and linguistics. The Danish expedition was itself a typical Enlightenment expedition, with its attempt to add reason and careful observation to what previously had been accepted on faith. But there were limits to what reason could achieve, and there was hardly a man in mid-eighteenth century Europe who dared call himself an atheist.[13] Beginning at least with Blaise Pascal (1623–62), Enlightenment thinkers had wrestled with the idea of God. Their conclusions were often diametrically opposed, but none went so far as to deny the existence of a Supreme Being. Pascal, the precocious French physicist, mathematician, and theologian, found a rational God too remote and academic, but was terrified of the void. Rene Descartes (1596–1650), the French mathematician and converted Catholic, insisted that the intellect could find God, and sought Him with the certainty of mathematics. He saw no contradiction between faith and reason. Baruch Spinoza (1632–77), a Dutch Jew of Spanish descent, was perhaps the prototype of a new secular outlook. He believed in God, but not in the God of the Bible. For his pains, in an extraordinary ceremony, he was excommunicated from the synagogue of Amsterdam. Symbolically, all the lights in the synagogue were turned out and so was Spinoza, with the imprecation "Let him be accursed by day and accursed by night." Isaac Newton (1642–1727), the English physicist and mathematician, saw a mechanical universe with God as the great watchmaker, the sole source of activity. The notion of gravitational force drew his system together, and proved the existence of God. Without an intelligent overseer, it could not work.

Later intellectual giants such as Milton, Kant, and Voltaire continued the struggle to reconcile belief with newfound knowledge. The God of Milton (1608–74) was cold and legalistic and Satan was the real hero of *Paradise Lost*. But without God, Satan was not possible. Immanuel Kant (1724–1804), the German philosopher who defined the Enlightenment as "man's exodus from self-imposed tutelage," found the way to God through conscience and

---

13. For the following survey I am indebted to Armstrong, *A History of God*.

reason, dismissing ritual and the authority of the Church. But he did not dismiss God. Even Francois-Marie de Voltaire (1694–1778), that epitome of anticlericalism, yearned for a simple religion that would make men just without making them absurd, that would not order them to believe things that were "impossible, contradictory, injurious to divinity, and pernicious to mankind; and which dared not menace with eternal punishment anyone possessing common sense." Voltaire may have been the embodiment of Enlightenment thought, but even he did not deny the existence of God.

Others struggled with the practical details, the human accretions of the religion of Jesus Christ. Gottfried Arnold's *History of the Churches from the Beginning of the New Testament to 1688* attempted to trace the historical manifestations of the institution back to the primitive church. Johann Lorenz von Mosheim (1694–1755), in his *Institutions of Ecclesiastical History*, recorded the development of theological doctrine. Johann Friedmann Mayer in his Wittenburg's *Innocence of Double Murder* attempted to reconcile the loving message of Christ with a vengeful God and centuries of Christian persecution and cruelty. In perhaps the most revolutionary development of all, Hermann Reimarus (1694–1768) attempted a critical biography of Christ, based on a careful analysis of Scripture. This may have truly represented the beginning of skepticism about the "truth" of Scripture, and would subject these previously unassailable, if not inaccessible, sources to the same analytical methods as secular texts.

In the midst of this ferment, much of it in Germany and much propounded by his contemporaries, Michaelis undertook his own critical analysis of Sacred Scripture. As we have seen, he was no revolutionary, but used his linguistic faculties to study the Bible as a document that, for all of its susceptibility to textual analysis, was still divinely-inspired. He appears to have accepted, at the outset, the factual truth of the document, and brought his erudition to bear on the building blocks of the text, the words and the context in which they were used. By implication, apparent difficulties were only the result of a lack of understanding of this original context. Much of what was not understood by Europeans about the Hebrew Bible was simply a lack of familiarity with the language, habits, practices, attitudes, flora, and fauna of the area in which the document originated. With research into the primitive meaning of words and the context in which they were used, the explication—the intensive scrutiny and interpretation of the interrelated details—of the Bible could be advanced.

Introduction

The Orient

European interest in the area where Judaism and then Christianity arose was, of course, centuries old. This was, broadly speaking, called the "Orient," although the term itself was difficult to define, being a part geographical, part linguistic, and part cultural and religious abstraction. Geographically, the Orient included the Near East—North Africa and the Levant; the Middle East—Arabia, Iran, and parts of Turan; and the Far East—everything else to the end of the Asian continent, and was not confined to our common acceptance of the word today as referring to northeast Asia. The geographical definition included most of the world of Islam, although the Muslim lands of North Africa were generally south, not east, of Europe and the question arose as to whether the Islamic portions of east Africa and the east Indian Archipelago, which were assuredly east of Europe, should be included. And the Muslim world was full of pockets of *dimmi*s—free, non-Muslim subjects living in Muslim countries, Armenians, Greeks, and Copts, to name only a few—who were certainly eastern and were often as remote and little understood as their Muslim neighbors. To Europe the Byzantine Greeks had been, of course, little better than the other Orientals, the term "Byzantine" itself, at least since Gibbon, a byword for intrigue and convoluted dealing. All these Easterners were treated by the West with equal disdain, and the sack of Constantinople in 1204 by the Venetians of the 4th Crusade was probably the greatest act of cultural and artistic pillage in history.

There were Oriental languages, and a linguistic definition of the Orient seemed to work as well as any other. The Orient was the place where certain Oriental languages were spoken. The most obvious were Arabic, Turkish, and Persian, but there were also the old languages of the Jewish and Christian scriptures, Hebrew, Syriac, Aramaic, and Chaldean. As we have seen, the study of these languages was a relatively recent phenomenon in eighteenth-century Europe, and they constituted a part of the Oriental branch of philology. There was also growing European interest in Sanskrit, Chinese, Japanese, Burmese, and other languages from the eastern regions of the globe. But to the Orientalists who set in motion the Danish expedition, these other languages were far afield from their concerns. Their exclusion only highlights the difficulty of the linguistic definition of the Orient: it simply depended on which languages were defined as Oriental, and was really no definition at all.

There was, however, a common thread that seemed to run through the concerns of these Orientalists, and that was religious. The Orient seemed to be for them that place where the religion of the Jews arose and gave rise to its successors, Christianity and then Islam. Again, there were other "Oriental" religions, but to eighteenth-century Orientalists the belief systems of the people in China, Japan, and India, for example, could hardly be called religions at all, however interesting they may have been as cultural phenomena. With the discovery of new worlds in the previous two centuries, European eyes had been opened to other peoples, most of whom shared nothing remotely close to European ideas about a supreme being. There may have been planted in European minds the idea that these people had beliefs that were not entirely contemptible, but we would be mistaken if we saw this as a form of eighteenth-century ecumenism.[14] There was simply no question in the minds of European merchants, travelers, scholars, and, especially, missionaries of the "truth" of Christianity. Judaism was accorded pride of place as the ultimate source of Christianity and, although the Jews had rejected the saving message of Jesus Christ, they had been the Chosen People. There was still hope that they could be shown the error of their ways.

Christian attitudes about Islam were less ambivalent. The claim of Mohammed to be the last in the prophetic tradition that began with Adam and proceeded through the Old Testament prophets to Jesus Christ, was dismissed as sheer imposture. Islam possessed a certain crude energy that had, admittedly, subjugated a good part of the known world in the first rush after its appearance, but its claims hardly deserved the attention of serious men. The acceptance by Muslims of the truth of, first, Judaism and then Christianity, was seen by Europeans as mere, shallow imitation. The study of Islam *had* undergone a change in the latter part of the seventeenth and early part of the eighteenth centuries, benefiting from the new openness as well as interest in the Orient as the source of the Christian past. Gone were the crude polemics with which Christians had approached Islam, and the unremitting hostility with which they had treated its founder. Edward Pococke (1604-91), J. J. Barthélémy d'Herbelot (1625-95), George Sale (1697-1736) and Simon Ockley (1678-1720) were among the leaders of this scholarly revolution, with their insistence on the importance of Arabic documents themselves as the sources of their research.[15] Even Gibbon, no friend of the East, recognized the exceptional character of the founder of Islam: "Conversation enriches the understanding," he said of the Prophet, "but solitude is the school of genius." The old polemicism was increasingly replaced by a more enlightened scholarship. But it still stopped short of

---

14. There were some of a more liberal bent, notably the French Heugenots Picart and Bernard writing in Amsterdam, that comparative hotbed of religious freedom. Their multi-volume *Religious Ceremonies of the World*, appearing between 1723 and 1737, was notably lacking in polemic. But they were out of the mainstream, being the exception that proved the rule.

15. Holt, *Studies in the History of the Near East*.

conceding to Islam a place with Judaism and Christianity in the galaxy of important religions, and only seemed to place the underlying hostility on a firmer scholarly footing. The translation of the Koran into English by Sale in 1734 was a vast improvement over the versions that preceded it. But in the *Preliminary Discourse* to the translation, itself a landmark in the European study of Islam, Sale finds hardly a manifestation of Islam that it did not owe ultimately to the Jews.

There was, of course, more to European feeling than resentment over the religious pretensions of Islam. The historic military conflict between Christianity and Islam was still fresh in the memory of both sides of the religious divide. The last large-scale European and Christian intrusion into the Arab and Muslim world, the Crusades, may have taken place centuries before, but the memory of that interregnum in what had previously been an uninterrupted string of Muslim successes had a remarkable life on both sides of the divide. The Muslims had eventually expelled the Crusaders from the Holy Land after a century of Christian rule, or misrule, and the unhappy cradle of Christianity had reverted to control of the infidels. But by the time the Danish expedition left Copenhagen, the conflict had reached the point where the two cultures could cautiously eye one another, if not accept each other openly.

But the memory of Islam as a threat was still too real and the behavior of the Turks was still too aggressive for Europe to take a dispassionate interest in the Muslim world. By the middle of the eighteenth century, the final expulsion of Muslims from the Iberian peninsula was already 250 years old, but parts of Spain had been Arab and Muslim far longer than they were currently European and Christian. And Catholic Europe was still alive to the threat in its midst of a kind of fifth column of *Marraños* and *Moriscos*, "new Christians" but still secretly Jewish and Muslim respectively. The Turks had begun the transformation of western Anatolia from a Greek and Slavic land to a Turkic one in the thirteenth century, but Constantinople had only fallen in the mid-fifteenth century. More recently, the Turks had subjugated southeastern Europe, decimating the flower of Serbian and Hungarian chivalry in the process. In their annual spring and summer campaigns, the Ottoman Turks regularly marched into the heart of Europe, and they had only been turned away for the last time from the gates of Vienna in 1683.

The Turks had inherited the mantle of the defenders of Islam from earlier Arab and Kurdish dynasties. For all of their status as foreigners and oppressors in many parts of the Muslim world, there was no question that the sympathies of Muslims lay with them in this increasingly unequal battle. For the insufferable air of superiority assumed by eighteenth-century Turks was the reaction of a culture under siege. In fact, for all of the importance of religion in the clash of cultures, the conflict itself was in the process of change.

Niebuhr in Egypt

The battle would increasingly be waged from the Western side by science, a kind of secular cousin to religion. In this, Niebuhr and his companions were unconscious and unwitting agents. But however the conflict was formulated, it was defined on Europe's terms. Europe set the stage for its incursions into the Orient, just as European scholars set the terms for this expedition.

The "Orient," then, in the minds of eighteenth-century scholars was an odd composite of equal parts geography, language, and religion. It represented a small part of the area lying to east and south of Europe, most specifically that area where the three "religions of the book" had arisen and were widely practiced. To attempt to define it more precisely would lend a specious clarity to what was, in the end, more a state of mind than a geographical location. It was more familiar than the other areas the preceding two centuries had opened to European eyes, and for that reason Europeans carried preconceived notions about what they would find there. It represented, through Judaism and the Judaic sources of Christianity, a putative cultural and religious past that was more important as it was less understood. Christian Europe had always maintained a curious nostalgia for this "Holy Land" and had maintained contact with it, with the exception of the two-century-long spasm of the Crusades, by means of pilgrims traveling to the holy sites. Their peregrinations took them to Palestine, to the Sinai peninsula, and often to Egypt, all part of the setting for the rise of Judaism, and then Christianity, as told in the Bible. Since the demise of the Crusader kingdoms these travels had always been at the sufferance of the Muslim authorities. North Africa and the Levant still represented a kind of debatable land in the ongoing conflict between Islam and Christianity, with the Mediterranean simultaneously separating the combatants and affording a medium for incursions into the territory of "the Other." As we will see, low-level naval warfare between the two sides still simmered in the "Middle White Sea" in the middle of the eighteenth century.

But it was not this narrowly–defined Orient, or the Muslim world, or the Arab world, or even the Holy Land that interested Michaelis. It was, instead, the Yemen or "Happy Arabia," the remote southwestern part of the Arabian peninsula, where a dialect of Arabic that differed from "western" Arabic was spoken, and whose habits, practices, and attitudes had been less corrupted by contact with outsiders. In his instructions to the members of the expedition, Michaelis makes this clear:

> ... the accounts we have of Happy Arabia are very small in number. Nature there has spread riches of which we are still entirely ignorant. Its history goes back to the highest Antiquity; we know that the idiom there differs from that of western Arabia; and as

the idiom has been the surest light to guide us to a knowledge of the Hebrew language, how will new illumination on that most important of books, I mean the Bible, be possible if we are not able to attain a knowledge of the dialect of Oriental Arabic to the same degree we understand that of the west?[16]

The area of this pristine manifestation of Oriental society was the laboratory in which the scholars of the expedition would work. It was an area deep in the heart of Islam and penetration by Europeans would be possible only with the greatest circumspection and care.

---

16. *Fragen*, V.

2

The Conceit

> "Conceit: an idea, thought, concept... a fanciful or witty expression or notion; often, specif., a striking and elaborate metaphor ... whimsical; fanciful." (*Webster's New World Dictionary*)

THE SERIOUS EUROPEAN STUDY of the Orient had begun with Spanish and Portuguese Jesuits, following in the wake of the fifteenth- and sixteenth-century conquests. But by the middle of the eighteenth century, Protestant Orientalism had overtaken the early lead of the Iberian Catholics, although the Protestants were still dependent on translations of early Arabic works made by Maronites in Rome. The next phase was distinctly Protestant but with equally religious roots. The new centers of Eastern studies were in Göttingen and Halle in Germany, Paris, Cambridge, and Oxford and, most importantly, Leiden in the Netherlands. There was a community of interest and a sharing of resources by these centers, a certain repetitiveness in their output, and the great Orientalists circulated among them. Michaelis made a seminal visit to England in 1741, visiting London and Oxford and experiencing, for the first time, an intellectual atmosphere not dominated by the dour tenets of Pietism. He also studied with the Arabic scholar Johann Reiske under the great Dutch Orientalist, Albert Schultens at Leiden.[1] These painstaking textual studies, undertaken by men who never saw the Orient about they had developed such erudition, represent something of a dead-end in the study of the Eastern world. Michaelis was, in many respects, the greatest representative of the type. The epitome of the eighteenth-century Protestant Orientalist, a man of prodigious intellect and a gift for Oriental languages, he was a giant in his field. But he never traveled to the Orient of

---

1. Reiske, whose irascibility was notorious, later turned on Schultens whose friendship Michaelis maintained. This was to influence Niebuhr's choice of a collaborator at a later date, as we will see below. Niebuhr's biographer says that Reiske was an early choice of Michaelis as the single scholar who would undertake the journey to the East.

his obsessions and his excursions were entirely textual ones into the world of Oriental scholarship. His published works were in the dozens, ranging from grammars and lexicographies to works of exegesis and civil law. During his lifetime, as we have seen, he laid the foundation of systematic Oriental studies in the universities of Europe.

Michaelis had a following on the continent far beyond his own circle in Göttingen, and he could draw on a common reservoir of scholarly resources. In 1756 he proposed to Count Bernstorff,[2] the Danish Foreign Minister, that King Friederik V sponsor an expedition to the Yemen or "Happy Arabia." Friederik was noted for his interest in science and was receptive to the idea of the expedition.[3] Its purpose would be biblical research, to do nothing less than fill in the gaps in European knowledge of the Bible by equipping a party of scholars to see and hear what the Oriental philologists in Europe could not. European scholars believed that a remnant of the Israelites still lived in the Arabian Peninsula. There *were* Jews in the Peninsula, of course, remembered for reasons more associated with Muslim than with Jewish history. It was known that the rejection of the Prophet's message by the Jews of Medina was responsible for much of the hostility of Islam to this first "People of the Book," and there were persistent rumors of Jews still living in Khaybar, north of Medina.[4]

But, more importantly, it was known that in the Yemen there was still a living Jewish community. A combination of its remoteness, the history of a brief period of Jewish rule in the southwestern part of the Peninsula, even the biblical story of the Queen of Sheba, lent the region importance in the minds of European scholars. The formulation in Michaelis's mind was straightforward: the Hebrew Bible was full of words and references to plants, animals and minerals that were unclear in Europe. If scholars were sent to the Yemen, where the living language was the closest to the original Hebrew, they could resolve much of the uncertainty. It was also clear to Michaelis that textual analysis was only a first step. There were contextual matters as well—the history, society, and natural phenomena in the area—that needed examination. The Yemen, unlike Palestine which had been corrupted by

---

2. Johann-Hartwig Ernest Bernstorff (1712–72), Minister of State in Denmark from 1750 to 1772. When Michaelis approached him with the proposal of an expedition to Arabia, he was receptive and, Niebuhr tells us, he was unstinting in his support.

3. He was succeeded by Christian VII in 1766 while the remnant of the expedition was still in the East. Christian VII was of a very different temper from Friederik and Niebuhr found little official favor on his return.

4. Doughty would look for them, unsuccessfully, in 1877. There is still a Muslim tradition that the end of the world will be presaged by, among other things, the return of the Jews to Khaybar.

centuries of contact with the West, represented a kind of time warp where, insulated from outside influences, it was thought that society and language were closest to the biblical originals. For that reason, it would serve as the laboratory where the necessary experiments could take place.

How this belief was to play itself out, how its purposes would be realized or frustrated, is the subject of the story that follows. As we will see, the expedition produced results that were not insignificant. They consisted of Niebuhr's *Travels in Arabia* and *Description of Arabia*; three volumes of botanical and biological studies, the posthumous work of Forsskal the botanist with illustrations by Baurenfeind, the artist; and scores of Oriental manuscripts that even today form the core of the collection in the Royal Library in Copenhagen, largely the work of von Haven, the philologist. But these contributions to an understanding of the Orient were incidental to the primary purpose of the expedition: research that would assist in the explication of the Hebrew Bible. Here, there also may have been incidental contributions to an understanding of that "Oriental" document. But the premise on which the expedition was based, that in the Yemen eighteenth-century Europeans would find a people speaking an idiom of "Oriental Arabic" closest to that of the Hebrew of the Old Testament and that, through its study, they would be able to shed light on linguistically-uncertain passages in the Bible, was fundamentally flawed. It is a measure of the reputation of Michaelis, of his powers of persuasion, and of European ignorance of the Orient, that a crowned head of state would have sponsored so quixotic a quest.

Enlightenment Expeditions

The idea of dispatching a body of scholars to a distant and little known land was in many respects a typical Enlightenment impulse, and the Danish expedition was typical of the state-sponsored expeditions of the middle of the century. Stories of the great sagas of exploration, for which there had long been an immense popular appetite,[5] married the new rationalism to create an interest in other peoples. They were now not merely exotics or grotesques, but fit subjects for scientific study. Moreover, Europe no longer seemed to have all the answers. To their curiosity was added the sense that other people had lessons from which even Europe could learn. Developments in science, alluded to above, also played a part in the background of the expedition. For the past hundred years, Europe had experienced the

---

5. Niebuhr used several of these compendia in the preparation of his works. Excerpts from his *Description of Arabia* and *Travels in Arabia* were, in turn, included in several of those that followed.

intellectual ferment, astonishing in its scope and vitality, known as the Enlightenment. Everything was now subjected to the test of reason. To the timid, the threat to accepted authority and traditional ways of thinking was unendurable. To bold thinkers such as Newton, himself a believing Christian and something of a Hebraist, the challenge was to forge a new synthesis between religion and science. In their heady optimism, nothing was—nothing could be—incompatible between the Bible as the word of God and the world as the manifestation of His creativity. The Bible was the inspired word of God, but it had been recorded by the hand of man, and its true meaning could be learned if it were subjected to the same kind of rigorous analysis as that applied to the physical universe.

The Danish expedition would include a philologist, a physicist (or botanist or natural philosopher), a mathematician or surveyor, a physician, and an artist. We have already seen the elaborate eighteenth-century conception of the discipline of philology. With philosophy, however, we find ourselves on less familiar ground. The natural sciences were also called "natural philosophy," and from the time of Aristotle had meant the study of all of "nature," that is, everything that was not created by the hand of man. It was a subject of immense scope and even at the time of the Danish expedition included the separate disciplines of botany, biology, chemistry, physics, geology, and even medicine. Its practitioners tended to range widely across these present-day boundaries. It is no accident that Forsskal was alternately called a "physicist," a "natural philosopher," and a "botanist" or that Kramer, the physician, is designated as his assistant in matters of natural philosophy. Many of the new discoveries in the sciences were, in fact, made by men trained in medicine. Recent advances in what we would today call botany and biology were largely due to the descriptive genius of Carl Linnaeus (1707–78), who was himself a medical doctor. Based on a system of sexual classification, Linnaeus almost single-handedly reorganized the study of the plant and the animal kingdoms. His *Systema Naturae* and *Species Plantarum* were the starting points of systematic classification, and we owe to him the binomial nomenclature so familiar to students of biology and botany.

The genius of the two men may have been different, but Linnaeus was to "natural history" what Newton was to the "exact sciences."[6] He was, in many respects, a figure like others we have already seen in the burst of Enlightenment science: a religious man and a believing, if unorthodox, Christian, at first destined for holy orders. His wide-ranging interests took him beyond the narrow confines of the classical and biblical learning typical in the preparation of the clergy of the day. His prose style was an odd amalgam

---

6. See Frangsmyr, *Linnaeus, The Man and His Work*.

of Ovid and the Old Testament, and he saw the hand of God everywhere in the brilliant manifestations of the natural world that were increasingly coming to Europe's attention. The task of the observer was to detect the divine plan in nature, and Linnaeus found order in what to others was chaos, a riot of genera and species. But his suggestion that plants had a sexual nature, critical to his system of classification derived from the numbers of pistols and stamens, shocked and scandalized the educators of the day with its suggestion of polygamy, polyandry, and free love, and the implication that God permitted such practices in the natural world. His new system would turn the teaching of natural history in Europe on its head, and resistance to the man and his revolution was immense. The debate in Europe over the Linnaean system, about artifice as opposed to nature, raged on. But however it raged, Linnaeus did a service to the methodology of botanical and biological taxonomy and his disciples, of whom a member of the Danish expedition, Petrus Forsskal was one, were equipped with a useful means of collecting and describing the "productions of nature."

If significant advances had been made in the study of languages and the macroscopic natural world, however, the same cannot be said of the microscopic world of medicine. It was unfortunate for the members of the expedition that physicians of the eighteenth century were as poorly equipped to understand the fundamental processes of nature as they were to treat the sick. The lack of understanding of these processes prevented the institution of systematic programs of treatment for most diseases. Such treatments as were effective tended to be accidental, and the "heroic remedies" of bleeding, cupping, and purging were prescribed well into the nineteenth century. Their justification was based on an incomplete understanding of physiology, which typically spoke of pressures of confined fluids, impacted blood driving even more impacted blood into smaller and smaller vessels, and of the need to relieve the pressure through drawing off the fluids. As we will see, the standard remedies probably hastened the end for several members of the expedition.

The world of eighteenth-century cities was one of filth, and Europe itself was no stranger to the phenomenon. While there were no longer the devastating outbreaks of disease that had characterized the Middle Ages, urban concentrations were still plagued by undrained marshes and stagnant waters, a lack of understanding of the importance of individual and collective hygiene, and imperfect systems to collect sewage and insulate the living from the dead.[7] But it was especially true of the cities of the eastern Mediterranean and the Orient where the members of the expedition would find themselves. The world described above bred the agents of disease and harbored the vectors

---

7. See Riley, *The Eighteenth-Century Campaign to Avoid Disease.*

of their spread. Contagion theory had dealt more or less effectively with the plague, a disease whose most virulent form was spread in its infectious stage by respiration. Even in the eastern Mediterranean, the practice of quarantine prevented the most serious outbreaks. But little was known of diseases spread by other causes and especially critical was the lack of knowledge of the cause of malaria. There is evidence to suggest that malaria was not just a disease of the tropics, but was endemic to Europe in the eighteenth century, included in the general afflictions called agues or fevers. Fevers were the great killers, causing about eight of every ten deaths in Europe during the century.[8] Programs to drain the swamps, and thus the standing water in which the mosquitoes carrying the disease bred, were effective in reducing the incidence of the disease in Europe. And the discovery of Peruvian bark, or quinine, had proved to be effective in the treatment of individual cases. However, in the case of the Danish expedition, as we will see below, even had the physician possessed the best medical knowledge of the day, there is probably little more he could have done for the stricken men.

European Rivalry and Cooperation

The eighteenth century was also a period of intense rivalry between the nations of Europe. The first half of the century had seen a succession of wars, often over dynastic arrangements, which preserved the balance between the powers. There were the War of the Spanish Succession (1701–13), the War of the Polish Succession (1733–38), the War of the Austrian Succession (1740–48), and the Seven Years War (1756–63). The last of these spanned the dispatch of the Danish expedition and, as we will see below, very nearly put an end to it before it had begun. But there was competition, as well, among the crowned heads for the laurel of the most enlightened sponsors of the new sciences. So when Bernstorff approached Friederik V, Michaelis's proposal fell on fertile ground.

The expedition also found favor for another reason, and this had more to do with cooperation than with competition. The transit of the planet of Venus across the face of the sun, occurring twice within eight years over intervals of, successively, 108 1/2 and 121 1/2 years, was due to occur again in 1761. To the astronomers of Europe it represented the chance of a lifetime to calculate "the astronomical unit," one of the fundamental building blocks of Newton's mechanical universe: the distance between the earth and the sun. In the seventeenth century Kepler had determined the proportions of the distances from the sun of the planets in

---

8. See King, *The Medical World of the Eighteenth Century*, 123.

the solar system. A combination of this knowledge and a careful measuring of the time it took for the planet Venus to cross the face of the solar disc would, by triangulation, permit calculation of the distance between the two bodies.[9] The dispatch of teams of astronomers by most of the states of Europe to locations as widely dispersed as India, the Cape of Good Hope, Newfoundland, China, and Siberia represented probably the first combined scientific undertaking in history. As it would be with Cook's first voyage to the South Seas eight years later, observation of the transit of Venus would be the first scientific task of the Danish expedition.[10]

Denmark had a factory at Tranquebar on the Coromandel Coast of India, founded in 1621 in the first rush of European mercantilist imperialism. To the factory was added a mission in 1705 and for the next 140 years it remained an outpost of Lutheran Protestantism and a profitable trading establishment. Missionaries were periodically sent to Tranquebar and this seemed an ideal means of fulfilling Michaelis's ambition. If the missionaries were suitably prepared in Europe to carry out their scientific enquiries, they could then proceed from Tranquebar to Yemen, that most remote part of Arabia, and there carry out their research. As we have seen, "Happy Arabia" had long held a fascination for the West. It had been the fabled source of incense, a commodity as important in the ancient world as pepper and petroleum would be to the medieval and modern worlds, and a source of enormous profit to the producers. The control of both its source and the transport network to the northern distribution points had bred, it was supposed, fabulous wealth in the kingdoms of South Arabia.

Nabataean trading colonies at Petra and Meda'in Saleh had led to direct contact, and eventually conflict, with the Mediterranean world. A Roman attempt to seize the sources of the incense, the military expedition of Aelius Gallus in 25 BC, had ended in disaster. The combination of its purported wealth and a general ignorance of its history—save for the biblical tale of the Queen of Sheba and knowledge of a later conflict involving Christians, Jews and Muslims—made it an especially intriguing subject. For some of the same reasons it was an attractive area for those interested in biblical research. Among eighteenth-century Oriental philologists it was known that a Jewish remnant still lived in the highlands of Yemen and it

---

9. See Woolf, *The Transits of Venus,* for a discussion of the expeditions that participated in the sightings, and their results.

10. It was hoped that they would be in the Yemen for the sighting. However, due to a series of delays, they were at sea off Marseilles and Niebuhr could not achieve the necessary precision on a moving ship. Cook's observation also failed in 1769, for different reasons.

was thought the inhabitants of that remote area still spoke a language—a kind of Eastern Arabic—that was close to Hebrew.

With the coming of Christianity and the end of the pagan world's demand for incense, the southwestern corner of the Arabian peninsula had fallen into a long period of slumber. But beginning in the seventeenth century a new South-Arabian product had reawakened Europe's interest: coffee. In the early eighteenth century the appetite for this commodity brought European ships in increasing numbers to Mocha. Coffee was traditionally shipped from Mocha to Jidda, and from there to Suez, whence it was carried overland to Cairo. From Cairo, it was distributed throughout the Mediterranean world. The obvious advantage of eliminating the many middlemen in the transaction attracted first the Dutch, then the French and English directly to the Yemen, at the time the only source of coffee in the world. There, they found the Arabs equally interested in a direct trading relationship. Some of the early accounts[11] of these contacts are surprisingly informed in their treatment of the Arabs and Islam, the history of the Yemen, and the cultivation and use of coffee. But they were the accounts of amateurs, and there remained much scholarly work to be done in this remote part of the Arab world.

Staffing the Expedition

A number of changes in the original plan took place in the five years from the time Michaelis first approached Bernstorff to the departure of the expedition from Copenhagen in early 1761. In place of the missionaries, it was decided to train scholars who would proceed directly to Arabia via the Red Sea instead of the long detour through India. It was shorter and they could improve their Arabic along the way. Amid acrimonious discussion,[12] much of it with nationalist overtones, the Danish authorities eventually settled on a party of five professional members[13], two Danes, two Germans, and a Swede. They were in many respects a typical eighteenth-century, state-supported expedition. Professor Frederik Christian von Haven, a Dane and a philologist,

---

11. See Laroque, *A Voyage to Arabia the Happy*. The account is written in the third person from the letters of the commander of the expedition of three French ships that sailed to Mocha in 1708 and returned to Europe via the Cape with cargoes of coffee. It is the account of an astute businessman who hopes, and succeeds with delightful ingenuousness, to reconcile "learning with traffick." As a matter of fact, it was cited widely by Sale in the *Preliminary Discourse* to his translation of the Koran.

12. See Hansen's *Arabia Felix*. The book contains a lively account of the politics surrounding the selection of the members of the expedition and the personality conflicts of those selected.

13. Lars Berggren, their Swedish servant, made up the sixth European in the party.

was responsible for inquiries into matters of language and for the purchase of manuscripts. He came from a family of travelers. His late brother had been chaplain to the Danish legation in Russia and was the author of books on his travels, Russian history, and a Danish grammar. Because of his nationality and training in the discipline closest to Michaelis's heart, von Haven was the member of the group of whom, perhaps, the authorities had the greatest expectation and in whom they made the greatest investment.

Professor Petrus Forsskal, Swedish by birth, a botanist and zoologist and student of the great Linnaeus, was responsible in general for "natural history" and, in particular, for botanical matters. There would later be much dissension over the rival claims of Linnaeus and the Danish authorities to Forsskal's work. He complemented his apprenticeship in natural history with the study of Oriental languages at Göttingen, Michaelis's own university. Michaelis had some concerns about Forsskal's political views, which the latter had expressed in a pamphlet called *Thoughts on Civil Liberty*. They were at odds with those of the establishment in Sweden.[14] But Forsskal told it like it was, a characteristic to be valued in an undertaking of this kind. For all of his attempts at managing the results of the expedition, Michaelis was not interested in credulous observers. It is probably, therefore, no surprise that much of Michaelis's expectations were invested in Forsskal and why his early death was such a blow to the older man. He is primarily known today for the volumes which Niebuhr edited on his return and for a genus of plants named by Linnaeus in his honor.

Dr. Christian Carl Kramer, a Dane and a medical doctor, was responsible for the health of the members of the expedition and was to assist Forsskal in his inquiries into natural history. In addition, it was hoped that his medical knowledge would be a means by which the expedition would gain the trust of the Arabs. Little remains as a memorial to Kramer. Herr Georg Wilhelm Baurenfeind, the German artist of the expedition, was to attach himself like a limpet to Forsskal and sketch the specimens collected by the other man. In addition, he was to record views of the inhabitants and plans and elevations of the cities. Niebuhr tells us that he was primarily an engraver in copper, and we know that he won the grand prize in engraving at the Academy of Arts in Copenhagen in 1759 with a work entitled, appropriately, "Moses in the Burning Bush." Herr Baurenfeind was a diligent worker, kept busy while he lived by his sketches of scenes, dress, musical instruments, hats etc., as well as fulfilling the constant requirement for recording Professor Forsskal's collections. We find in Baurenfeind's sketches of people a characteristic eighteenth-century inability to see natives as anything but

---

14. See Hansen, *Arabia Felix*, 24–29.

differently dressed Europeans, much like the earliest sketches of American Indians. The Niebuhr and Forsskal books contain many copper engravings of his sketches and it is ironic that it was left to others to carry out the work for which he would have been so well suited.

Finally, Carsten Niebuhr, mathematician and surveyor to the expedition, was primarily responsible for maps of the areas visited. No professor or doctor, and only recently a university graduate, Niebuhr had been pressed to take a title suitable to his position in the expedition. He had reluctantly chosen the simple "lieutenant des ingénieurs," too diffident to take a title more grandiose. For all of his intellectual quickness and tenacity, Niebuhr struck observers as timid and cautious, as might be expected of a man of his humble origins. He spent the time after his appointment acquiring the tools he would need to carry out his duties as cartographer. He studied with Professor Johann Tobias Mayer, the most eminent German astronomer of the eighteenth century, under whose tutelage he mastered the techniques of position finding through celestial sightings, so necessary to his mapmaking. Niebuhr tells us in the introduction to the *Travels* that Mayer himself calibrated the quadrant with a radius of four feet that was to play such a large part in Niebuhr's work on land. Niebuhr also studied Arabic with Professor Michaelis. While he was not a skilled linguist in the academic sense, and soon became impatient with the complicated grammar of classical Arabic, the preparation proved to be of great value. The later estrangement of Niebuhr and Michaelis complicated Niebuhr's work of recording the results of the expedition. There is even the suspicion that Niebuhr's impatience with these early grammatical studies was a sin, however venial, for which Michaelis never forgave him. The two men represent opposite poles of "Orientalism," the one a man of prodigious erudition but little practical knowledge of the subjects with which he dealt, and the other a man whose knowledge was nothing if not practical. As we will see, Michaelis would dissociate himself completely from Niebuhr's later work of publication.

While the members of the party were selected for their particular areas of expertise, there was some degree of overlap. Forsskal, for example, was also a linguist and philologist and had studied both subjects under Professor Michaelis. And, as we have seen, in addition to tutoring in position-finding Niebuhr had also taken lessons in Arabic from Michaelis. It was fortunate for the world of scholarship that Niebuhr concerned himself with more than his astronomical sightings. In fact, all the members of the expedition were instructed to keep journals recording everything they saw and heard, and the several accounts would be compared in Copenhagen after their return. This five-sided view would be more conducive to the emergence of the truth than the testimony of a single man, however eminent. But the two professors, von Haven and Forsskal, were clearly the most important members of the party, two eminences around whom the others orbited like lesser bodies. Both men rather despised the others, and they intensely disliked one another. This would have major repercussions on the work of the expedition.

The *Fragen*

Michaelis had advertised in the gazettes of Europe that the expedition would accept questions from scholars prior to its departure. He was aware that the expedition could make a contribution to secular as well as religious knowledge. There were other vines and silkworms out there, and he dismissed the contentions of those who argued that there was nothing more Europe could learn from Asia. Only a few questions had been received by the time the party left Copenhagen. As they came in, however, they were bound with a set of instructions for the members of the expedition and later published as *Fragen an eine Gesellschaft Gelehrter Männer, die auf Befehl ihro Majestät des Königes von Dannemark nach Arabien reisen*, or "Questions directed to a Company of Scholars on a Journey to Arabia by Order of His Majesty the King of Denmark" (hereafter called the *Fragen*). The *Fragen* is the key to understanding the purposes of the expedition. In the Preface, Michaelis gives a brief history of the conception and dispatch of the expedition, with suitable blandishments to his Danish Majesty and the statement that if he had been aware of the interest of this sovereign in the progress of science, he would have been even more bold in his requests. It was of considerable satisfaction to Michaelis that the original, single member of the expedition, the philologist von Haven, was a Dane. As the expedition grew in Michaelis's mind, His Majesty graciously consented to the inclusion of a naturalist and a mathematician. Finally, a physician and an artist were added.

The Preface continues with some very sensible advice on the subject of travel. The first thing travelers need is a rudimentary knowledge of the language of the lands through which they will travel, as learned through a study of the principles of grammar. Michaelis very wisely deprecates the services of translators, who often act as filters, screening out what they judge to be superfluous or unnecessary, often telling the foreigner only what he thinks the foreigner wants to hear. And a critical need in the recording of geographical names is their original language, often transformed beyond recognition as they pass from one language, and one script, to another. European travelers

> write these things in Latin characters, and in as many variants of spelling as there are persons to write them. These variations will be influenced by the dialect or gross pronunciation of the Arab

> who utters them as much by the European who hears them . . . It will not be easy to conform the reports of a German traveler, for example, with those of an Englishman.[15]

Science, to whose purposes the expedition was ultimately devoted, required a common language among its practitioners to be truly useful. In Europe, that common language for centuries had been Latin. But travel to lands outside Europe presented problems of a different kind, where place names consisted of often wholly-unknown sounds in several Oriental dialects, recorded through the filter of several European languages. As we will see, Niebuhr provides his listing of villages in the Egyptian Delta in German characters and orthography, while Forsskal's list of the same villages appears in Swedish. In the end, Michaelis even understood that a grammatical knowledge of Arabic acquired in Europe would be insufficient to the purposes of the travelers. Familiarity with the colloquial languages of the areas through which the expedition passed would, alone, yield satisfactory results. But, if anything, he understated the difficulty of language, if only because there often did not exist standard spellings in Arabic in the same country, much less from one country to the next.[16]

A second problem was that travelers were inundated with impressions and did not know what to look for. With his customary thoroughness, Michaelis supplies this want. Instead of abandoning the travelers to their own curiosity, he will prepare questions "indicating to them matters about which we desire clarification."[17] Here, Michaelis attempts unequivocally to put his stamp on the expedition:

> They see an infinite number of objects, without paying them the proper attention; but they would pause if they knew precisely which were the objects about which Europe wanted to know, and which of their contributions might serve to lift the obscurity. So to render their travels truly useful, questions should be proposed for them: without guidance, they may be able to make good observations, but only of things that have already been seen by other travelers without enlightening us at all . . . The traveler passes only a week or a month in these places to conduct his research: in so short a time, and amidst so many distractions, he will not have the advantage of all the resources

---

15. *Fragen,* Preface, ii. My translation.

16. Even in Cairo in the twenty-first century, street names are sometimes spelled differently, the "s" sometimes with a "saad" sometimes with a "seen." To take the difficulty a step farther, the lands where Arabic was spoken were not broadly literate in the eighteenth century.

17. *Fragen,* Preface, IV.

> that would facilitate the enterprise, resources which a European scholar has only in the leisure of his study. He, surrounded by a compendious library, can collect ten facts, the eleventh of which is missing to properly make his discovery; this eleventh is seen by the traveler, but he has not been instructed to look for it . . . *But if a European scholar takes the time to furnish the traveler with a list of his questions, complete and detailed, then he will be able to provide what other travelers cannot.*[18]

As mentioned above, Michaelis had advertised in the academic gazettes of Europe for questions other scholars would pose to the travelers. But behind the tepid expression of gratitude in the Preface, Michaelis is unable to conceal his disappointment in the response:

> Although in truth I expected more assistance of this kind than I have received, I must also say that among the works which were sent me, were found some of the greatest importance. These had nothing in common with the questions proposed in this Work, but the originals have been sent to the travelers, so that nothing would be lost in passing through the hands of a translator.[19]

So the bulk of the questions are from Michaelis himself, although they include the contributions of friends and scholars who constituted a kind of scientific society in Göttingen. The questions number only 100 in the table of contents, but the time they are elaborated in the body of the text, they represent nearly 1,000 specific queries. Michaelis is under no illusion that all will be answered satisfactorily. But the more the travelers are aware of what is unknown, the more attentive they will be, and the better able to filter through the multifarious sensations that will assault their senses. Nearly all the questions relate to the explication of Sacred Scripture. Michaelis makes no apology for the theological nature of the questions:

> I must state that I believe that all of them are important, which will explain a book on which our religion is founded . . . In a word, as long as we occupy ourselves with a knowledge of *this most*

---

18. Ibid., IV. My italics.

19. They included queries from Mr. von Helm, Counselor of the Chancery of Oldenberg, Mr. Pastor Pagendarm, Mr. Rust, Keeper of the Registers of Baerenbourg, Mr. de Navarre, living in Amsterdam, Mr. Doctor Thiery of Paris, and Mr. Doctor John Collet of London. Finally, the Academie des Belles Lettres in Paris sent "an excellent memoire, full of sagacious questions regarding the History, Geography, and the language of the Arabs." It also included a list of the kings of the Yemen dating from 1817 BC to the end of the reign of the Jewish Dhu Nawas in 502 AD when the country seemed to pass out of view. See *Fragen*, Preface, IX.

> *ancient of books*,[20] we must at the same time pursue the study of Natural History and the customs of the Orient, matters of which one would not be able to dream if the cause had not been furnished by so memorable a monument of oriental antiquity. I can name no other book, at least any whose subject is moral, which is capable of rendering these services to the sciences.[21]

Michaelis's intellectual frame of reference is, again, clear for all to see: biblical and scientific truth are one and cannot be contradictory; textual difficulties with the Hebrew Bible derive from an imperfect knowledge of the language of those who wrote it; the Bible itself is the "most ancient of books," capable of rendering signal service to science; and Happy Arabia will be the laboratory in which this service is rendered. It is not being wise after the fact to recognize the limitations of this framework. If nothing else, it illustrates the limitations that religious preconceptions placed on the study of the Orient.

As we will see below, European scholars were just beginning to make the discoveries that would lead to the deciphering of the hieroglyphs of ancient Egypt, a body of writing two millennia older that the Hebrew Bible. But, even with these advances, serious scholars were still attempting to view the history of Egypt through the prism of Hebrew Scripture. It is not too much to say that another hundred years would pass before Europe finally freed itself from the baleful influence of the Bible as a purportedly historical document. It probably *is* being wise after the fact to find fault with Michaelis for his lack of curiosity about the writing of one of the most ancient of the Oriental peoples, the Egyptians. But appearing in the Judeo-Christian tradition only as accessories to the great biblical drama, they apparently did not matter.

The flavor of the questions in the *Fragen* can be appreciated by an extract from Question XXXIII, *Of certain Insects that are commonly mistaken for Locusts*:

> It is commonly believed that Joel (ch. I, 4) listed four species of Locusts. From the little that I know of them, it appears incredible to me that one species of Locust would hatch later than another; nevertheless, it is this which must be admitted if the Prophet speaks of four different species, since the following must consume what the preceding swarm has left, and that the Prophet would have the supernatural, or better, the unprecedented singularity of the scourge of Judea consist in the appearance of four successive destructive swarms in the country, where ordinarily they would

---

20. My italics. The Hebrew Bible is in fact a comparatively recent work, having been codified in its final Hebrew form after the beginning of the Christian era.

21. *Fragen*, Preface, XV.

> not be seen in the same year ... The ילק must also feed on what the locust has left behind. The Travelers will also find this word in Nahum, III, 15:16; Jerem. LI, 14:27; and Ps. CV 34 ... Will all this be confirmed or refuted by the Arabic Language?[22]

The 100 questions themselves were devoted to subjects as varied as the tidal movements in northern reaches of the Red Sea, leprosy, the medicinal use of dill, the various sorts of Arab *manna*, the occurrence of gold and other precious metals in Arabia, the usefulness of circumcision, marriage customs, the venom of serpents, the maladies of wheat, and unclean animals and birds. A single thread, however, connected them: mention in the Hebrew Bible, or relation to an account in the Bible, where a better understanding would explicate the Sacred text.[23]

Attached to these questions were a set of instructions to the travelers, incorporated in forty-three articles. They were drafted by Michaelis[24] for the King's signature. They were an attempt, with his customary thoroughness, to specify in detail the responsibilities of the expedition in general and of the individual members in particular. We have already seen the outlines of the individual responsibilities. There were also admonitions to the party in general:

> The above mentioned travelers shall proceed together to Happy Arabia, never losing from their view the purpose which we have so graciously proposed, that is to say, to advance the Sciences and Letters; they will make in that country as many discoveries as to them shall be possible.[25]

They were to keep the goal of Happy Arabia ever before them and not be distracted by other matters, however attractive they might seem. As we will see, they were diverted from their goal at an early stage, and spent a year in Egypt. It would be the longest period of time they were together as a group in the Orient. From Egypt, they should proceed to *Gebel el Mokatab*, or the written Mountain, in the Sinai peninsula (Article Three) to examine what Michaelis suspected were the original Hebrew characters, scratched on the rocks by the Israelites during their sojourn in the wilderness.

---

22. Ibid., 64. For a list of the questions see Appendix A.

23. A second series of questions, which continued to trickle in after the departure of the expedition, were later bound and sent, in several sets by three different routes, to the expedition *en route*. Niebuhr tells us that the first set only reached him in Bombay, after the deaths of the two professors for whom they were primarily intended.

24. They also included contributions by a number of professors at the University of Copenhagen, reproduced as *Beylage I–III* in Michaelis's *Literarischer Briefwechsel*.

25. *Fragen*, Article 1, XX.

Article Four was an exhortation to learn the Arabic language and Article Five urged that they "penetrate into the interior of Arabia as soon as possible," and not content themselves with the coasts. It also included the most impractical of the instructions: they should decide the places they would visit by "a plurality of votes." As we will see, this democracy of professional academics was the first casualty of the journey, and the bitter feud between two camps threatened the progress of the expedition. Article Six elaborated on this "perfect equality among the Scholars:" the "maintenance of peace and harmony shall be their constant and principal study." Unfortunately, their constant preoccupation became the protection of themselves against threats, real or imagined, from within the expedition itself.

Article Seven required that each traveler keep a journal, entrusting nothing to memory. These independent journals would then be reviewed in Europe so that a balanced view of their reports could be achieved. This article ensured that the results of the expedition would be preserved for posterity and the *Tagebuch* of Niebuhr was the basis of his *Travels*. But his were not the only journals that have survived: von Haven's diary is kept in the Manuscript Department of the Royal Library in Copenhagen and was recently published.[26] That of Forsskal was published in 1950.[27] Copies of the journals were to be sent periodically to Europe, addressed to the Privy Counselor von Moltke, as a kind of back-up, to ensure that nothing would be lost along the way.

Article Ten enjoined the travelers to have "the greatest regard for the inhabitants of Arabia," with particular respect for their religion. The language of the instruction expresses some old—and some not so old—fears and suspicions that attended travel of Europeans in this part of the world. The travelers should

> abstain from everything which would cause these people the least discomfort: they shall use an extreme delicacy in researches which can demonstrate the ignorance of the Mohammedans, in making them suspect that their treasures will be disinterred, to serve the art of magic, or to suggest the activity of a spy... They shall carry on this research in a manner the least perceptible,

---

26. Von Haven left a handwritten diary of almost 400 pages, written in his "fluid and animated pen," as well as a small diary covering the period of the journey to Mt. Sinai, published by Michaelis in the *Literarischer Briefweschel*. The former is in the Manuscript Department of the Royal Library in Copenhagen and was recently published as *Frederik Christian von Havens Rejesjournal fra Den Arabiske Rejse 1760–1763*, edited by Anne Haslund Hansen and Stig T. Rasmussen. In addition, in the Danish state archives, there remain records of the correspondence between the members of the expedition, Herr von Gahler (the Danish Consul in Istanbul) and Count Bernstorff. These sources are quoted extensively in the Hansen and Rasmussen books.

27. See Forsskal, *Resa till Lycklige Arabien. Petrus Forsskal's Dagbuk 1761–1763*.

and give to the people an impression the most favorable. They shall be careful to guard against exciting the jealousy and cruel vengeance of the Arabs, either by amorous intrigues, or by taking with women the kinds of liberties tolerated in Europe.[28]

This delicacy would particularly be put to the test in Egypt, still actively hostile to Europeans.

Articles Eleven through Fifteen dealt with the logistics of the purchase and disposition of manuscripts and the copies of inscriptions, sketches, charts, and plans they would collect. They should all be sent to von Moltke in Copenhagen. Like the efforts of all consultants, their work would belong to the client, in this case the King of Denmark. This is followed by individual instructions to Forsskal (Articles Sixteen through Twenty-two) regarding the "natural productions" of Arabia, including their collection, a determination of their Arabic names, and their dispatch to Copenhagen. Where it was not possible to collect specimens themselves, he was to ensure that they were drawn, if possible in color. Articles Twenty-three through Twenty-six were addressed to "Physician Kramer" specifying that he should act as an assistant to Forsskal—and this assistance should be graciously requested by Forsskal, not demanded!—as well as to use his medical knowledge to secure the favor of the Orientals among whom they traveled. He was not the last to travel in the Orient exercising the honorable profession of a healer.

To the mathematician Niebuhr (Articles Twenty-seven through Thirty-four) was entrusted responsibility for geography, especially latitudes and longitudes of the places they would visit. Without prejudice to the principal object of the journey, he should place himself in a favorable position to observe the transit of Venus on June 6th, 1761. Niebuhr should also use his mathematical knowledge to guard against "the sophistry which has deluded so many travelers who were not Mathematicians."[29] This last admonition was issued with respect to a delicate mathematical task that lay beyond his geographical duties: a study of the increase in population, with particular regard to the effect of polygamy on populations in the Orient. This pre-Malthusian[30] interest in populations was in some respects mathematical, but it veered into the highly-charged areas of sex, reproduction, and marriage, against too great an interest in which they were all simultaneously warned. Among the most common of human impulses was to find fault with other people for their differences in these sensitive matters and Niebuhr's research would have to be very careful indeed.

28. *Fragen*, Preface, XXV.
29. Ibid., XXXVII.
30. Malthus was born in 1766.

Professor von Haven was the addressee of articles Thirty-five through Forty-two. He should apply himself especially to those philological subjects which would "serve to clarify points from Sacred Scripture and the Law of Moses."[31] He should also devote himself to the study of Arabic, to the various dialects they would find, and to the differences between the common dialects and the grammatical rules of the language taught in Europe. His was also the primary responsibility for the purchase of manuscripts, dictionaries, evidence of the paleography of the Arabs, and copies of inscriptions they should find along the way. In one of the more intriguing entries Michaelis refers to the "mysterious writings" of the Sabaeans of the Yemen, about which Kaempfer[32] had provided recent enlightenment. It was one of the few evidences of Michaelis's interest in languages other than the purported cognates of Hebrew. And finally, in article Forty-three, we learn the responsibilities of the artist Baurenfeind. It was to assist Forsskal and von Haven to make copies of the things, whether living organism or example of scripts, that would be of interest to the scholars of Europe.

The Instructions were sealed with the hand of the King in the Castle of Jaegersbourg on the 15th of December, 1760. We learn in Article Twenty that His Danish Majesty had instructed that the expedition should expect the assistance of the Danish East Indies Company and the captains of His Majesty's ships. This part of the instructions would be put to immediate test when the five, with their servant and baggage, embarked early in the next month on His Majesty's warship *Grönland* for the voyage to Constantinople.

---

31. *Fragen*, p. XXXVIII.
32. See Kaempfer, *Amoenitatum exoticarum*.

3

To the Orient

> I have been careful in all my travels to take astronomical sightings not only at noon but also during the night, by means of a good Hadley's octant ... Many geographical and marine charts could be corrected if the positions of the principal features were known to within one or two minutes, and I believe that I have been successful in establishing nearly all my sights with at least this degree of precision. (Niebuhr, *Travels,* vol. I, 2)

THE SHIP SET SAIL on January 4, 1761. The highest levels in the Danish government had been involved in the selection of the members of the expedition and it was said that the King himself had taken a personal interest in the preparations. The results were eagerly awaited in scholarly circles throughout Europe. But there were deep rifts within the group, the residue of the haggling over the selection process and a still-simmering conflict over authority.[1] The *Fragen* decreed, with more wishful thinking than good sense, that the five members would be equals and that decisions would be taken democratically. Human nature, even in the absence of the difficult personalities involved, would have quickly made this little more than a pious hope. Von Haven was imperious in his demands, could not endure life aboard ship, and he quickly received permission to disembark and proceed overland to Marseilles. He felt that he, as a Dane, should be the leader of the expedition. Leaving aside the question of nationality, Forsskål's claim was probably greater, given the wider range of his interests and training.

Surprisingly—or perhaps not surprisingly, given the clash of the two professors—Niebuhr was made the expedition's treasurer, to the disgust of von Haven who coveted the position for himself. One biographer[2] states that Niebuhr had already acquired the instruments necessary for the journey out

---

1. See Hansen, *Arabia Felix.*
2. See *Biogrpahie Universelle Moderne,* vol 30.

of the stipend provided by the Danish government and was surprised when Bernstorff offered to reimburse him on his arrival in Copenhagen. Bernstorff was equally astonished at such disinterestedness and decided on the spot that Niebuhr was the man to manage the expedition's funds. The account of the incident ends laconically with the observation that "never was such a confidence better placed." Niebuhr, whose account alone was published in his lifetime, ignores these details, we suspect out of his lack of interest in matters of precedence, or embarrassment at the unseemly nature of the conflict. This kind of thing simply did not belong in print. Besides, he had work to do.

Their track from Copenhagen would take them slightly east of north in the Kattegat, the narrow waters between Denmark and Sweden, before turning southwest into the Skagerrak that exited into the open sea. But the attempt to clear the Kattegat in the dead of winter proved abortive, again and again. They were at nearly fifty-seven degrees north latitude and there were squalls and snowfalls. Only after a two-month agony, involving much heavy weather, backtracking, and loss of life among the ship's company, was the *Grönland* able to reach the North Sea. Life aboard ship was not easy in the eighteenth century, particularly in the winter in the northern latitudes. But Niebuhr seemed to have been born to a life at sea:

> I never felt the least effect of sea sickness, properly so-called, even during the most violent of storms. For the rest, I placed myself entirely in the hands of the Most-High Providence: and as I was, moreover, confident of the skill of our officers and men, I slept tranquilly through every storm, however uncomfortable the wind, rain, and cold, while they saw to the safety of the ship.[3]

He spent the time taking celestial sightings when the weather permitted, comparing his positions with those of the navigator, and taking interest in nearly everything that went on aboard ship. He clearly felt that the sooner he understood navigation at sea, the better prepared he would be for his cartographic duties on land. An incident early in the voyage showed Niebuhr's mettle and was a harbinger of things to come. In examining the log-book, he noticed that in the hour-glass, half-minutes were calibrated at only twenty-nine seconds, before learning that it was a method used by the pilot to correct for a lag in marking the line when the ship was moving rapidly. This kind of attention to detail was technically beyond his brief—he had no responsibility for navigation of the ship. But over the next seven years we will see it again and again, and come to appreciate it as a characteristic of the man. He was a serial overachiever and, as much as any other trait, this is responsible for the repute that Niebuhr earned among those who followed him.

3. *Travels*, vol. I, 5.

Niebuhr in Egypt

The problem of latitude had already been solved, and for centuries navigators had taken measurements of celestial bodies—the pole star, the sun, or some other body—above the horizon to assist in the determination of a ship's position. The instruments could be as crude as the latitude hook or the *kamal* (an early Arab device) or as sophisticated as the quadrants, sextants, and octants with which Europe was now equipping its navigators. In addition to an instrument, the method required an almanac[4], which listed the positions of celestial bodies as a function of time and the position of the observer on the earth's surface. Niebuhr probably carried the *Connaissance des Temps*, published since 1678–79 by the Paris Observatory. Early latitude sailing called for the mariner to sail to the appropriate latitude, a relatively straightforward calculation, before turning north or south to reach the desired destination. The voyages through the North Sea and, particularly, the Red Sea, were primarily transits in a north-south direction, suitable for this kind of direction finding. While he does not generally burden the reader with his astronomical sightings, Niebuhr lists them all, to within a second of accuracy, in an appendix ("Niebuhr's astronmische Beobachtungen") to vol. III of the *Travels*. The precision of his sightings was such that, a century later, positions in the Greek archipelago were still plotted using Niebuhr's figures.

The Longitude Problem

The determination of longitude at sea represented a far more complex problem and by the time the Danish expedition set sail it had not yet been satisfactorily solved. Longitude did not involve a straightforward relationship of an observer's north-south position to that of a celestial body, and it was a research matter of considerable importance to Europe. In England, the Board of Longitude had been established in 1714 at the Royal Observatory at Greenwich to encourage work on the problem, and the considerable sum of 20,000 sterling pounds had been offered to the man who developed of a solution workable for mariners. A large reward had also been offered in France in 1716. The solution seemed to be close at hand in not one, but two, competing methods. The lunar distances method made use of the passage of the moon through the celestial sphere, and used this passage as a kind of celestial clock from which the position of the observer could be calculated. It was the far more elegant solution and the one favored by astronomers. Prof. Mayer was its foremost proponent. The chronometer method, on the other hand, was favored by instrument makers. It required a chronometer,

---

4. The word is simply a transliteration of the Arabic *al-manakh*, or "climate."

or clock, of sufficient precision that the local time at sea, anywhere in the world, could be related to the standard time at an accepted position along a north-south band of longitude passing around the earth's surface.[5] The method would then make use of comparison of the times at the two locations to establish the longitude of the second, or unknown, position.

The chronometer method seemed mathematically crude, and it was rather despised by astronomers. The lunar distances method, on the other hand, seemed based on sound astronomical and mathematical principles. Instead of measuring differences in time, it made use of the moon as a reference point and measured the varying distances between it and other bodies as they moved through the celestial sphere. These measurements, varying with time and the position of an observer on the surface of the earth, would then be used to calculate the position of the observer. But the method required instruments of a precision, calculations of a sophistication, and an observing platform of a stability that were impossible for the average navigator at sea. Not surprisingly, having studied with Mayer, Niebuhr was a firm believer in the lunar distances method. In spite of the fact that it represented a dead-end in the search for a longitude solution, his careful observations on the voyage from Copenhagen to Gibraltar had an immediate and very practical consequence:

> On this voyage I made various observations that I sent from Marseilles to Professor *Mayer*, who was so satisfied with them that, on his deathbed, he ordered that they be sent to England as evidence of the usefulness of the lunar tables, and the English had them printed along with the corrected tables of Mr. *Mayer*.[6]

Mayer's tables had been first sent to the British Admiralty in 1755 through the intercession of one Johann David Michaelis, whose cousin was a private secretary to George III. The Cambridge astronomer Nevil Meskelyne added the corrected tables, which Mayer's widow sent in 1763. It was on the basis of these that she was awarded 3,000 pounds of the reward set aside for those who made significant contributions to the solution of the longitude problem.[7]

---

5. The early standard had been Paris. Later, when Harrison's chronometer—developed by John Harrison (1683–1776)—pointed the way to the solution, the standard became Greenwich, a borough south of London.

6. *Travels*, vol. I, 7. Presumably, he means the lunar tables published by Meskelyne in the *British Mariners Guide*.

7. See Forbes, *Tobias Mayer (1723–62)*. Mayer died at the age of 39, worn out says Niebuhr's biographer, by "his incessant exertions in the cause of science."

To Marseilles and Malta

The *Grönland* was now in open water, but it was still winter and they were still subject to the icy blasts from the Arctic Ocean and the Norwegian Sea. In early March a contrary wind pushed them north to nearly sixty-four degrees north latitude, just below the Arctic Circle, and west to the vicinity of Iceland. Spring finally brought an improvement in the weather. But it also brought a perfect calm and, on the 31st of March they found themselves still north of sixty degrees latitude and making little headway. The calm allowed Niebuhr to observe St. Elmo's Fire, an electrical phenomenon originally attributed to the apparitions of saints. However, the overcast was still too great to permit a sighting of the aurora borealis, or northern lights. But the wind suddenly shifted on the 6th of April and they began their southward passage towards the Mediterranean. They sighted Cape St. Vincent, the most westerly point of Europe, on the 21st of April and passed through the Pillars of Hercules in early May. Baurenfiend made a sketch of Gibraltar to the north and Ceuta to the south as they passed through the straits, and it was later engraved and included as Plate I of the *Travels*.

The Mediterranean has a reputation for violence in the winter, but in the spring it was beautiful, although the calm made progress slow. Finally, on the 14th of May they dropped the anchor near St. Eustace, a league south of Marseilles. This was the first opportunity for the men to plant their feet on dry land since early January, and they were all sick to death of the cramped, smelly quarters and the poor water on the ship. While the weather may have become calm, however, their own relationships now became more tempestuous. Von Haven rejoined the party on their arrival in Marseilles and the conflict between him and Forsskal—indeed, between him and everyone else in the party—flared up anew. At a dinner given by the captain in honor of the now–reunited expedition, an exchange of insults between Forsskal and von Haven was to have serious ramifications for the future equanimity of the group.[8]

In Marseilles they visited libraries, the collections of natural history "amateurs," and an astronomical observatory. After a pleasant and productive stay of nearly three weeks, the *Grönland* again prepared to set sail. But there was another threat to their onward progress. The Seven Years War (1756–63) was at its height and, although on land it involved nearly every nation in Europe, it was largely between France and England for command of the seas.[9] The Mediterranean was a major theater of war. The French were still involved

---

8. See Hansen, *Arabia Felix*, 72.
9. Its end would put an end to French ambitions in India and the Americas.

To the Orient

in the Levant trade, although the presence of an English squadron in the Mediterranean had reduced the traffic to a fraction of its previous level. But even neutral shipping was now subject to boarding and seizure, especially if it was proceeding from a French port. In Marseilles they found three Danish merchantmen waiting for an escort and the *Grönland* was pressed into duty. On the 3rd of June the little convoy set sail for Izmir. The Danes had maintained a profitable neutrality in the war, but any ship leaving a French port was still suspect. Three times in the next week they would be challenged by English warships, but the captain of the *Grönland* resolutely refused to permit inspection of his charges, several times preparing the ship for battle. Niebuhr says that English finally retired, although grudgingly.

On the 6th of June Niebuhr made preparations to observe the transit of Venus. As we have seen, it was hoped that the Danish expedition would be in the Yemen for the sighting, where it would occupy a place in what would be worldwide coverage of the event. However, on the 6th of June 1761 they were at sea off Marseilles, still so near to other observation points in Europe that Niebuhr held out little hope for the value of his contribution. The sea was calm but a heavy shower had obscured the entry of the leading edge of the planet across the solar disc and Niebuhr could observe only the entry of the trailing edge. Nevertheless, the exercise was useful if only to his own developing skill with the quadrant, and he completed the observation:

> These observations gave me the apparent transit of the trailing edge of Venus in front of the solar disc at actual time 9 hours, 3' 53" at a latitude of 40° 6', and a little east, although not far from the meridian of Marseilles.[10]

The sighting does not appear in the list of 120 official sightings recorded in 1761.

In fact, the observations in that year generated as much heat as light. In the first place, the entry and the exit of the planet were visible only in a narrow band across the earth's surface. The rest of the sightings were either of the entry or the exit. In the absence of a standard time or reliable chronometers, the comparison of these results was immensely complicated. In addition, the observations themselves were very difficult with the optics then available. It was almost impossible to determine precisely when the planet exited or entered the edge of the solar disc. The phenomenon where a tag of the planet appeared to hover on the edge of the disc made the timing almost a matter of judgment rather than scientific precision. So, the timings differed widely, as did the ultimate calculations of the astronomical unit. They ranged from

---

10. *Travels*, vol. I,15.

as little as 82 million miles to as much as 102 million miles, bracketing the accepted mean value today of about 93 million miles. And we now know that the calculation is not the function of a single variable, but must consider other factors including precession and nutation, the lunar parallax, and the masses of the earth and moon.[11] But there were some solid results from what has been described as the first combined scientific undertaking in history. Observers had been sent to locations from Siberia to the west coast of the Americas and several of the expeditions, including those of the Frenchmen Chappe d'Autoroche in Siberia and Le Gentil de la Galassiere in the far-east, produced multi-volume compendia like those of Niebuhr.

On the 14th of June, the convoy arrived in Malta. In some respects, it was the introduction of the members of the expedition to the Orient. The houses were built in the eastern style, flat-roofed and of made of the limestone that seemed to be all that covered the rocky prominences of the main island and the little island of Gozo to the north. The Maltese spoke a language that was a blend of Italian and a residue of Arabic from the period of the Muslim domination, and the names of the towns were an odd combination of the two—Mdina, Marsa, Zebbug, Gzira. More importantly, these northern Europeans were now entering that debatable land between the worlds of Islam and Christianity where warfare had surged back and forth intermittently for a millennium. It was an area of shifting alliances, where the enemy of today might be the ally of tomorrow, and where the old certainties seemed less certain. The warfare had traditionally been between Islam and Catholic Europe—the Empire, the Pope, the Venetians, and the Genoese—and the emperor lately led it with a zeal that mirrored his prosecution of the counter-Reformation.

But shifting alliances and competition among the parties for trade with the East complicated this latter-day crusade and their willingness to cut separate deals with the Turks diluted the power of Christianity. They had briefly come together in 1570 at Lepanto and inflicted an historic naval defeat on the Turks. But centrifugal forces ever drew them apart. The Venetians and the Genoese vied for trading privileges with the Turks, and the Emperor, the Spanish, the Kingdom of Naples, and the Pope all pursued their independent interests. The Protestants had traditionally been on the sidelines in this great, ongoing clash of civilizations, taking advantage when it presented itself. They were considered by the Catholics to be little better than heathens themselves. And the French, as always, pursued their commercial interests. They may have been nominally Catholic but the Levant trade was too important to sacrifice to the vagaries of religious contention.

---

11. See Woolf, *The Transits of Venus*, 195.

To the Orient

The wealth of Malta was largely booty acquired during centuries of warfare with the Turks. And the Maltese, located squarely in the middle of this maritime cockpit, may have been Christians, but the Grand Masters of the Order of the Knights of St. John owed their allegiance to the Pope in Rome. To the northern European Protestants of the expedition, he might as well have been just another Oriental potentate. There is a hint of scandal and condescension in Niebuhr's remarks on the wealth of the Cathedral of St. John in Valetta:

> In short, from the description I was given, the riches of the *Kaba* of *Mecca* hardly equal those of this church, and they may even surpass those of Mohammed's tomb in *Medina*.[12]

The island had been taken by the Arabs in the ninth century, before being freed by the Norman Roger Guiscard in 1091. But its fame rested on the celebrated sieges in 1565 and again in 1614, both of which ended disastrously for the Turks. To this day a common saying in Arabic refers to *al-uzun fi Malta*, or when "the call to prayer sounds in Malta," expressing the Muslim idea of its formidable reputation. It is the equivalent of our "when hell freezes over."

The island was still a fortress in 1761, and well it might have been. Low-level naval warfare still simmered between Christians and Muslims in the Mediterranean, and the presence in the harbor of a captured Turkish warship was an introduction to the odd relationship that had prevailed for centuries between the two civilizations in this inland sea. Niebuhr tells us that rebellious Christian slaves had captured the ship near the island of Stanchio in September of 1760 and had sailed it to the vicinity of Malta. Slaves were still being used as a kind of currency in the Mediterranean in the middle of the eighteenth century, being taken and ransomed in large numbers by both sides. These men had apparently cut their own deal: they offered the ship to the Maltese authorities provided they were able to divide the booty among themselves. The deal having being struck, Niebuhr dryly remarks, the wisest returned to their own countries with their share of the spoils while the others were still in Malta, "engaged in dissipation." The ship was built in the Turkish fashion, with small galleys amidships where the officers could recline, "for the Turks like to walk as little aboard ship as on land."[13] Niebuhr later learned that the French had purchased the ship and returned it to the Sultan as a gift, an act unthinkable to this northerner, even for the French of the Levant. But the

---

12. *Travels*, vol. I, 16.
13. Ibid., 17.

Sultan's good graces were the key to the Levant trade and here, as was often the case, commerce took precedence over religion.

The Maltese had three warships manned by galley slaves chained to their oars. They were part common criminals and part Muslims captured from "Barbary"—Algeria, Tunisia, and Tripolitania. Technologically, this was a different world from that of northern Europe with its Atlantic outlook. Nearly 300 years after the Iberians had rounded the Cape in sailing ships, oared galleys were still in use in the Mediterranean.[14] A treaty between the sultan and the king of Naples now prevented the Maltese from entering the Greek archipelago. But flags of convenience were used by privateers and the official ban was widely flouted. Niebuhr suspects that these Maltese were Christians "whose desire for gain was greater than their zeal for Christianity."[15] And he saw, for the first time, the other side of the coin: the Maltese were blamed by the Turks for the same kinds of depredations that the Barbary pirates committed against the Europeans. Early in the next century the young American republic would be drawn into the conflict, and the action against the Barbary pirates (1801–5) would be its first foreign war.

There were other sights in Malta and Niebuhr and Forsskal visited the salt-pans, the country houses of the Grand Master, a statue of count Roger, and a hill where the apostle Paul was supposed to have preached. Von Haven was not idle and began his purchases of manuscripts in Malta. He was one beneficiary of the naval action described above. His diary for June 18th, 1761 includes the following entry:

> At noon I had lunch on board . . . and was able to purchase two Arabic books. One was a Turkish history from 1704 to 1728, printed in Constantinople in 1740 . . . the other was a Koran, beautifully written. Both were taken from the large aforementioned Turkish warship. I bought the Koran for a Venetian sequin, or 5 Maltese scudi.[16]

To Constantinople

They set sail from Malta on the 20th of June and entered the Greek archipelago on the 26th. The islands appeared to Niebuhr to be considerably north

---

14. For a discussion of the unique space occupied by the Mediterranean, see Braudel's *The Mediterranean and the Mediterranean World in the Age of Philip II*.

15. *Travels*, vol. I, 18.

16. From von Haven's diary, quoted in *Den Arabiske Rejse 1761–1767*.

of the positions as indicated on the chart of d'Anville[17] published in 1756. He lists his own observations in the *Travels* in an attempt to correct the errors: there were latitudes of the islands of Sapienza, Serigotto, and Ovo, and of Cape St. Angelo. But the position- finding soon came to an end with an attack of dysentery that nearly cost him his life:

> I wish I had been in a condition to continue my observations in these waters. But so violent a flux of blood overcame me that I soon lost all hope of seeing Constantinople, much less Arabia. However, thanks to Providence . . . I lacked here nothing in the way of comforts and conveniences.[18]

On the 3rd of July, the ship reached Izmir. Niebuhr was prostrate throughout the week's stay, and was scarcely able to raise himself from his bunk to catch a glimpse of the famous merchant city through the little window in his cabin. They left on the 10th and proceeded to the vicinity of the island of Tenedos, where they would transfer to a Turkish vessel for the journey through the Dardanelles to Constantinople. While they waited, they were exposed for the first time to the Orientals they had come to see. A Turkish gentleman and his suite waited with them:

> Their language, their dress, and their whole bearing appeared to us so strange that we conceived little hope of finding very much agreeable among the Orientals.[19]

They arrived in the old capital of the Eastern Empire, its name long-since Turkified to Istanbul, on the 30th of July. There, they were welcomed by his Danish Majesty's Ambassador Extraordinaire to the Porte, Sigismund von Gahler, and settled down for a stay of nearly six weeks. Their education in the manners and customs of the Orient was about to begin.

It was also an opportunity for the authorities to review the progress of the expedition since it had left Copenhagen, nearly seven months before.

---

17. Jean-Baptiste Bourguignon d'Anville (1697–1782), Royal Geographer and a giant among early cartographers. His particular interest was in the geography of antiquity, particularly in Egypt. His *Memoires sur l'Egypte, Ancienne et Moderne, suivi d'une description du golfe arabe* was one of his few written works. The bulk of his immense output was of maps. See *Notice des Ouvrages d'Anville*. The quality of his work was attested by the scholars who accompanied Bonaparte, who were astonished by the accuracy of the maps made by a man who never visited the country. He may have had assistance from Niebuhr: a map entitled *Carte Physique et Politique de l'Egypte*, bearing his name and published posthumously in Paris in 1796, bears a striking resemblance to the painstaking map of the Delta prepared by Niebuhr and published as Plate X in vol. I of the *Travels*. See Chapter 10 below.

18. *Travels*, vol. I, 21.

19. Ibid, 22.

Niebuhr in Egypt

Von Gahler was not pleased, nor was his Majesty, when he learned of the unseemly state of affairs between von Haven and Forsskal. There were serious matters involved here, including the expenditure of royal funds and the reputation of the crown in a widely advertised scientific undertaking. Von Gahler was not about to see Danish treasure and reputation forfeited to the petty animosities of two men. So the ambassador brought them together, and while he took note of their grievances, he insisted that the two reconcile their differences in the greater interests of the expedition. But it was clear to the other members of the expedition, if not to von Gahler, that the state of relations between von Haven and Forsskal had reached a point of no return. The reconciliation was cosmetic and the feud, now decidedly not petty, would soon manifest itself in a more deadly form.

Meanwhile, there was much to do in the city: bookshops for von Haven to visit, botanical trips for Forsskal in the surrounding areas, provisions to lay in for the journey that lay ahead of them, and personal matters to discuss with the ambassador. Von Haven found an intermediary to assist in the purchase of manuscripts, one Mr. Barukh. This man at first brought collections to the Dane and charged his own fee. Von Haven bought several manuscripts and arranged for them to be sent to the Royal Library in Copenhagen. Later, he decided to dispense with the middleman and plunge into the *suq* himself. It was an introduction to the difficulties of religion and culture that would plague them throughout their stay in the East. He found the street of the booksellers in the *Bezestan* and dealt with the dealers directly, although with Barukh along as advisor and a janissary[20] along for protection. However, the effort ended in a brawl, a crowd having gathered and objected to the fact that an unbeliever was buying Arabic manuscripts. Von Haven escaped intact, although, as he informs us, his hat was trodden underfoot. He eventually bought thirty-four manuscripts in Istanbul, but there clearly was Muslim sensitivity to one of his primary duties.

There were already many descriptions of Istanbul available in Europe, and Niebuhr was still recovering from his bout with dysentery. But, as his health returned, he took his first tentative steps in the exploration an Oriental city. For some explorers to be first is everything and to be second nothing. But, throughout his travels Niebuhr makes clear that this was not his philosophy. He rarely claimed precedence, but suggested that if each traveler contributed his own observations to a steadily growing body of knowledge, the progress of science would be advanced. He speculated on the differences that made it difficult to compare Oriental with European cities: in general,

---

20. In Turkish, *yeni-cheri*, "new troops" members of the permanent military corps of the Ottoman Empire until its abolition in 1826. We will see janissaries later in Egypt, as a kind of military-commercial class trading with Suez and the Gulf.

they were more spread out in the Orient, with lower buildings and much more open space. Consequently, they contained fewer people in the same area than in a European city of the same size. In addition, the narrow streets, the number of transients in the *suqs,* or commercial districts, and the lack of reliable census figures, made it difficult to estimate their populations. There already were maps of Istanbul, but Niebuhr found them unsatisfactory. So he made his own,[21] using the instruments that would be his standbys in the years ahead: a small pocket compass and his own two feet. Both had to be used with circumspection. A European who took too great an interest in the details of an Oriental city was immediately suspect, if for no other reason than that maps had military uses. This suspicion was not out of place. As we have seen, the relations between Ottoman Turkey and Europe were still strained and it would be years before a military solution to the "eastern question" was achieved by Europe. From the walls of Istanbul itself to the fortifications in the Dardanelles, to the islands of the archipelago, it was clear to Niebuhr that the military infrastructure of the Ottomans had fallen into disrepair. The walls could no longer withstand a siege by a modern enemy, and the cannons, their carriages rotted and their bores fouled with debris, could no longer be fired safely. The Sultan's fleet lay anchored in line near the arsenal and although it was "prettily painted" it was in a very poor state of repair. Niebuhr was not the first European—nor would he be the last—to comment on the lack of military readiness of the Turks. Increasingly, they would be armed only with their arrogance against the growing challenge of the Europeans.

We suspect that, in his mapmaking, Niebuhr used the compass only to establish general headings, and only when he was not being watched. He probably paced off distances, using as a counter the *subha,* or Islamic rosary, to keep track of the hundreds. That left the problem of notes and transcription of the notes onto the finished map. We can imagine Niebuhr pacing the enceinte, the old Byzantine walls to the west of the city, before crossing over the Golden Horn to Galata. From Galata he would have crossed by ferry to Scudar on the Asiatic shore. Once the general outlines of the city were established, he could then fill in the spaces with the principal buildings. It was a technique he would use over and over again in the next six years, from Cairo to Bombay, Diarbakr, Jerusalem, and Brusa. Much of the finished product remained a matter of speculation, since residential districts were generally closed and the residents were suspicious of outsiders, especially outsiders wearing tight pants.

21. The map that appears as Plate III of vol. I of the *Travels* was probably the product of two visits to the city, this early one and another on the return journey in February of 1767.

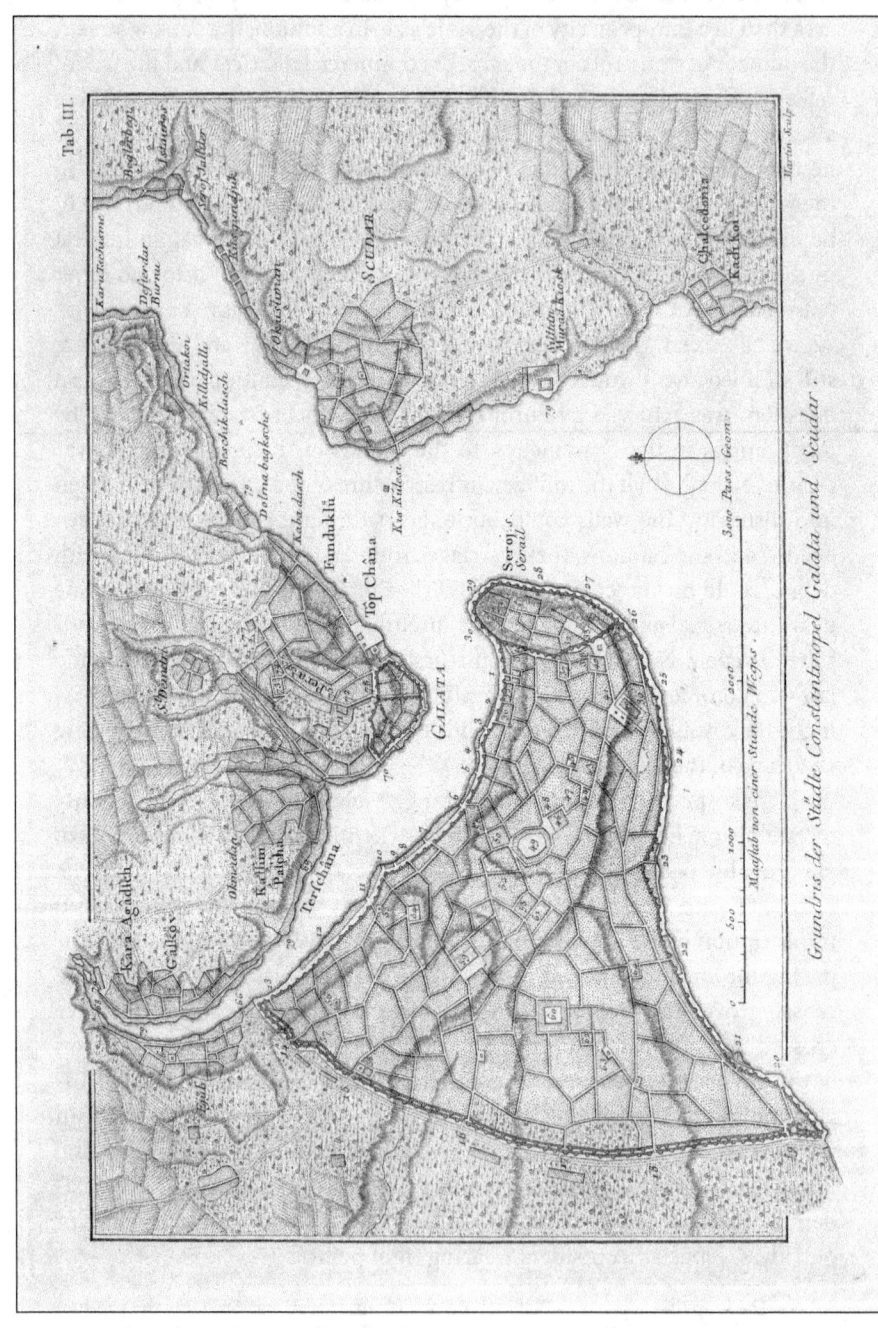

Nonetheless, the map of Istanbul (Niebuhr calls it Constantinople throughout) that appears as Plate III in vol. I of the *Travels* is tolerably detailed, with seventy-nine specific notations, including the major districts, mosques, palaces, embassies, and churches. Many of the locations are listed in Turkish in the original Arabic script, of great use to the student of the old language.[22]

The major landmarks were similar to those he would later list in Cairo where he prepared the first detailed map of that city by anyone, European or otherwise. The sixth-century church of St. Sophia, built by the Emperor Justinian and turned into a mosque after the fall of the city, served as the model for other mosques of similar size. Many of them had been built by dowager sultanas, unmarried members of the royal family, and still others were converted Christian churches. The city was full of these and other public buildings, including baths, *sabils* where water was distributed freely to the populace, public hostelries or *karwanserays*,[23] and the covered market districts where crafts and trades of a similar nature were gathered together in a single location. The arrangement of the commercial cities in the Islamic world into quarters or streets, where a particular trade or guild conducted its business, facilitated the close control exercised by the guild over the members and prevented the intrusion of outside influences. The arrangements governing the *suq* were, in some respects, the equivalent of the municipal institutions in large cities in the West.[24] Exposure to the medieval arrangement of a typical Muslim city was an excellent preparation for Cairo. In spite of the grand appearance of the public buildings, however, the residences were meanly built of stone or wood, the latter looking like birdcages. This made them susceptible to the twin scourges of earthquake and fire, not uncommon occurrences in the city.

Even after the coming of the Ottomans, the capital of the eastern emperors remained a great cosmopolitan city. Niebuhr reports that, in addition to the mosques, there were twenty-three Greek Orthodox and three Armenian Orthodox churches, as well as facilities of six Roman Catholic orders, each under the protection of a different European ambassador. The Jews had numerous synagogues in the city and the surrounding villages. Although the representatives of non-Islamic religions were now forbidden to build new churches and synagogues, they freely met in Galata, the European quarter, without the authorities taking undue notice. The city and suburbs were full of gardens, and in the latter were many country homes

---

22. For a fuller discussion of Niebuhr's mapmaking, see Chapter 10 below.

23. The English caravansary, from the Turkish compound of *karwan*, or caravan, with *seray*, or house, from the old Tatar capital on the Volga. See Redhouse, *A Turkish and English Lexicon*.

24. See Lewis "The Islamic Guilds."

Tab. IV.

of the sultan, although the incumbent in 1761 went only to *Kara Agadsche*, a gloomy and melancholy place that, Niebuhr remarks, suited his equally melancholy humor. It was clear to this particular European observer that the sick man of Europe was in the advanced stages of his affliction.

In the Hippodrome, or At Meidan, stood an ancient Egyptian obelisk whose hieroglyphs Niebuhr painstakingly copied. It was his introduction to a subject that he would later take up with enthusiasm in Egypt, the copying of hieroglyphs, and a first attempt at arranging them in some sort of logical order. This particular obelisk was one of a pair erected by Tutmosis III of the 18th dynasty. It originally stood south of the seventh pylon of the Great Temple at Karnak[25] and was removed by the Romans to Alexandria. It was probably the Emperor Julian (332–63 AD) who had it shipped from Alexandria to Constantinople, although the raising of the obelisk is credited to his successor, Theodosius. In 1761, it was missing an unknown portion of the lower half, although it still stood nearly sixty-five feet high. Originally, it was probably a third higher. We now know that it contains scenes of the Pharaoh making offerings to the god Amon-Re[26], as well as texts describing important incidents in his reign, including a celebration of his successful crossing of the Euphrates in northern Syria.

On this first visit to an Oriental city Niebuhr confesses to his initial fears of the reaction of the populace to his curiosity. He also suggests that he gradually laid them to rest:

> I had such fearful ideas of the Mohammedans on my first visit to Constantinople that I hardly dared look carefully at this bit of antiquity. But on my return, I copied all the hieroglyphs on this obelisk (Plate IV) without the slightest fear of more than one hundred and fifty Turks, spectators who watched me work.[27]

There would be many experiences like these in the six years between the two visits. Niebuhr was the cartographer on the expedition but, as we will

---

25. Portions of the lower part, as well as fragments of its companion piece, still lie in the original setting.

26. The fact that the god remained and was not defaced suggests that it had already fallen at the time of the iconoclasm of Akhenaten and, so, survived intact. See Habachi, *The Obelisks of Egypt*.

27. *Travels*, vol. I, 31. To Niebuhr, they were "Mohammedans." We should resist the temptation to be wise after the fact. Even Richard Burton, that most obsessively learned of men, was still using the term a century later. But if anything could epitomize the lingering misunderstandings between Christianity and Islam, it was the use of the word "Mohammedan." It suggested that Muslims were the followers of a man, rather than of God. And he was a man for whom Europe's intense hostility was only beginning to change under the influence of more dispassionate scholarship.

see, he gradually assumed other duties as well. Not the least of these was his interest in inscriptions, whether Arabic, hieroglyphic, Himyaritic, or cuneiform. It was the first time these hieroglyphs had been copied in their entirety by a European and the copy is shown as Plate IV of the first volume of the *Travels*.

To Alexandria

In Constantinople the Danish expedition had entered the Oriental world, although they were still able to live as Europeans in its relatively cosmopolitan atmosphere. But it was clear that they must soon adapt themselves to the dress and customs of the local population if they were not to be the objects of curiosity or, worse, hostility. Cairo was not Istanbul, and the Cairenes were notorious for their dislike of outsiders, particularly if the outsiders were Franks. So the members of the expedition spent the days before their departure from Istanbul outfitting themselves for the onward journey. They visited tailors and acquired the long, loose-fitting garments that were common in the East, replacing the tight-fitting European fashions that were not only uncomfortable in a culture without chairs, but made their wearers objects of derision. The ever-helpful von Gahler procured passports, letters of recommendation, and bills of exchange for Egypt.

They left the city on the 8th of September on a ship from Dulcigno, a trading city on the Adriatic not far from Ragusa.[28] Their destination was Alexandria. The transit of the Dardanelles was a reverse of their earlier journey, but at a more leisurely pace. And Niebuhr, who tells the story, was now healthy and able to devote himself fully to his duties. The ship was boarded by the customs authorities at Kum Kalla, to look for contraband and fugitive slaves. The inspection took a day, and that allowed Niebuhr to inspect the fortifications, which were in an advanced state of decay.[29] On the 18th they passed through the entrance to the straits and again entered the archipelago. They sailed nearly due south, and on the 19th at noon were in the vicinity of Samos, at 37° 46' latitude, based on Niebuhr's sighting of the sun at that hour. On the 21st the ship dropped anchor in the roadstead at Rhodes and the Europeans prepared to go ashore.

28. Now called Dubrovnik in Croatia. Dulcigno is now Ulcinj.

29. Europeans, however, were consistently to underestimate the difficulty of forcing the Dardanelles. In 1808 Admiral Duckworth advanced to within eight miles of Istanbul before contrary winds forced him to retire. But the British discovered in 1915 during the Dardanelles campaign how difficult the combination of topography, mines, and guns could be. They lost three battleships in the attempt to force the narrows before they brought in the army. The agony of Gallipoli followed.

The island had fallen to Suleiman the Magnificent in 1522 after a lengthy siege and heroic defense by the Knights of St. John, who were permitted to decamp for Malta. It had once been a byword for impregnability and the Turks, remembering the price they had paid, thought it still impregnable. But the fortifications, here too, were in no state to withstand a modern siege. Niebuhr speculated on the location of the colossus, one of the seven wonders of the ancient world and found the span where it was supposed to have bestrode the harbor too great to be believable.

The depredations of the Turkish sailors (probably Greeks, since the Turks traditionally manned their fleet with natives of the archipelago) were so feared by the populace that the city considered itself, once again, under siege: the shops were shuttered and the streets were deserted. Niebuhr, Baurenfeind, and Forsskal, now dressed as Turks, went ashore and found the residence of the French consul closed. But, by accident, they met a Capuchin friar who returned with them to the consulate and they were ushered into the presence of the consul. He lent them his *dragoman*[30] to escort them around the city. They had a filling but primitive Turkish meal, eating with their hands, before paying visit to a Jew who made a business out of serving wine to European visitors. There were still many Greeks on the island but they were not permitted to live in the city. The next day von Haven and Kramer paid a visit to bishop in a nearby Greek village where, according to the Oriental custom, they left their shoes at the door of the residence. When they left, von Haven discovered that his brand-new slippers had been exchanged for an old pair.

However difficult their relations with the inhabitants of the island, their own relations now were of greater concern. Rhodes was where the simmering discontent among members of the party appears to have come to a head.[31] In spite of the surface reconciliation that had taken place in Istanbul, the unsettled state of relations between Forsskal and von Haven still prevailed, and the expedition had by now divided itself into two camps: Forsskal, Niebuhr, and Baurenfeind on one side, and the two Danes, von Haven and Kramer, on the other. On the last day in Constantinople, Kramer had noticed that von Haven bought a large quantity of white and yellow arsenic. As a physician he knew that it was an amount large enough to carry off a regiment. In Rhodes, in spite of his national loyalties, he disclosed his fears to the other members of the party. The information produced the darkest suspicions in Forsskal, Niebuhr, and Baurenfeind, and together they penned a letter to von Gahler in which they confessed their fears:

---

30. A kind of general factotem. From the Arabic *targama*, "to translate."
31. See Hansen, *Arabia Felix*, 83–87.

von Haven intended to poison one or all of them, and requested that he be separated from the group. The letter, dated April 21st, still preserved in the Danish State archives, was carried back to Constantinople by a gentleman returning from Rhodes. Whether or not the suspicions were justified, the letter and the incident were to poison the atmosphere of the expedition for remainder of their time together. Needless to say, it also produced the profoundest dismay among the Danish authorities, whose response would reach the travelers later in Cairo.

They left on the 22nd for Alexandria. After several days of coasting, or sailing within sight of land, they now entered the open sea where modern aids to navigation might have been useful. Niebuhr busied himself with his observations and found the latitude of Rhodes to be 36° 26'. But the position was useless in the absence of a chart and this the master lacked, along with a logbook and compass. He had probably made the descent from Rhodes to Alexandria a hundred times, and could find the harbor in his sleep. The Turks *did* understand the use of the technology and privateers were sometimes suspected of passing the latest European aids to navigation to the enemy. But if the master was unconcerned about navigation, he was in mortal fear of pirates. The ship was heavily laden, and the master feared attack by corsairs who, probably flying the flag of an Italian prince, made it their habit to lie in wait off the coast of Egypt. The ship carried guns, but most were not in condition to be fired. The cargo also included a group of female slaves[32] who were quartered immediately above the Europeans.

In spite of Michaelis's stern warning, the slaves were a source of innocent dalliance by the members of the expedition after the initial shock of discovery. By the end of the four-day passage they were exchanging little gifts of fruit and refined sugar, passed by means of kerchiefs lowered from the window above. They learned a few Turkish words, which, they found, were those of a warning to be careful lest they be discovered. They also had an exchange about religion with the clerk, or first mate. Niebuhr remarks that Catholics had filled the man's mind with absurdities about the Protestants, such as the fact that there were still many pagans in Germany and Hungary, and they were called Lutherans. The clerk listened patiently as Niebuhr tried to correct these notions. But he would tolerate no discussion of the relative merits of Christianity and Islam, calling all polytheists (which these trinitarian Christians were) "cattle and asses" before stalking off. He taught Niebuhr an early lesson in this most sensitive of issues:

32. The use of female slaves, typically Circassian, in good Turkish households, lasted into the twentieth century. The practice was so widespread, and so absent the horrors normally associated with the institution in the West, that it was among the reasons the Turks resisted to the end the anti-slavery conventions sponsored by the British.

The good man thereby reminded us to leave each man's belief to himself, and that his own religion was the best as long as he had no doubts about it. I don't consider it my calling to make proselytes. But, when, afterwards, I asked sensible Mohammedans about the principles of their religion, I occasionally told them something different about Christianity, without asserting that it was preferable to the teachings in the Koran, and no one took it amiss.[33]

They reached the vicinity of Alexandria early on the fourth day out of Rhodes, although a contrary wind kept them at sea until nearly nightfall. They anchored in the western harbor, reserved for local ships, on September 26th, 1761. As they entered the harbor in the failing light, they would have seen the low skyline of the city, certainly the outline of Qaytbey's fort, and perhaps, in the distance, the misnamed "Needle of Cleopatra" or the similarly misnamed "Pompeii's Pillar." Other than those landmarks, there was little that remained of a city that had once been the queen of the Mediterranean. The Danish expedition had now truly entered the Orient, and it was probably appropriate that they should have done so in a city that had long-since lost its foreign patina. Alexandria was now an Egyptian city, and it was not even the first Egyptian city on the Mediterranean.

---

33. *Travels*, vol. I, 40.

4

Alexandria

> There is no more impressive and majestic reflection of the achievement of the Greeks in Egypt than the great city which bore the Alexander's name. It dominated the Eastern Mediterranean world politically, culturally and economically for six-and-a-half centuries and rivaled the new eastern capital of the Byzantine empire, Constantinople, for another three ... by the middle of the first century BC Diodorus of Sicily could describe it as 'the first city of the civilized world' ... Materially enriched by the exploitation of its enormous potential for maritime trade and culturally unrivaled as the fountainhead of the Greek literary and intellectual tradition for more than a millennium, Alexandria was truly the queen of the Mediterranean. (Bowman, *Egypt after the Pharaohs*, 204).

ALEXANDRIA MAY HAVE ONCE been the first city of the civilized world but in 1761 it was, to use Niebuhr's phrase, little more than "heaps of ruins." Arrian tells the story of its founding by the Macedonian Greek, Alexander the Great, in 332–31 BC, of his personal choice of the position and establishment of the principal outlines of the city. The excellence of the site was immediately obvious to him, from the mild climate, to the natural bay west of the mouth of the Nile, to the propitious rise of land between the lake to the south and the sea to the north. The eponymous city quickly arose as an outward-looking, intellectually vital, commercial hub. It was the reverse of what it had been under the Pharaohs, traditionally a remote outpost guarding against the intrusion of foreign goods and influences from the north, most particularly Greek. Too much may be made of the insularity of the ancient Egyptians, of their self-sufficiency and lack of interest in other civilizations. They certainly were interested in projecting their military power into Palestine, if only to guard against the incursions of the Levantine peoples.

Alexandria

But for two and a-half thousand years, from the founding of the Old Kingdom to the rise of Classical Greece, other than a few evidences of Aegean polychrome ware, there was little that demonstrated their interest in the north and the seaward approaches of the Mediterranean.

All this changed when the Greeks came in the second third of the first millennium BC. Alexandria became the capital and the Mediterranean world became the focus of the new ruling class. The city waxed and then waned, depending on the outlook of its rulers. It grew into the cosmopolitan metropolis described above, dominated by a foreign elite, largely Greek, then Greek and Jewish, and then Roman. It retained that outward orientation through nearly a millennium of uneven development. Without laying the destruction of much of the city to the Arabs,[1] the triumph of Amr ibn al-Aas in 642 AD ended the focus on the Mediterranean and ushered in a period of more than another millennium of Alexandria's contraction into a cultural and commercial backwater. The capital under the Arabs was moved south again, to what eventually became Cairo. Too much may also be made of Arab responsibility—if blame is to be laid—for the end of the brilliant growth that Alexandria had once been. There were other factors, including earthquakes and natural subsidence, which contributed to the decline after the coming of the Arabs. But long before they arrived, the unnatural phenomenon of Alexandria had begun to collapse under the weight of its own contradictions.

At its height under the Ptolemies, the city reportedly had a population of a million souls, although the figure is probably inflated. In addition to Egyptians, it was a city of cosmopolites, of cutthroats, brigands and freebooters from Greece, Cilicia, and Syria. There were also exiles, fugitive slaves from the mercenary army, Hellenized Jews, and immigrants from Syria and Asia Minor, Italy, Syracuse, Libya, Carthage, and Massalia.[2] When the Danish expedition arrived in September 1761, the population of the city had fallen to about 8,000. According to Strabo[3], the most reliable source on the physical layout of the city in the first century BC, the city had been a rectangle of thirty stades from east to west by seven to eight from north to south. A stade was approximately 185 meters, so the city would have been just over five and a half kilometers in extent from east to west, and a quarter of that from north to south. By 1761, and indeed through most of the Middle Ages, the inhabited area had shrunk to the land bridge to the

---

1. The native, Coptic Egyptians were responsible for much of the damage, some on religious and some on cultural grounds.

2. See Saad El Din, Mursi et al, *Alexandria: The Site and the History*.

3. Strabo arrived in 24 BC with the Roman expedition of Aelius Gallus, a first, ill-fated attempt to conquer *Arabia Felix*. He accompanied the expedition as far south as Philae before returning to Alexandria, where he remained until 20 BC.

island of Pharos, the old seven-stade Heptastadion and a small part of the old western half of the city. Niebuhr did not make what would later become his customary city map. As the reason, he cited the danger to himself and his instruments from thieves. It should be remembered that he probably made the map of Constantinople only after his return to that city in February of 1767, and this was his first real exposure to a city in the East. But the layout cannot have looked too different from that in the map made by the cartographers of the French expedition forty years later. The coming of the French in 1798 would be followed by that of another Macedonian, Mohammed Ali, who set the stage for the restoration of both the commercial importance of the city and the domination by foreign elites, again Greek, Italian, and Jewish, with French and, to a lesser extent, English influences. The coming of the foreigners also led to a modest recrudescence of the intellectual pretensions, if not the importance, of Alexandria in the twentieth century, if E. M. Forster, Constantine Cafavy, and Lawrence Durrell can be compared with such giants as Euclid, Archimedes, and Claudius Ptolemy. But all this lay in the future, and the map prepared by the French shows a warren of dwellings on the causeway and little else.

In 1761 Alexandria was no longer the commercial center it had been, or would become again in the nineteenth century. It was a sleepy provincial town, given to chronic, low-level warfare between the few residents and the Arabs, or Bedouins, in the outlying areas. The authorities—the Mamluk beys in Cairo—were more interested in the commercial importance of Rashid, or Rosetta, as we will see below. So, Niebuhr's fears were probably justified. And Michaelis's admonition must have seemed very fresh in his mind: the goal of the expedition was Happy Arabia, a prospect that must have seemed very near now that they had finally arrived in the Orient. There were sights to be seen in and around Alexandria, but Niebuhr could not justify the possible loss of his instruments, much less his life, for such stuff.

Long gone was the lighthouse, one of the seven wonders of the ancient world which, after centuries of decay, had finally succumbed to an earthquake in the fourteenth century. On its former site, at the eastern end of the peninsula now called Ras at-Tin, was Qaytbey's fort, built in 1480, with its garrison of 500 janissaries. The fort had also fallen into disrepair and the Jews were blamed for the ruin.[4] It was the first of many times Niebuhr would see the works of this greatest of the Mamluk builders. In Cairo there would be the mosque, the mausoleum complex (the only one Niebuhr identifies among the tombs of the northern cemetery in his map

---

4. They had been ordered to live in the fort towards the end of the Mamluk period, and were later accused of having abandoned it and allowing it to deteriorate. See Winter, *Egyptian Society under Ottoman Rule*, 218–19.

Alexandria

of the city); the Bab al-Qarafa, or gate of the same name; the *madrasa*, or school; the *sabil-kuttab*, where water was distributed gratis to the poor, and the *wikala*, or caravansary. In Giza, there were two bridges over the Nile with inscriptions whose text and translation he provides,[5] both attributed to Qaytbey. In Tanta, there would be the tomb of Ahmed al-Badawi, that most Egyptian of saints, whose expansion Qaytbey had financed in 1483. This greatest of *burghi* (see Chapter 7 below) Mamluk sultans (1468–95) aspired to be the first prince of Islam and did not confine his attention to Egypt. Other buildings in the feverish activity that defined his reign included the citadel at Aleppo and the rebuilding of the Prophet's tomb in Medina after it had been destroyed by lightning in 1475.

The description of Qaytbey's method of raising funds for his building program, largely through extortion and exaction, was not untypical of the Mamluks:

> Although Kaitbai refused the customary succession largesses, money was needed to repel Bedouin ravages, as well as meet dangers threatening Asia Minor; and the mode of raising it was fit prelude for that which was to come. The President of the Council being held responsible, was robbed of everything he possessed, and then was assessed in a sum which on his declaring his inability to pay, he was flogged in the Royal presence. On this producing no effect, the Sultan himself took the cudgels till the wretched Emir's blood besprinkled the by-standers, At last, on agreeing to pay down 200,000 dinars, the bleeding courtier was not only set at liberty, but clothed in a robe of honor. Such was the rude and versatile barbarity of Kaitbai's court.[6]

Not a great deal had changed in 1761, as we will see below.

Even longer gone in Alexandria were the palaces, temples, stadia, and the famous library. The Ptolemaic walls surrounding the Great City—as it was known—had been succeeded by the Roman walls, and these in turn by Arab walls. In the sixth and seventh centuries AD, the Roman walls had been formidable enough that they on two occasions would have withstood sieges had there not been treachery from within. The story of their fall to the Persians in 617 AD was, in many respects, the tortured history of the ethnic conflicts that continually beset the city:

> The siege lasted some time, and with all their skill the Persians were unable to force their way into the great fortress. Indeed its defenses were so strong as to be virtually impregnable . . . the

5. *Travels*, vol. I, 193–94.
6. Muir, *The Mameluke or Slave Dynasty of Egypt*, 173.

long lines of bulwarks and towers were as formidable as ever; and a united and resolute garrison, drawing endless resources from the sea, which the Empire still commanded, would have wearied out the besiegers . . . But union had long been impossible to the turbulent population of Copts, Romans, Syrians, Jews, and students and refugees from all parts of the Empire. The Copts and the Syrians hated the Romans and the Jews hated the Christians, with an enmity on which no common peril could act as solvent; while all would have laughed to scorn the idea that between the different races, classes, and creeds there could have been any bond of patriotism, which alone might have given them cohesion. It is therefore not surprising to learn that the city fell through treason.[7]

The fall of the city to the Arabs barely twenty-four years later was also achieved without a fight, for many of the same reasons.

Even allowing for the characteristic hyperbole of the untutored conquerors, the Great City still left an indelible impression on the Arab army. Many of them had seen large cities before, including Edessa, Damascus, Palmyra, Antioch, and Jerusalem. But nothing can have prepared them for Alexandria, the first city of the civilized world. The white marble with which the city was paved was so brilliant, said an Arab source, that no one could enter the city without a covering over his eyes to shield them from the glare. At the full moon the reflected light was still enough to allow a tailor to thread a needle without a lamp.[8] The buildings were so many and so magnificent that they could only have been raised by the *jinn*, invisible beings who either helped or hindered humanity, but typically excited the baser passions.

Standing out among the thousands of edifices there had been the great church of St. Mark, the church of the Caesarion, the two standing obelisks, and the Serapaeum in the Acropolis that had once housed the Library. Most extraordinary of all, there was the Pharos or lighthouse. Built in four sections of decreasing diameter,[9] it rose to a height that is still disputed, but must have been several hundred feet. At the time, Cairo was little more than a few miserable settlements surrounding the Roman fort of Babylon. But the decision had been taken by the Caliph Omar that the future capital would be south at the beginning of the Delta, not on the sea. Amr's description of a seagoing vessel was a measure of the Arabs' initial lack of knowledge and fear of this new medium:

7. Butler, *The Arab Conquest of Egypt*, 72–73.
8. Ibid., 369.
9. The sections were, beginning at the bottom, square, octagonal, round and open, perhaps prefiguring the classical Mamluk minaret configuration.

> Verily, I saw a huge construction upon which diminutive creatures mounted. If the vessel lies still, it rends the heart; if it moves, it terrifies the imagination. Upon it a man's power ever diminishes and calamity increases. Those within the ship are like worms in a log.[10]

So, the victorious army of Amr began to lay out the Arab capital at Fustat, beginning with a fort across the Nile at Giza.

In 1761, even the Arab walls were no longer intact. They had been high enough 300 years before the arrival of the Danish expedition that Bernhard von Breitenbach[11] had never seen a city so well fortified as Alexandria. The dual harbor, with its eastern and western portions separated by the causeway, was still in use although it, too, was in a sad state of disrepair. The safer, western harbor, protected from the strong easterly winds, was reserved for local ships such as the one in which the Danish expedition arrived. European ships were forced to use the eastern, or old royal, harbor. Niebuhr reports that the bottom of the latter was so covered with debris that it was difficult to anchor safely without a series of booms and cables attached to the anchor. A recent archeological survey of the eastern harbor has shown why this was the case: the bottom was strewn with columns, capitals, statues, lead ingots, slabs of rock, red granite blocks bearing hieroglyphs, and the remains of Ptolemaic palaces and the Temple of Poseidon. Over 2,000 architectural pieces, mostly of Aswan granite, have recently been located, in addition to a dozen sphinxes, three obelisks, and a series of gigantic blocks, weighing on average 75 tons, which probably belonged to the lighthouse.[12]

The members of the expedition stayed in Alexandria for a month, temporarily accommodated in the home of the French consul. The absence of a direct link between Alexandria and the Nile meant that Rashid was now the most important entrepôt for trade, as it communicated directly with the river. Although the city had not recovered its maritime or commercial importance, Alexandria was still home to a few European merchants as well as the consulates of France, Venice, Holland, and Ragusa. The consuls of these countries also represented the interests of Denmark, Sweden, Tuscany, and Naples. The Europeans generally spoke Italian among themselves. The natives who dealt with the Europeans had learned their several languages

---

10. Quoted in Butler, 467.

11. Breitenbach was a German from Mainz who traveled to Egypt in 1484 and whose Latin account appeared soon thereafter. Niebuhr referred to it in the preparation of his own account of the city.

12. See *Egyptian Archaeology*, No. 9, 1996. Makrizi suggests that a twelfth-century governor purposely had the columns broken up and thrown into the sea as a kind of breakwater. See Butler, 388.

quickly, and Niebuhr remarks that there were "native Mohammedans who spoke Provençal, Danish, or Swedish almost as well as if they had been born in France, Denmark, or Sweden."[13] The month gave the members of the expedition time to see the sights. Forsskal was busy from dawn to dusk, botanizing in the environs of the city, and the absence of classical remains was no drawback in this exciting world of unfamiliar flora. For Niebuhr, the onetime queen of the Mediterranean held few attractions. Aside from two churches, a mosque, a Franciscan convent, and a few mean houses, it was a desert within what remained of the walls of the Arab city.

At the Coptic Church of St. Mark he was exposed to an example of the unedifying internecine squabbles that seemed to characterize Christianity in the Orient. In spite of a common belief in the saving message of Jesus Christ, the Latin and Orthodox Churches still regarded each other with the suspicion and contempt normally reserved for unbelievers. Alexandria had entered with gusto into the doctrinal disputes of the early Church, and over the years much blood had been shed over so abstruse a matter as the nature of the trinity. More recently the quarrels seemed petty, but from such pettiness dissension and even massacre arose. Niebuhr's Coptic informant told him that the body of St. Mark was no longer on display since the Venetians had removed the head. The Catholics, on the other hand, claimed that they had spirited away the entire body of the saint to Venice, dismembered and packed as that of a swine. The customs authority in the city was in the hands of the Jews and the body, identified as that of an unclean animal, had been passed without inspection. This was only one example of the ongoing conflict between Eastern and Western Christianity in Egypt. As we will see below, Franciscan and Capuchin brothers worked quietly in this largely Muslim country to bring Roman Catholicism to the heathen. They were tolerated by the Mamluk and Ottoman authorities as long as they confined their proselytizing to the Copts. And the frequent quarrels between the *dimmi*s, or non–Muslim subjects, created opportunities for extortion by the beys. The activities of these religious orders excited the liveliest resentment of the Latin Church among the native, Coptic Christians. And the brothers, conspicuous in the western habits and open sandals that were unsuited to an eastern city, had a reputation among the natives for a lack of personal hygiene. Of such superficial characteristics was religious contention made.

But even in the wasteland of eighteenth-century Alexandria, Niebuhr found a few monuments that appealed to his surveyor's bent. Still within the enceinte were two red granite obelisks, one standing and the other fallen and shattered. They were a pair originally erected by Tutmosis III before the

13. *Travels*, vol. I, 53.

Tab. V.

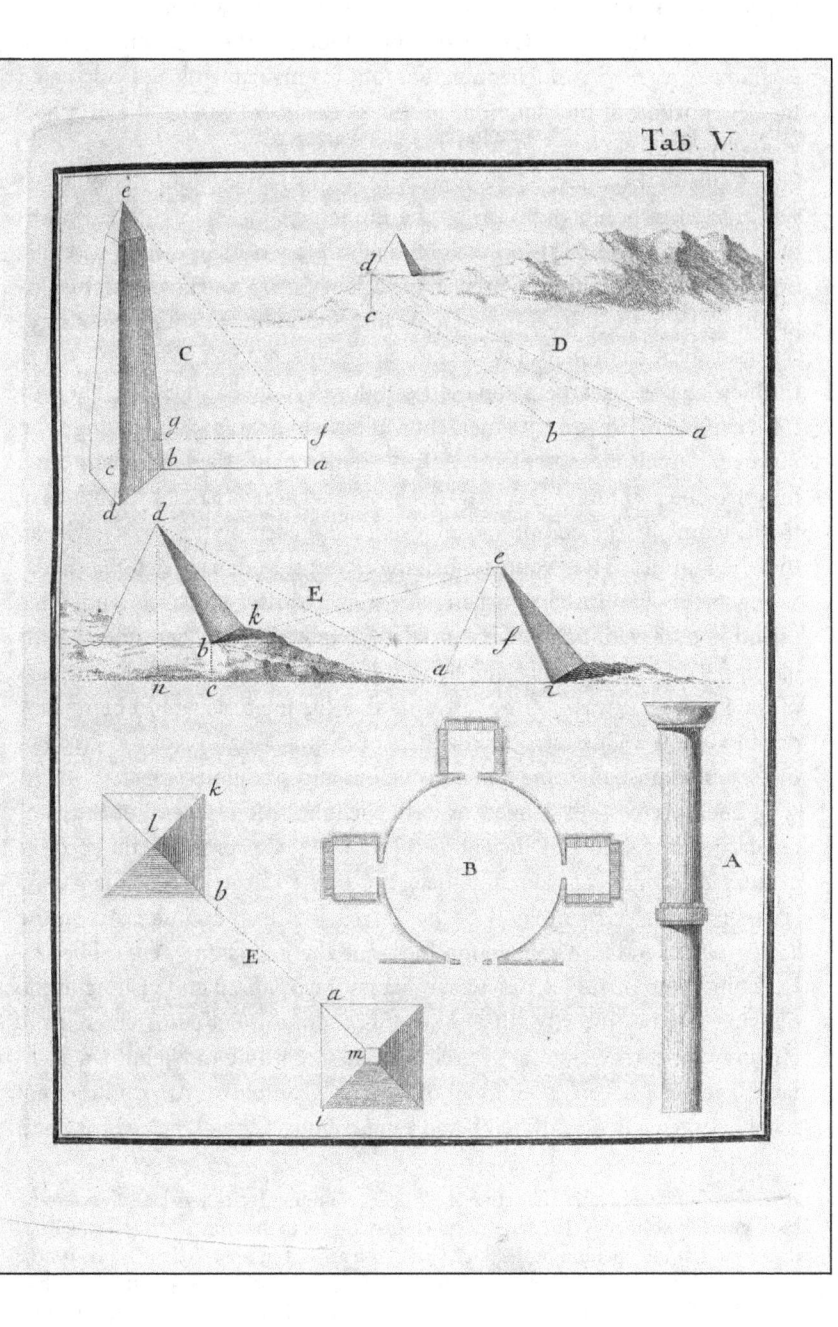

Temple of the Sun at Heliopolis in 1468 BC. They were taken to Alexandria in about 10 BC and set up in front of the Caesareum, the temple devoted to the worship of the deified Caesar and they stood together until one fell in an earthquake in 1301 AD. Niebuhr took out his instruments and calculated that the portion of the standing "*obelisk of Cleopatra*," above the ground[14] was sixty-one feet eleven inches in height.

Outside the walls, Niebuhr found the misnamed Pompeii's pillar (it was erected in honor of a visit by the Emperor Diocletian and had nothing to do with Pompeii) to be eighty-eight feet ten inches high, carefully explaining the geometry he used in his calculation. Von Haven put his linguistic skills to the test with the Greek inscription on the southwest side of the base of the column. But it was indecipherable, not being as deeply cut as the hieroglyphs. And, being on the weather side, it had also suffered heavily from exposure. However, an incident while Niebuhr was triangulating soon put an end to his measurements. A Turkish merchant asked to look through his astrolabe and found the image reversed, a characteristic of the optics of the instrument. The rumor spread, eventually reaching the governor's ear, that the Europeans had come to turn the city upside-down. Niebuhr's janissary guard now refused to accompany him and this effectively ended his scientific observations in Alexandria, although he began here the records of rainfall and temperature that he would keep throughout the seven years of the journey. It was not the last time that his interests would be misunderstood, and it illustrated the dangers of mapmaking—or of too great an interest in almost anything—in a suspicious and credulous world.

There were other dangers as well. Niebuhr visited several of the catacombs in the vicinity of Diocletian's column. One was probably the second-century AD tombs of Kom al-Saqafa, according to the plan of the site he shows in Plate V (B) in vol. I of the *Travels*. But in the same area, on the 11th of October, two Venetian mariners and their janissary were robbed by Bedouins, who made it a practice to harass the *fellaheen* and visitors in the outlying areas of the city. That same afternoon, the Bedouins entered the city to make their customary purchases and, according to Niebuhr, a spectacle "such as I haven't seen in all my travels"[15] unfolded. After a disagreement between a Bedouin sheikh and a merchant, a general melee broke out.

---

14. Its full height, after the removal of the surrounding earth, was later found to be sixty-nine feet six inches. The standing obelisk was given to the United States by the Khedive Isma'il at the opening of the Suez Canal and arrived in New York in 1881. It now stands in Central Park. The fallen twin was given to the British in 1875 (although it had been officially presented much earlier by Mohammed Ali) and erected on the Thames Embankment in London, where it still stands. See Habachi, *The Obelisks of Egypt*.

15. *Travels*, vol. I, 53.

The Europeans retired to the roof of their house to watch. The townspeople, "perhaps as misbehaved as any in the entire Ottoman Empire,"[16] were armed with stones, the Arabs with firearms, and the struggle surged back and forth in the narrow streets. Finally, greatly outnumbered, the Arabs withdrew after losing fifteen men and several horses. They then seized all the animals that were pasturing in the outlying area, thereby bringing the *fellaheen* into the fray. But, after a few days, peace was restored, as it was in the interest of all sides to come to an accommodation.

The Bedouins in Egypt

The depredations of the Bedouins had been a continuing problem since the arrival of the Turks in 1517. At first resolving to drive the usurpers out, the Bedouins had nearly exhausted the Ottoman armies in the early days of Turkish rule.[17] They greatly outnumbered the Turks, and it was only their characteristic independence and lack of unity that allowed a small number of the latter to maintain control. But as the Mamluks reasserted their control,[18] the Bedouins became their allies against the Ottomans and they waxed increasingly arrogant. By the time the Danish expedition arrived in Egypt, they had been integrated into the Ottoman administrative structure, and Bedouin sheikhs were tax farmers and governors. Istanbul conferred Imperial patents and bestowed robes of honor on important sheikhs, as much to pacify them as to ensure good governance. But, even though many had settled on the land and were indistinguishable from the *fellaheen*, they had not lost their predatory ways, and the above incident was typical of the low-level civil strife that prevailed throughout the country. When the central government was stronger, the Bedouins were weaker, and vice-versa. The middle of the eighteenth century represented a watershed and, if it represented the high water mark of Bedouin power and influence, it was about to change, as we will see below.

Problems with the Bedouins may have also kept the Danish Expedition in Egypt for the better part of a year. Hansen suggests that the reason for the delay was the turmoil caused by von Haven's purchase of the arsenic, as the members of the expedition waited for guidance from Copenhagen before proceeding to Arabia. But there was another reason as well, and it may have been the deciding factor. Caravans between Cairo and Suez, where

---

16. Ibid, 54.
17. See Winter, *Egyptian Society under Ottoman Rule*.
18. See Chapter 7 below.

Abbildung der Araber in Egÿpten.

the expedition would take ship for Jidda, were held up for most of 1762 while the consequences of a violent clash between the Egyptians and the Bedouins of the Hejaz played themselves out. Every year the Egyptians sent the pilgrimage caravan, under the command of one of the Mamluk beys, to the Holy Cities. The way was paved by gifts to the Bedouins through whose territories they passed, and in this particular year the authorities decided that they would no longer accede to their extortionate demands. Jabarti tells the story and it is worth recounting in full:

> Command of the pilgrimage caravan had been given to Hussein Bey Kashkasha, who led the caravan in 1174 (mid-1761 AD, my note). The Bedouins lay in wait along the route and when, suspecting nothing, Hussein Bey entered a narrow gorge, the Bedouins barred the way. Their chiefs presented themselves to him and demanded the usual gifts . . . The Bedouins were invited to present themselves at the next station to collect their money. The caravan then set out, but as soon as it emerged from the narrow gorge, Hussein Bey had his troops drawn up in a wide space and when the Bedouins and their chiefs presented themselves met them with blows of the saber. More than twenty chiefs were killed, including Hazzah. Hussein Bey then ordered the caravan to set out again; the cannon shot announcing departure was sounded, the caravan moved off and the wives of the men who were killed dispersed, vowing vengeance. The Bedouin tribes rushed from every quarter and barred the way of the caravan. They prepared ambushes at the turnings of the roads and in the gorges. Hussein Bey was faced with fierce combat, both at the head and the rear of the caravan. He was everywhere and charged at the head of his Mamluks. He inflicted great loss on the enemy. Finally, he arrived safe and sound in Cairo with the pilgrims. He brought back on camels the heads of the Bedouins put to death. The caravan and its commander made a triumphal entry into the city of Cairo. The emirs, nevertheless, found fault with the conduct of Hussein Bey, and Ali Bey Ballout Kabban told him: "You have turned the Bedouins against us and so barred the route of the caravan. Who will lead the caravan next year to the Holy Places? I will, answered Hussein Bey, I will deal with these Bedouins. Don't worry about anything . . . But in proximity to Egypt there were Bedouins who never passed up the opportunity to plunder travelers and fellaheen. Hussein Bey decided to bring them to their senses; he led several expeditions that cowed the thieves. The taking of their flocks, the death of their warriors whose heads were carried to Cairo,

and the terrible reprisals exacted by Hussein Bey had excellent results. The Bedouins renounced their criminal behavior. Roads became safe, danger and fear were banished, and the report of Hussein Bey resounded far and wide."[19]

The splendid results outlined by Jabarti were not immediately apparent in late 1761, and it was not until August of 1762 that the Danish party was able to find a caravan to accompany them from Cairo to Suez.

Doctrinal Disputes

Alexandria may have been a ruin in 1761 but its early history might have been of interest to the practitioners of sacred philology in the eighteenth century. We have already seen that the city was the epicenter of contention in the early Christian era. But even before the triumph of Christianity there had been ethnic and religious conflicts that would presage later controversies. And there had even been mooted an issue that, so far from being raised, would have been unthinkable to Michaelis: a challenge to the historicity of the Hebrew Scriptures. The city had always been dominated by foreigners and in its Golden Age, the three centuries before the beginning of the Christian era, the intellectual lights had not even been citizens of the city, but resident alien Greeks. The body of citizens was itself tiny, ruling over a polyglot population of Greek non-citizens and non-Greeks, of whom the least privileged were the Egyptians who increasingly flocked to the urban center from the outlying areas. By the middle of the second century BC, the mix of the lower class Greeks and Egyptians had become combustible, and the next two centuries were characterized by constant outbreaks of mob violence. The Egyptians were not happy under Greek rule, and they were even less happy when, after the revolt of the Maccabees in 60 BC, Ptolemy Philometer admitted large numbers of Palestinian Jews into Egypt, largely into Alexandria and the Fayoum. The Jews grew into a privileged and officially protected elite, their status provoking a fierce, inevitable spasm of anti-Jewish sentiment in first-century AD Roman Alexandria. The always-flammable combination of ethnicity and religion had already produced

---

19. Jabarti, *Merveilles Biographiques et Historiques,* vol. II, 215–18. My translation is from the French. Jabarti lists the Hijra year 1174 as beginning on 13th of August, 1760. So, the Haj ceremonies in 1174, from the 10th to the 13th of *Dhu al-Hijja,* would have occurred in mid-July 1761. The caravan would have set out several months before that. It was the return of the caravan in late August of the following year that was the signal to the Europeans that it was safe to depart. See Chapter 13 below.

outbursts of violence that wracked the city, and the contention had spilled from the streets into the libraries and research centers.

Philosophical schools had always been important in Alexandria and, even before the rise of Christianity, philosophical disputes mirrored the ethnic fault lines in the city. Epicureanism, Stoicism, Gnosticism, Neoplatonism, and many eclectic schools competed for the allegiance of the educated. Christianity arrived in the person of St. Mark in the middle of the first century AD, and added its own strongly felt tenets to the mix. At first practiced secretly, by the end of the next century it had become a powerful movement that challenged the pagan philosophical schools. At the outset of the third century, its increasing prominence provoked the first official program of persecution by the Emperor Severus and prisoners were brought from all over Egypt to be martyred in Alexandria. Fifty years later another wave of persecution aimed at eliminating Christianity from the Empire. In 284 AD arrived the Emperor Diocletian, who ushered in the period of persecution that still marks the beginning of the Coptic calendar. Believing Christianity to be an insidious force in the empire and threat to the old gods, Diocletian launched a ferocious program of persecution of the Christians. But its growth was inexorable. The victory of Christianity in 312 AD, however, did not put an end to doctrinal disputes. If anything, they now came to the fore with increased strength, leavened by the always-present and fierce chauvinism of the Alexandrians.

The Arian controversy may have turned on the nature of Christ, but the excommunication of Arius by Alexander, the Bishop of Alexandria took a decidedly nationalist turn when the ban was opposed by Eusebius, Bishop of Nicomedia. The Emperor Constantine was moved to intervene to bring about unity in the Church. The Council of Nicea in 325 AD, brought together 300 bishops from throughout Christendom, but the creed that issued from the council only papered over the differences. Egypt, in the person of Alexander followed by Athanasius, stood in opposition to Constantinople. In 391 AD the Emperor Theodosius made Constantinople second only to Rome in the Empire and first in the East, forever condemning Alexandria to a minor status in the Church.

But if opposition to the Christians had been fierce, their own efforts to eradicate paganism were equally savage. In the same year, 391, a decree was issued against paganism, and the ensuing destruction included many of the remnants of the secular heritage. The pagans took refuge in the Serapaeum in the Egyptian quarter, but Christians ran amok in the temple and demolished it. Theophilius ordered a church to be built on the sight. The same fate was shared by other libraries and temples in the city, and attacks on philosophical schools were redoubled. The famous library, the Mouseion or "Temple of the Muses," had been founded in 295 BC by Ptolemy Philadelphus, who appointed Demetrius of Phaleron, a pupil of Aristotle, to furnish it with "all books in the world." It had been destroyed in 48 BC, under circumstances that are unclear. But a daughter library in the Serapeum shared the fate of the temple in 391 AD. From a city that had once been the intellectual center of the world, its descent into the depths of religious intolerance and fanaticism was nearly complete.

As might be expected, the Hebrew Scriptures had come under scrutiny as a product of this religious and philosophical ferment. Long before the Bible was subjected to critical analysis by the philologists of the Enlightenment, and indeed before the rise of Christianity, the subject of the Hebrew Scripture as history had been hotly debated. It was an emotional subject and had already leant itself to accusations and counter-accusations of bad faith and bad history.

Two products of Ptolemaic Alexandria had a direct bearing on sacred philology, although their influence may have long been forgotten. The first was the Septuagint,[20] the translation into Greek of the Hebrew Scriptures made by seventy (hence the name) Jewish scholars for Ptolemy II in about 250 BC. Later Christian tradition held that the translation was made to provide a copy for the library of Alexandria. But it probably had more prosaic purposes as well. The first was to provide an understandable text of the Scriptures for the Greek-speaking, Jewish population of Alexandria. Working independently the experts were said to have produced versions that agreed with one another. But, in fact, the extant texts of the Septuagint seem to have been produced as a result of the practice of oral translation in the synagogues in Alexandria, and they differ from one another, with each in its own way differing from the masoretic Hebrew text. In spite of the several versions, the Septuagint would become the Greek standard for the Eastern Church. And, in spite of misreadings, mistranslations, and internal corruption of the text, it was subsequently translated into Latin, Coptic, Ethiopic, Arabic, Armenian, Georgian, and Slavonic.[21]

A second purpose of the translation may have been to educate the ruling Greeks about the details of Jewish history, just as a later work, Aristobolus's *Explanation of the Book of Moses*, was an attempt in the mid-second century BC to explain Hebrew Scripture to Ptolemy Philometer. But the Jews were not the only barbarians clamoring for the attention of historians in an attempt to place the histories of their peoples within the canon of Greek learning. Also in the third century BC, the Chaldean Berossus

---

20. It is said there were supposed to be six representing each of the twelve tribes of Israel—Reuben, Simeon, Levi, Judah, Issachar, Zebulun, Joseph, Benjamin, Dan, Naphtali, Gad, and Asher—but the number fell short.

21. See *The New Standard Jewish Encyclopedia*.

Alexandria

prepared an account of Babylonian history for the Seleucid king, Antiochus. And in the latter part of the third century BC, some say in response to Berossus, Manetho, a priest and a native of Sebennytus in the Delta, prepared his *Aegyptiaca*, a history of Egypt written by a native Egyptian.

Written in Greek and making use of Greek method and literary form, but taken from Egyptian priestly sources, it was a chronology of the Egyptian kings from the beginning through the 30th Dynasty and the second Persian period. With allowance for the fact that Manetho has come down to us in fragments and only through the filter of Jewish and Christian commentators, it would still become the key to an understanding of the outlines of ancient Egyptian history. If he did nothing else, Manetho bequeathed Egyptologists the system of dynasties that remains in use today, although his importance would not become apparent before the deciphering of the hieroglyphs. Then, his chronology, the names often corrupted beyond recognition in their passage through other languages, would be correlated with other evidence such as the Turin Hieratic Papyrus, the Palermo Stone, and the king list at Abydos to establish the basic soundness and consonance of the several lists.[22]

But in the two millennia that would separate Manetho from Champollion and Lepsius,[23] the *Aegyptiaca* was put to another and more tendentious use. Originally, it had been written from an Egyptian point of view with the primary purpose of correcting the errors of Herodotus. It was felt that Herodotus had correctly reported the broad outlines of Egyptian history. But with his penchant for tall tales and for reporting, unfiltered, popular beliefs, he had gotten many of the details wrong. The suggestion that the pyramids had been built with slave labor and that prostitution was widespread in Egypt seemed particularly egregious errors. In fact, Herodotus may have represented a first western attempt to describe the mysterious and inscrutable East. But, to some, the most controversial part of Manetho's account was not about Herodotus but about what it said—or did not say—about the part of Egyptian history of most interest to Jews and what would become

---

22. This will be discussed more fully in Chapter 12, "The Antiquities of Egypt," below.

23. Richard Lepsius was the embodiment of the term "Egyptologist." Champollion's *Grammar* had advanced the study of the language of the ancient Egyptians but Lepsius was the man who put the first, tentative principles of Champollion into practice, correcting, correlating, and expanding the knowledge of the hieroglyphs. His brief description of the temples in *Egypt, Ethiopia and the Peninsula of Sinai*, his account of the Prussian expedition to Egypt in 1842–1845 reads like just another dry recitation of remote facts—Pharaohs, gods, and texts—until one realizes that he was correctly identifying them—by name, provenance, and purpose—for the first time. The work of Lepsius was an extraordinary exercise in practical scholarship and the 12 immense volumes of the Prussian expedition, published in 1858 as the *Denkmaler aus Ägypten und Äethiopien*, are probably the greatest work of Egyptology ever produced.

the Christian West: the purported sojourn of the Israelites in Egypt and the account of the Exodus as preserved in the Bible. For, quite simply, there was no mention in Egyptian history, as recorded by Manetho, that a large subject population such as the Israelites ever lived and worked in Egypt and, after a period of servitude and a visitation of plagues in the country, were allowed to emigrate by the reigning Pharaoh.

So, Jewish history and its place in the history of Egypt had clearly become an issue in Ptolemaic Egypt, no less an issue than it would become two millennia later. The problem with Manetho's history was not immediately apparent to the Jews of Alexandria, although an Alexandrine Jew, Demetrius, may have written an account of Jewish history as a counterpoint to Manetho. But the gauntlet was thrown down and when Josephus would pick it up in the first century AD, the field was virtually his alone. Flavius Josephus (37–100 AD) was the Jewish historian and author of *Jewish Antiquities*, a twenty-volume history of the Jews, from the beginning to the close of Nero's reign. Born in Jerusalem of a priestly family, he was governor of Galilee when war broke out between Jews and Rome in 66 AD. He went to Rome in 70 AD, where he combined loyalty to the Empire with loyalty to his own people. In his *Contra Apion*,[24] Josephus would take on Apion the Grammarian, called by Tiberias "cymbalum mundi," and the leader of the anti-Jewish movement in Rome. But it was as interpreter of Manetho to the world that Josephus is of interest here. He would find the Egyptian historian full of calumnies against the Jews, and undertook to provide what he thought to be the necessary editorial corrective. The editorializing has sparked its own critical commentary:

> The Jews of the three centuries following the time of Manetho were naturally keenly interested in his *History* because of the connexion of their ancestors with Egypt—Abraham, Joseph, and Moses the leader of the Exodus; and they sought to base their theories of the origin and antiquity of the Jews securely upon the authentic traditions of Egypt. In Manetho indeed they found an unwelcome statement of the descent of the Jews from lepers; but they were able to identify their ancestors with the Hyksos, and the Exodus with the expulsion of these invaders. The efforts of Jewish apologists account for much re-handling, enlargement, and corruption of Manetho's text, and the result may be seen in the treatise of Josephus, *Contra Apionem*.[25]

---

24. Born in Assiut in Upper Egypt, Apion lived at Rome during the reigns of Tiberius, Caligula, and Claudius (14–54 AD).

25. Waddell, *Manetho*, xv.

Alexandria

The controversy—if such it can be called between a living spokesman and a chronicler long-since dead—broke out anew and the cudgels were taken up in Rome three hundred years after Manetho. The language of the commentaries by, first, Jewish and, later, Christian chroniclers, indicates the strength of feeling against those who did not support the account as related in the Hebrew scriptures. First, Josephus:

> Thus Manetho has given us evidence from Egyptian records upon two very important points: first, upon our coming to Egypt from elsewhere; and secondly, upon our departure from Egypt at a date so remote that it preceded the Trojan war (traditionally 1192–1183 BC, my note) by well nigh a thousand years. As for the additions which Manetho has made, not from the Egyptian records, but, as he has himself admitted, from anonymous legendary tales, *I shall later refute them in detail, and show the improbability of his lying stories.*[26]

Manetho presents problems to chroniclers because of obvious inconsistencies, ambiguities, and absurdities, although it is impossible to determine how much these were his own and how much those of the interpreters through which his history passed. But to Josephus, Manetho's sin was not that he was inconsistent, or that his regnal years are often suspect, but that the *Aegyptiaca* did not corroborate the account of events as told in Genesis and Exodus. Instead, it mentions two stories that bear on the possible sojourn of the Israelites in Egypt, one later accepted as historical and the other impossible to verify. The first, the coming of the Hyksos, represented the infiltration and gradual seizure of power in Egypt by a Levantine people in the seventeenth century BC. The Hyksos reigned for just over 100 years in what is commonly accepted as the 2nd Intermediate Period between the Middle and New Kingdoms. But they were hated overlords and oppressors and were expelled militarily by the resurgent Egyptians, hardly the stuff of the stirring account of bondage, faith, and escape told in the Bible.

The second, the story of the expulsion of the lepers, seemed a more probable basis for the Biblical account. There, a group of lepers, led by a charismatic revolutionary and banished to work in the limestone quarries of Turah, had revolted before being expelled from the country. But the small scale of the uprising, the fact that the principals were diseased, and the fact that they did not seek to leave but were expelled made it an unlikely source for the account in the Bible. It is impossible to say whether Manetho, in his reaction against things foreign, unconsciously changed or embellished two events that clearly stood out negatively in the collective memory of the Egyptians.

26. Ibid., 107; my italics.

Niebuhr in Egypt

There is the possibility that he did so purposely, that his omission of details of the sojourn of the Israelites in Egypt was intentional, that he was consciously imparting an "Egyptian" interpretation to a commonly held body of knowledge. But this seems improbable in a third century BC priestly historian for whom the history of the Jews was probably not an issue at all. More probably, the lack of corroboration reflected his ignorance of what would become the "sacred history" of the Jews and the Christians. That is, the biblical account was not mentioned because it was *not* commonly accepted history.

Later Christian chronographers would find Manetho equally ignorant of the "truth" as it appeared in Jewish and then Christian tradition. He would subsequently be brought to us through the filters of Antiochus (third century), Eusebius (early fourth century), and George the Monk, otherwise known as Syncellus who, at the beginning of the ninth century AD, wrote a *History of the World from Creation to Diocletian:*

> Manetho of Sebennytus, chief priest of *the accursed temples of Egypt*, who lived later than Berossos, in the time of Ptolemy Philadelphus, writes to this Ptolemy, with *the same utterance of lies* as Berossos" . . . They imagine that they have attained a striking result, but one must rather say that it is *a ludicrous falsehood which they have tried to pit against the Truth.*[27]

Again, the strength of feeling against Manetho is obvious, and the basest motives are imputed to the Egyptian. The absence of mention or corroboration of the biblical account is the same problem for Syncellus that it had been for Josephus. The omission can only be the result of a willful attempt to ignore or pervert the truth. Manetho's account clearly did not suit the taste of these commentators, and they were forced to tailor the Manetho they found to suit their needs.

The fact of the matter appeared to be that there was no textual support in Egyptian records for the biblical account: no record of northerners other than, briefly, as hated overlords, no mention of a mass exodus of a servile population. Indeed, if the numbers making up the Exodus in Numbers 1:46 were to be believed—a total of 603,550 men, not including women and children—the event would have been a catastrophe in a country of only a few million people. The number so concerned Flinders Petrie, a believing Christian as well as probably the greatest Egyptologist of them all, that he went to great pains to rationalize the numbers of the host and the capacity of the Sinai to support such a number. The departure of a third of the population would surely have

---

27. See Waddell; again, my italics.

merited mention in Pharaonic annals. Instead, there was only silence.[28] We will explore Petrie's reasoning more fully in Chapter 12 below.

Subsequent medieval attempts to tie the history of Egypt to the biblical account, long after Manetho was lost or forgotten, consisted largely of attempts to find in the ancient monuments evidence of the sojourn of the Israelites in Egypt. An early effort of this kind was that of Rabbi Benjamin of Tudela in Navarre, who made a tour of synagogues in Central Europe, Greece, Palestine, Mesopotamia, India, Ethiopia, and Egypt between the years 1165–73.[29] Benjamin saw the monuments of Pharaonic Egypt through the filter of Jewish history and appears to be the author of the notion that the pyramids of Giza had been built as the corn storehouses of Joseph (Exod 47:14). But the notion that Egypt has importance only as a bit player in the great biblical drama of mankind has had remarkable staying power. It is a religious and cultural provincialism that has lasted, even among rigorous European scientists, well into modern times. Even so otherwise-sensible a man as Charles Piazzi Smyth, the national astronomer of Scotland, would blithely opine in the late nineteenth century that the pyramids could only have been built by the Israelites under divine guidance, the ancient Egyptians, in his opinion, being incapable of the necessary technical sophistication.

Not only was Smyth wrong, but he was obstinately and repeatedly wrong, as the increasing refinement of the measurements of the pyramids demonstrated the errors of his "sacred cubit" and "pyramid inch," among others. But his book *Our Inheritance in the Great Pyramid* found a receptive audience and went through multiple issues in the late nineteenth century. The struggle of Flinders Petrie, who was responsible for much new knowledge, to reconcile what he knew of Egyptian history with the details of the account in the Bible, has already been mentioned. There was one area however, where the Bible did make a contribution, albeit mistaken, to the science of Egyptology: the term "Pharaoh" comes to us from the Old Testament. The Egyptian original simply meant "Great House" and referred to the royal palace, not to the

---

28. See Redford, *Egypt, Canaan and Israel in Ancient Times*, for a fuller discussion of the record as it eventually came to be known. Redford is unambiguous on the influence of the Bible on history: "In fact, the biblical writers are wholly and blissfully unaware of the colossal discrepancy to which their 'history' and 'chronology' have given rise" (259). And "Biblical scholarship has of late been bedeviled by acceptance, either tacitly or explicitly, of a number of arbitrary preconceptions that fail to be honestly acknowledged, and by a tedious tendency to rationalize the conservative views learned at the feet of priest, preacher, or rabbi as a philosophy of history" (261).

29. His account, originally written in Hebrew, was printed in Constantinople in 1543 and was later translated into Latin and French, which latter version was included in Bergeron's *Recueil de Voyages*.

individual who occupied it.[30] Our continued use of it today is an anachronism, but a useful one, and we would as profitably discard it as we would the names "Cairo" or "Damascus" for obscure, but more scholarly correct, versions.

And what of the attitudes of these particular European travelers in the Danish expedition, that embodiment of Enlightenment science? As we have seen, Johann David Michaelis had effectively given Egypt the back of his hand, although his dismissal had more to do with the quality of the accounts of the country than with any lack of intrinsic interest:

> How many accounts of Palestine and Egypt do we have, all full of repetition and useless notions about the supposed holy places? It is not that I would discourage those who are inclined to make excursions into these countries. There remain assuredly many things to discover there. Most of their predecessors have neglected those things most worthy of being noted.[31]

Had he known that the Danish expedition would be detained in Egypt for nearly a year, we suspect that Michaelis would have attempted to fill the void with questions on Egyptian flora and fauna, the incidence of famine in the land, the probable location of the Land of Goshen, and the treasure cities of Pi-'thom and Ra-am-ses, and the Egyptian method of making bricks. As it was, the members of the expedition were left to their own devices. But even left untutored, as we will see, Niebuhr produced results that would represent an important contribution to the nascent discipline of Egyptology. The lack of a specifically biblical focus undoubtedly had a great deal to do with his success.

30. See Gardiner, *Egypt of the Pharaohs*, 52.
31. *Fragen*, Preface, V.

5

To Cairo

> The ancient descriptions of Egypt abound with the names of cities, of which most have undergone such considerable change that whoever tries to locate them would think that the country is deserted . . . so the ancient Egyptian cities have little by little disappeared, and have been replaced by others, many of which have, in their turn, fallen into oblivion. If the cities of Egypt appear to us to have undergone greater change than those of other countries, it is simply because we have more ancient accounts of them. (*Travels*, vol. I, 94–95)

As they made preparations to depart Alexandria, the members of the Danish expedition were about to enter the timeless land of Egypt. As we have seen, Alexandria had always been an anomaly, an outward-looking window on the Mediterranean world and the creation of outsiders. With the coming of the Arabs in the middle of the seventh century the window had effectively closed, and the vast majority of the inhabitants of the country reverted to their unchanging preoccupation with the river and the fertile land it had created. But there was a new group of outsiders who ruled this timeless Egypt, as there had been continuously at least since the coming of the Persians in the sixth century BC. The Persians had been followed by the Greeks, the Romans, the Byzantines, and then by the Arabs. The Arabs had ruled through the Umayyad and Abbassid Caliphates until the middle of the ninth century. But they had been succeeded by the Fatimids from the west and then by a series of usurpers, first Turkish and afterwards Circassian, Georgian and Mingrelian who had given the ruling classes of the country a Central Asian then a Cuacasian character. These had been followed in the sixteenth century by Ottoman Turks who brought the country under the nominal control of the greatest empire of its day. By the time the Danish expedition arrived in 1761, this latest domination by foreigners had already lasted for nearly a thousand years. And, as it had been for the previous suzerains—Persians, Greeks, Romans, and Arabs—so it was for these latecomers: Egypt was a wealthy land, to be plundered of its resources. If there was a constant in Egypt, amid all this superficial change, it was this pillage and Niebuhr sees it at once:

> The Persians, Greeks, Romans, Arabs and, finally, the Turks, all foreigners, who have ruled in succession in Egypt, and who seem to have dedicated themselves to ruining this fertile country by their misgovernment, annually extracted such considerable sums, and thereby reduced the means of subsistence of the inhabitants, that the country has necessarily become continually more depopulated and the number of cities reduced.[1]

The country was remote and unruly, and this gave ample scope to the unscrupulous representatives of the ruling powers, as well as the most disorderly of the locals. We have already seen the state of relations between the Bedouins and the settled population. We will see below, in greater detail, the relationships between the other factions in the country.

The members of the expedition would have preferred to travel overland to Cairo, if only to see the Delta. But, after the tumultuous events they had witnessed in Alexandria, it was not a risk they were willing to take:

> The Europeans who have published accounts of their journeys from Alexandria to Cairo all took one route, going first to Rashid, and from there to Cairo on the Nile. We would have preferred to travel overland, to see parts of Egypt that were still very little known. But, from the above, the reader can surely see that the nomadic Arabs made this course impossible ...[2]

They left Alexandria by sea for Rashid (or Rosetta) on the 31st of October 1761. However, after leaving the city they had traveled only four leagues[3] before they were delayed by an unfavorable wind and some of the party decided to proceed overland with a group of Turks. Niebuhr waited and kept to the original plan, covering the remaining six leagues between the two cities and arriving in Rashid on the 2nd of November, about the same time as the others. With the effective end of European trade through Alexandria after the Ottoman conquest of Egypt in 1517, riverine traffic had been

---

1. *Travels*, vol. I, 96.
2. Ibid., 54–55.
3. Niebuhr means throughout a German league, each of which was equal to 3.25 English miles or 5.23 of today's kilometers. For a full discussion of the various measures used in the eighteenth century, see the chapter on the Delta.

diverted to Rashid, and the latter city had become the entrepôt for merchandise moving between Cairo and the Mediterranean. There were consuls representing France and Venice in the city, as well as European merchants to see their merchandise onward to Alexandria or the Levant. The Europeans living in the city believed it was the site of the ancient seaport of Canopus, and twenty marble columns had been unearthed nearby that year and sent to Cairo. Niebuhr, able to take sightings again, determined that Rashid was at latitude 31° 24' 21". Perched on a height on the western bank of the river, it had a "charming view of the *Nile* and the Delta."[4] Baurenfeind sketched a view of the city that appears as Plate VI in vol. I of the *Travels*.

The members of the expedition stayed with the Franciscans, the first of many such sojourns with Catholic religious orders in the East. They found the inhabitants of the city to be friendly, unusual enough in Egypt that Niebuhr made a note of it, and remarked that they wished that their stay could have been longer. But they were in some haste to reach Cairo, and on the 6th of November they left Rashid in a small flat-bottomed boat. Niebuhr took one sighting ashore near Deirut, but for the rest of the time was content to note the changes in the course of the river with his pocket compass and the elapsed time from bend to bend. But in this preliminary gathering of information the boatman was not cooperative, and "couldn't, or perhaps wouldn't" remember the names of the villages they passed. Pirates infested this stretch of the river and they posted an armed sentry at night. The obvious presence of Europeans in the boat was both a temptation and a deterrent, the prospect of plunder not being outweighed by the probability that the passengers were armed. The thieves were as at home in the water as the Europeans were on land, and would slip aboard unnoticed and even steal items from under the heads of passengers while they slept. So they often sailed at night, which made careful observation impossible and Niebuhr's detailed map of this part of the river was made only after several short trips he made while they were in Cairo.

Niebuhr's Sources on Egypt

In spite of Michaelis's comment above, Niebuhr suggests that the country was "very little known," at least to Europeans. What was the information about Egypt available to Europe in the middle of the eighteenth century? A review of the sources Niebuhr himself used in the preparation of the *Travels* will help to answer the question. A first group are the ancient authors, most of them Greek and Roman, including Strabo, Arrien, Pliny, Herodotus,

4. *Travels*, vol. I, 57.

Ptolemy, Agatharcides, Aristophanes, Diodorus Siculus, Curtius, Xenophon, Nearchus, and the anonymous author of the *Periplus of the Erythrian Sea*. The Greek Strabo (63 BC to 21 AD, est.) was the most catholic of the ancient geographers in terms of the breadth of his interests, which included history, ethnology, anthropology, and botany as well as geography. His seventeen-volume *Geography* included an account of Egypt, based on his residence there in 25–24 BC, during which time he ascended the Nile as far as the first cataract at Aswan. With regard to the location of certain of the ancient cities of the country, Strabo had stated that the Pyramids could be seen from Babylon and that Memphis was opposite that city, details that were important in establishing the site of the former capital of the country. Pliny the Elder, or Gaius Plinius Secundus, (23–79 AD), the Roman historian whose *Natural History* also survived, is cited by Niebuhr for his testimony that the Pyramids of Giza lay between Memphis and the Delta. As we will see below, the location of Memphis was still a matter of controversy in the middle of the eighteenth century.

Claudius Ptolemaeus (100–165? AD), or Ptolemy, was the Egyptian astronomer and geographer, who left a thirteen-part *Geography*, including maps and latitudes and longitudes, with a part devoted to Egypt. Ptolemy placed Memphis at latitude 29° 50',[5] important in that it showed the city to be well south of the bifurcation of the Nile. Flavius Arrianus, or Arrian, of Nicodemia (96–180 AD) whose most important work was the *Anabasis of Alexander*, was the most trustworthy source on Alexander the Great. We have already seen above his contributions to the history of Alexandria. Another contributor was Diodorus Siculus, a Roman and contemporary of Julius Caesar, who traveled to Egypt in 59 BC and, among other things, identified forty-seven tombs in the Valley of the Kings. His *Bibliotheca* included a description of Egyptian ruins. There were other sources as well, including Plutarch and Ammianus Marcellinus, who had contributed to Western knowledge of the ancient Egyptians and all these sources became available to western scholars of the Renaissance when the works of the ancients were opened again to critical examination. They were also available to Niebuhr in the preparation of his *Travels* and *Description*, some in the small traveling library the expedition carried with it. Most, we suspect, he found in the Royal Library in Copenhagen after his return.

---

5. Niebuhr himself, in spite of the fact that he correctly placed Memphis to the south of Cairo in the neighborhood of Saqqara, did not visit the site and so could not take his own sighting. However, he determined that the latitude of Cairo as 30° 2' 57" and this geographical evidence buttressed his argument that Memphis was not located in the neighborhood of Giza as some maintained.

To Cairo

But the greatest resource available to scholars was, surprisingly, Herodotus of Halicarnassus. Herodotus (484–25 BC) was the earliest of the classical sources, although even his account, dating from the mid-fifth century BC during the Persian period in Egypt, was written two thousand years after the building of the pyramids at Giza. The "father of history" has always been controversial, and he remains controversial to this day. A reading of his account of Egypt, largely comprising Book Two in *The Histories*, where he covers plant and animal life, the rise of the Nile, religion, the hieroglyphs, the major monuments, and quasi-historical anecdotes, demonstrates why. The mixture of fact with what appears to be patent fiction, involves him and the reader in a little conspiracy: it is difficult to believe that he doesn't know that he is putting the reader on. It is equally hard to believe that Herodotus doesn't know that the reader knows it too. He is careful to couch the most fantastic of his tales in the language of a reporter, claiming that he really doesn't believe them himself, but is only relating what others have said. This allows him to weave his characteristic tapestry, part fact, part fiction, part folk-tale, without being answerable for the result.

Credulous he is not. His claim to have ascended the Nile as far as Aswan appears dubious, if only because he doesn't describe the monuments at Thebes. But many details in his description of Egypt we now know to be accurate: his careful description of the sacred ibis, the widespread use of barley beer, the eleven fathoms of muddy bottom a day's sail from the mouth of the Nile, the "black and friable" earth of the alluvium, the "sacred and common writing," the lower half of the pyramid of Mycerinus being cased with "Ethiopian stone" (red granite), among many others. These details could have been gleaned from other sources, and there are those who suggest that Herodotus never visited Egypt at all. But they are enough for some to substantiate his claim that he saw the country with his own eyes, sometime between 450 and 430 BC.

His history—probably learned from the priests of Hephaestus at Memphis—is confused, although he has the broad outline and many details right: his identification of the first king with Min is commonly accepted today in its variations as *the* 1st Dynasty Pharaoh. But his chronology is convoluted, covering "three hundred and thirty monarchs in the same number of generations," before arriving at "Sesostris, who succeeded them."[6] He correctly identifies the 4th dynasty Pharaohs Cheops (Khufu), Khephren (Khaphre) and Mycerinus (Menkaure) as the builders of the pyramids at Giza. But before reaching the 4th dynasty he has passed through Pheros, Proteus,

---

6. Sesostris appears to be an amalgamation of the 12th-dynasty pharaohs Senrowset I and Senrowset II with 19th-dynasty Ramesses II. Herodotus, *The Histories*, trans. Aubrey de Selincourt.

and Rhampsinitus, all Greco-Egyptian folkloric blends with no historical basis in fact. He has also digressed into a discussion of Helen and Homer, and the impulse to weave into the history of Egypt details from his own national epic appeared as irresistible to this fifth-century BC Asiatic Greek as it would later to Jewish and Christian apologists. Herodotus also placed Memphis in the narrowest part of Egypt, lending further support to the contention that it lay south of Giza.

Niebuhr cites Herodotus for his statement that Egyptians made yearly pilgrimages to the cities of Heliopolis, and Busbastis and Busiris in the Delta.[7] The Delta seemed to Niebuhr to be less favored than Upper Egypt for the preservation of ruins, if only because of the greater population and the natural tendency to re-use the building materials from the ancient monuments. However, in his map of the Delta (see Chapter 10 below) he has indicated where there are "unmistakable monuments of ancient cities,"[8] and he suggests that the upper Delta is a rich vein for those interested in the location of the ancient cities of the country. Also interesting from our point of view, Niebuhr suggests that the modern city of Samanud is the old city of Sebennytos, the birthplace of Manetho. These citations generally have to do with geography, not surprising given Niebuhr's interests.

A second group of Niebuhr's sources is made up of the medieval historians and travelers, Arab and otherwise. The earliest appears to be the Egyptian Sa'id bin al-Bitraq (died 939 AD), alias the Patriarch Eutychius of Alexandria. His *Eutychii Annales*, or the "Annals of Eutychius,"[9] was an early purportedly historical text, beginning with creation and ending with the reign of the Caliph al-Radi in the tenth century AD, to which an Arabic history of Sicily was appended. Along with the chronologies it included lives of the patriarchs of Alexandria, Rome, Jerusalem, Constantinople, and Antioch, a chapter on "the Knowledge of Our Lord Christ," and chapters on the periods of Diocletian and Eutychius himself. The complete document was translated into Latin by Edward Pococke (see below) as *Contexio Gemmarium*, published in two volumes in 1658. It was a source document widely used by early Orientalists. Niebuhr cites it as evidence in his discussion of the relative location of the cities of Babylon and Memphis. Eutychius says that one crossed the Nile in going from Babylon to Alexandria and thus, Niebuhr reasoned, Babylon must have been on the east bank of the river. It could not therefore, be the same location as that of the ancient city of Memphis, which was on the west bank. We now know this to be true.

7. *Travels*, vol. I, 98.
8. Ibid., 96.
9. In Arabic known as the *Nizam al-Jawhar*, or the "System of the Essential."

Abu 'Abdullah Mohammed b. Idris al-Hamudi, or Sherif Idrisi, was another medieval historian and geographer. He was born in Ceuta in Morocco in 493 AH (1100 AD) and spent many years at the court of Roger II of Sicily. There, in 1154 he completed the *Kitab Rudjar*, or the "Book of Roger," a description of the world and probably the most important medieval work of geography. It was synopsized in Rome in 1619 and translated into imperfect Latin as *Geographia Nubiensis* by the Maronites Gabriel Sionita and Johannes Hesronita. Although incomplete and inaccurate, it was used widely by early Orientalists and was one of their most important early sources. Even in the middle of the eighteenth century, after the rise of Protestant Orientalism, Rome remained an important center for Oriental studies. As we have seen, von Haven spent time in the Eternal City preparing for the expedition, studying Syriac and Arabic at the *Collegeo Maronitico*. The Church still had important contacts in eastern societies through its missionary and educational work, and Rome still had lessons to teach those interested in Oriental studies.

Interestingly, Rome was also home to the greatest collection of Egyptian obelisks in the world, Egypt not excepted. They were thirteen in number and had been removed in Roman times. Most had fallen during the centuries of turmoil following the fall of the Empire, and laid buried for centuries before renewed interest in antiquities beginning in the sixteenth century had rescued them. The obelisk in Piazza San Giovanni in the Lateran, for example, the largest of all surviving obelisks, was removed from Karnak and taken to Rome in 357 AD. It fell at some unknown date and was only rediscovered in the sixteenth century, lying seven meters beneath the surface in the marshes near the Circus Maximus. It was re-erected in 1588. The story helps to explain why some of the early European attempts at reproducing the hieroglyphs were so clumsy: these prodigious masses, most of them covered with hieroglyphs, were unseen, hidden beneath the debris of centuries. Niebuhr cites the *Geographia Nubiensis* as incontestable evidence of the location of the city of Heliopolis.

Another medieval Arab source was Isma'il Imad ad-Din al-Aiyubi (1273–1331), or Abulfeda, a prince, historian, and geographer. Born in Damascus of a branch of the Egyptian Aiyubides, where the family had fled after their overthrow by the first of the Mamluks, he served in various capacities including that of governor of Hama, eventually acquiring the titles al-Malik and al-Saleh and the hereditary title of Sultan. But he is primarily known as a scholar and his geography, *Takwim al-Buldan* (or "Survey of the Countries") was completed in 1321 AD. Latin translations of the geography appeared in Europe as early as 1650,[10] and one of the early chapters on Egypt is cited by Niebuhr as evidence of the location of Memphis.

We have already seen the dubious contributions of Benjamin of Tudela, or Rabbi Benjamin, the twelfth century traveler from Tudela in Navarre. Although Tudela was an early prize of the *reconquista* and had been in Christian hands since the ninth century, it had been a dependency of the Cordoba Caliphate and retained a large Moorish quarter. It is probably there whence Rabbi Benjamin came. He traveled with the intention of seeing synagogues all over the world and his itinerary suggested that he visited central Europe, Greece, Palestine, Mesopotamia, Ethiopia, India, and Egypt, although there is some question as to whether he saw many of these places with his own eyes. He wrote his account in Hebrew in 1160. It was printed in Constantinople only in 1543 and translated into Latin in 1575 by Arius Montanus.[11] There appears to be no doubt that Rabbi Benjamin actually was in Egypt and he provides valuable information on the location of Jewish monuments in Cairo. His contribution to ancient Egyptian history was more problematic.

Leo Africanus, or al-Hasan al-Wazzan al-Ziyati, was an Arab geographer of the sixteenth century. He was born in Grenada in 1496, of a distinguished Moorish family, shortly after completion of the re-conquest. He was taken to Africa as an infant and educated at Fez. He later accompanied an uncle as the envoy of the sovereign of Fez to Timbuktu, beginning a life of travel that took him throughout the Muslim world. He visited most of North Africa, Arabia, Persia, Armenia, Syria, and Egypt in various capacities, private as well as diplomatic and official. In 1517 he was captured by Christian corsairs and taken to Rome where Pope Leo X recognized him as a man of learning and settled on him a considerable pension. He learned Latin and Italian and converted to Christianity, taking the Pope's name as his own. His *Description of Africa*, written in Arabic, secured him the title "the African." He translated it himself into Italian in 1526 and, although it was full of grammatical errors, it came to the attention of Ramusio, the compiler of one of the great compendia of travels, who included it in his *Navigationi*. It was translated into French as the *Description de l'Afrique* in 1536. This is surely the version Niebuhr used. The editor advertised that no one had yet described that part of the world with so much detail, accuracy, and truth. It was a judgment widely shared and the work was translated into Latin, Dutch, English, and German and heavily used by European scholars. However, after the death of Leo X, the African fell from favor. He returned

---

10. It is probably the version printed by Gagnier in Oxford in 1723 to which Niebuhr refers.

11. The Latin was translated into French and included in Bergeron's *Recueil de voyages*, and it is probably this translation to which Niebuhr refers.

to Tunis and died a Muslim in 1585. Niebuhr refers to him for evidence of the development of the city of Cairo.

Another Oriental source was the Latin translation of the life of Salah ad-Din by Baha ad-Din Yusuf Ibn Shaddad, an Arab biographer born in Mosul in 1145 AD. He was appointed Qadi al-Askar[12] of Jerusalem in 1190 by Salah ad-Din, shortly after the city was recaptured from the Crusaders. He died in 1235. His chief work was this biography which was translated into Latin by the Dutch Orientalist Albert Schultens[13] as *Index geogr. in vitam Salidini*. Niebuhr cites the work as evidence of the location of the city of Heliopolis and of the building program of Salah ad-Din in Cairo. Finally, there was Ibn Yousef al-Mukadessi Marai, whose *histoire des Soverains de l'Egypte*, was translated into German by Reiske, whom we will see later as one of Niebuhr's collaborators in his own works. It was cited by Niebuhr as a source for the Fatimid building period in Cairo.

The members of the Danish expedition were, of course, only the latest in a long line of Europeans that had sojourned in Egypt. There had been many travelers to the country between Rabbi Benjamin and the Danish expedition in the mid-eighteenth century, and many had left interesting accounts. Many of these accounts were also cited by Niebuhr as he prepared his *Travels*. We have already referred to Bernhard von Breitenbach, the German from Mainz who traveled to Egypt in 1484 and whose account appeared soon thereafter in Latin and German. The Polish Prince Radziwil, the Duke of Olica and Nieswitz, was a famous sixteenth century traveler who traveled to the Holy Land and then Egypt in 1583. His *Voyage to Jerusalem* was published in Polish, which was afterwards translated into Latin in 1601. These early Europeans, with their religious and cultural provincialism, tended to see Egypt through the lens of the biblical story of the bondage, a distortion rather like viewing an object through the wrong end of a telescope: it diminished rather than enlarged the object. The Arabs, whose tradition of history was not dominated by the story in the Hebrew Scriptures (although Muslims accepted its broad outlines), were more matter-of-fact about Egypt. However, beginning with the seventeenth century, this European view had begun to change and much valuable work was done in the few European sources that did exist. John Greaves, Thomas Hyde and Edward Pococke, in particular, made major contributions to a scientific understanding of the monuments of ancient Egypt.

12. Literally, "the judge of the army," the position later became important in the Ottoman times as an effective vice-chancellor of the Empire. Traditionally a Qadi al-Askar, or *Qazasker*, was appointed for each of the Asian and European parts of the Empire, respectively Anatolia and Rumelia. Rumelia was the senior position.

13. See Introduction.

Niebuhr in Egypt

Greaves (1602–1652) was an English mathematician who, in addition to his scientific pursuits, studied Oriental languages and was able to read works on astronomy in the original Arabic, Persian, and Greek. He was a contemporary of Archbishop Laud (see below) who became a supporter, and in 1638 he traveled to the East, including a visit to Egypt in his itinerary. Greaves spent his time in Cairo measuring the pyramids and collecting Oriental manuscripts. He was a precursor to Niebuhr in that he viewed the monuments in a spirit of scientific detachment and mathematical precision, carrying with him "a radius of ten Feet most accurately divided into Ten thousand Parts." His survey of the pyramids was the most thorough to date. He returned to England in 1640 and was chosen Sullivan Professor of Astronomy at Oxford. In 1646 he published his *Pyramidographica*. However in the general religious ferment of the age, he suffered from the opprobrium attached to those with too latitudinarian an approach to the subject of the Orient. His rooms were ransacked and many of his manuscripts were lost. The *Miscellaneous Works of John Greaves* was published posthumously in 1737 and was used by Niebuhr in the preparation of his own works.

Thomas Hyde (1636–1703) was another English Orientalist who brought a spirit of scientific detachment to the subject of the Orient. His career was like many of those we have already seen, beginning with an interest in religious matters before expanding into an interest in a more secular Orient. He matriculated at Oxford, devoting himself to Persian and assisting in the publication of Persian and Syriac versions of the Bible. In 1658 he became a reader in Hebrew at Queens College, received his MA in 1659 and was subsequently appointed keeper of the Bodleian Library. After taking orders, he occupied a series of religious posts before succeeding Edward Pococke as Laudian Professor of Arabic at Oxford in 1691. He then held the post of secretary and interpreter in Oriental languages to the government under Charles II, James II, and William II and was regarded as the greatest expert on Oriental subjects in Europe. In 1700 he published his most important work, *Historia religionis veterum Persarum eurumque majorum*, the first attempt to treat the subject of the ancient Persian religion in a scholarly way. He also published the text and translation of an Arabic treatise on astronomy by Ulugh Beg bin Shah Rukh[14] on the celestial latitude and lon-

---

14. He was the fifteenth century ruler of Turkestan and Transoxiana. Born in 1393, he was an astronomer, artist and theologian, and he is credited with making of Samarkand what Timur had only dreamed of. His astronomical work was particularly valuable, developing powerful new instruments and building an observatory in Samarkand, correcting the observations of Ptolemy in the process. He was less successful in statecraft and suffered from the ingratitude and eventual rebellion of a son who had him overthrown and executed in 1449.

gitude of the fixed stars. Hyde intended to publish a later compendium of various Oriental subjects. Death cut short the intention although Gregory Sharpe gathered together some of the parts and had them published in 1767 as *Syntagma Dissertationum et Opuscula*.

Another of the early pioneers was Edward Pococke (1604–1691) the great English Orientalist whose research, in the words of P. M. Holt, "marked the emancipation of scholarship from bigotry and who, with the other great Orientalists of his time, laid the foundations of the modern understanding of Islam, its history and its culture."[15] Pococke was ordained in 1629 and shortly after was posted as chaplain to the Levant Company in Aleppo. He returned in 1636 to the newly established Chair of Arabic at Oxford, created especially for him by William Laud, the Archbishop of Canterbury and a great supporter of Oriental studies. Laud would later be tried and executed by the Long Parliament, in 1645, which put the whole subject of Oriental studies in jeopardy. The attitudes of the traditionalist, broad-minded clergy were increasingly out of favor with the Presbyterians and Puritans. Pococke survived the purge and devoted the remainder of his life to scholarship and publication. Like many other early Orientalists, he was a clergyman and he survived the vicissitudes both of the religious controversies that swirled about Oriental studies, as well as the increasing focus on science, which began to eclipse the traditional interest in languages. Pococke was widely quoted by Sale in the latter's *Preliminary Discourse* to the Koran. The *Specimen historiae Arabum* for which Pococke is primarily known, an excerpt from the Arabic *al-Muktassar fi'l dowal* of Abu'l Faraj b. al-Ibri, or *Bar Hebraeus* was, again in Holt's words, "profoundly erudite in content and noncontroversial in tone."

The above travelers and scholars had all contributed to a growing body of knowledge about Egypt in the middle of the eighteenth century. But the most detailed were the accounts of Norden and Richard Pococke, and to the latter Niebuhr owed his greatest debt. It is probably no accident that they were also the least credulous of travelers. We have already mentioned Frederick Ludwig Norden (1708–1742), the officer in the Royal Danish Navy who was dispatched on the first Royal Danish Expedition to Egypt in 1737. He sketched the antiquities of Alexandria before proceeding to Cairo where he arrived on July 7th, 1737. In the vicinity of the capital he examined the pyramid complexes of Giza and Dahshur, and he may have stumbled on the truth when he pronounced that the pyramids of "Dagjour" (Dahshur) had "suffered more, since they are more damaged: from whence one may presume that they are more ancient" than those of Giza. Built by Sneferu,

---

15. Holt, *Studies in the History of the Near East*.

the first pharaoh of the 4th dynasty and the immediate predecessor of Cheops, they were, in fact, earlier than those of Giza and inaugurated the age of the great pyramids. Norden quoted Herodotus on his (correct) assertion that the three major pyramids at Giza were built by Cheops, Khephren and Mycerinus.[16] He also offered the opinion that the age of the pyramids was much greater than commonly accepted:

> We must absolutely throw back the first epocha of the pyramids into times of so remote antiquity that the vulgar chronology would find a difficulty to fix the aera of them.[17]

Norden left for Upper Egypt in November of 1737, where he continued his mapmaking and sketching, eventually reaching Aswan in December of that year. He kept a detailed journal and filled it with impressions, maps and sketches of Meydum, Assiut, Karnak, Luxor, Esna, Edfu, and Aswan. At "Lukoreen" (Luxor) he was certain, correctly, that he saw the remains of ancient Thebes. The expedition returned to Cairo in February 1738, passing Pococke at Esna on the return journey down the Nile, and to Copenhagen later in the same year. His Egyptian material remained unpublished after his return to Denmark although he translated his journal into French. He was later assigned to England as part of his naval duties and was elected a member of the Royal Society in 1741. Norden attracted considerable attention with his publication in English in 1742 of *Drawings of some Ruins and Colossal Statues at Thebes in Egypt*. He died of consumption in February 1742 at the age of just thirty-four. But his unfinished work was entrusted by King Frederick V to the Danish Academy of Letters and Sciences and published in 1755 in French as *Voyage d'Egypte et de Nubie*. It was translated into English as *Travels in Egypt and Nubia*, appearing in 1757. In spite of the fact that Norden was a Dane, sent to Egypt on another Danish expedition, there is surprisingly little reference to Norden in Niebuhr's account. Niebuhr refers to Norden's description of the facing of the "second pyramid," where, as we have seen, Norden himself, having read his Herodotus, refers to the three pyramids at Giza by name.

But Niebuhr owed his greatest debt of all to Richard Pococke (1704–65), the English traveler and divine who is not to be confused with, or apparently related to, the Edward Pococke we saw above. Niebuhr remarks that Pococke "examined everything with so much care and intelligence, and

---

16. See Herodotus, *The Histories*, Book 2, 131–33. The first direct proof, other than Herodotus, of the names of the builders came only in 1837. In that year Howard Vyse blasted out entrances to three relieving chambers of the first pyramid. Hieroglyphics left by workmen in one of the chambers included the royal name of Cheops and this confirmed that he had built the pyramid. See Lehner, *The Complete Pyramids*.

17. Norden, *Travels in Egypt and Nubia*, vol. I, 109

of all of them left the most complete account."[18] As mentioned above, he was in Egypt in 1737, and made a tour of the country, examining Alexandria, "Grand Cairo," and the Giza complex. He, too, names the builders of the great pyramids as reported by Herodotus. He visited Memphis, correctly identifying its location on the west bank of the Nile, the subject of an ongoing controversy. He also saw Saqqara, Dahshur, and Fayoum where he searched for the famous labyrinth. As mentioned above, he passed Norden in Upper Egypt late in the year and ascended the Nile as far as Ethiopia, examining the granite quarries, Philae, and the cataracts. Along the way he stopped at Thebes and measured some of the ruins at Karnak. From Upper Egypt he returned to Cairo before proceeding to Arabia Petraea and Mount Sinai, where he speculated on the journey of the children of Israel during the Exodus. He returned to England in 1742 via, among other places, Jerusalem, Baalbeck, and Cyprus. In 1743 he published *A Description of the East*, which was far superior to the usual traveler's fare on the Near East. No less a critic than Gibbon described it as "of superior learning and dignity."[19] Pococke's works would be Niebuhr's constant companion during the sojourn in Egypt, perhaps a part of small traveling library the expedition carried.

The many accounts of Egypt occasionally sparked controversy, and Pococke and Thomas Shaw (1694–1751) tilted in print over the location of Memphis. Shaw was an English traveler and antiquary who began his career in the East, like so many others we have seen, as a man of the cloth. He was made chaplain to the English factory at Algiers in 1720 and traveled widely from that base, going to Egypt, Sinai, and Cyprus in 1721 and afterwards to Jerusalem, Palestine, and North Africa. Returning to England in 1733, he was made a fellow of Queens College, Oxford and a member of the Royal Society. He published his *Travels or Observations Relating to Several Parts of Barbary and the Levant* in 1738. Parts of it were challenged by Richard Pococke, and Shaw issued supplements in 1746 and 1747 which were included in the second edition of the book. Niebuhr refers to the dispute between Shaw and Pococke over the location of the ancient Memphis. Pococke was closer to the truth, placing Memphis on the west bank of the Nile near Mitrahina while Shaw maintained that it was located at Giza. That did not prevent the authors of *An Universal History from the Earliest Accounts to the Present Time*[20] from coming down on the side of Shaw:

18. Pococke, *A Description of the East*, 43.
19. Gibbon, *The Decline and Fall, of the Roman Empire*, Chapter li, n. 69.
20. The work of many contributors, the *Universal History* was a multi-volume compendium first published in England in 1736–44 and afterwards updated frequently. The first eighteen volumes were of ancient history from the Flood to the early Christian period, "drawn from the most authentic documents of every nation." The modern portion

> The city of Memphis was in the same place as the present village of *Geeza*. This we learn from Dr. Shaw, whose observations on Egypt and Arabia Petrea are more worthy of praise, and preferable to those of all modern accounts for their truth, at least plausbility, erudition, accuracy, and judgment ... in a word, his book still stands up after all the envious and malicious attacks, whose authors have taken up their pens to imitate or discredit him, and have sunk into the oblivion, or at least are read with the contempt that they so rightly deserve.[21]

Niebuhr is plainly scandalized by the dispute and the strength of the emotions it excited. Greater men than Shaw have maintained equally wrongheaded notions, and one is reminded of Richard Burton's dispute with Speake over the source of the Nile. It was Sieur de Villamont, who arrived in Egypt in 1590, who is credited with first placing Memphis at its correct site to the west of the river and south of Giza. But this did not end the controversy and in the eighteenth century we find Pococke and Shaw still at each other's throats over the issue. After reviewing the available evidence, Niebuhr casts his vote, correctly in this case, for Pococke and Villamont.

Finally, with Jacob Bryant, we will complete our review of the near-contemporaries of Niebuhr and their contributions to the nascent discipline of Egyptology. It will represent a return from the secularists to a man who attempted to reconcile what he knew of ancient history with the Hebrew Scriptures. Bryant is a relative modern, and his syncretic effort demonstrates that such attempts depended not on the century or the state of knowledge or the scientific temper of the age, but on the predilections of the author. The effort to reconcile secular with what purported to be sacred history would continue in the nineteenth century with a figure like Piazzi Smyth who as we have seen, was the national astronomer of Scotland, and the twentieth with Flinders Petrie, probably the greatest of all Egyptologists. This was long after advances in Egyptology and geology, to name only two disciplines, had armed scholars with the tools to better disentangle historical fact from legend.

Bryant was a Fellow of Kings College, Cambridge and former Secretary to the Duke of Marlborough. His *Observations Relating to the Various Parts of Ancient History* published in Cambridge in 1767, is probably as

---

in 42 volumes, *The Modern Part of a Universal Modern History*, covered the period from the rise of Mohammed to the then present day. The complete sixty-volume set contained an astonishing 128 million words, and served as a kind of eighteenth-century precursor to the Internet. Niebuhr probably used the German translation *Algemeine Welt-historie von Angebum der Welt bis auf gegenwaertige zeit* (Halle, 1744–91). Like the English editions it had a modern supplement, *Historie der Neuern Zeiten*.

21. Quoted by Niebuhr in *Travels*, vol. I, 104.

To Cairo

close as any of the works we have seen in attempting to reconcile science with the Bible. It would be difficult to imagine a more rigorous proponent of the type. After animadversions on the lack of rigor of those attempting to establish the place of residence of the Israelites in Egypt, Bryant goes on to state in the preface that there *are* sources that have never been contested, one of those being the Bible.

But Bryant exceeded even himself with his later syncretic work, *A New System, or, an Analysis of Ancient Mythology* published in 1774–76. A sampling of the part on Egypt is probably not untypical of a certain kind of contemporary Egyptology:

> The Mizraim seem to have retired to their place of allotment a long time before these occurrences (note: dispersion of the Cushites) . . . The country, of which they were seized, was that which in aftertimes had the name of Upper Egypt. They called it the Land of Mezor . . . The lower region was a great morass, and little occupied . . . In the process of time, the Mizraim were divided into several great families, such as the Napthuhim, Lehabim, Ludum, Pathrusin, and others . . . At last, the Titanic brood, the Cushites, being driven from Babylonia, fled to different parts: and one very large body of them betook themselves to Egypt . . . They took Memphis with ease, which was then the frontier town in Egypt . . . There are many fragments of ancient history, which mention the coming of the Cushites from Babylonia into the land of the Mizraim . . . An account of this sort is to be found in Suidas. He tells us that Ramesses, the son of Belus (of Babylonia) who was the son of Zeuth, came into the region called Mestraea, and gained sovereignty over the people of the country. He was the person whom they afterwards called Aegyptus and the region was denominated from him.[22]

The conflation of Hebrew, Babylonian, Greek, and Egyptian myth and its repackaging as historical truth was typical of the effort, although it was not without a hint of truth. Egyptian prehistory shows unmistakable, although still puzzling, signs of Mesopotamian influence.[23] Needless to say, Bryant was not a traveler but a compiler and the work had the decided odor of the study, not the field. He did use sources that were legitimate in spite of his penchant for wrong-headed synthesis. The history of Egypt, by far the largest of the treatises, is largely devoted to the period when the Israelites were purportedly

---

22. Bryant, vol. II., 233–35.
23. See Gardiner, *Egypt of the Pharaohs*, 397–98.

in the country. Niebuhr obviously read the book after his return. It does not appear to have colored his views of ancient Egypt, as we will see below.

We left Niebuhr and his companions adrift on the river in early November, the north wind in the winter an ally as they beat against the strong current of the Nile. Occasionally it was necessary for the boatmen to land and drag the flat-bottomed boat with ropes when the wind died down. On the 7th of the month in the vicinity of Deirout, finding the inhabitants friendly, Niebuhr took out his instruments and established the latitude of the place as 31° 13′. It was the last observation he would take before they reached Cairo, and he reverted to the practice of simply noting the bearing of the bends in the river and elapsed time between landmarks. He was not the first to notice the lush verdure on either side of the river, the flat-roofed huts made of unbaked brick, and the date groves and pigeon cotes rising above the level surface of the Delta. They arrived at Bulaq, the port of Cairo, on the evening of the 10th of November, 1761 and settled down for an enforced stay of nearly a year. Egypt was an afterthought in the expedition to Happy Arabia. As we have seen, had it not been for Hussein Bey Kashkasha's decision to brook no further extortion from the Bedouins of the Hejaz, Europe might have been deprived of Carsten Niebuhr's careful observations in the country. The world of scholarship can be grateful for Hussein Bey's truculence.

6

The Mother of the World

> It is not possible to decide for certain whether the city of Cairo has grown or shrunk in size in recent centuries, since until now, we have not had a plan of it . . . So in Plate XII I have prepared a map of *Cairo* and the adjoining cities of *Bulak, Masr el-Atik,* and *Djise.* This was so difficult, and the well-known insolence of the populace of Cairo towards all those of a religion different from their own made the work so dangerous, that as yet no European has undertaken . . . such a thing. (*Travels,* vol. I, 108).

When the members of the Danish expedition arrived in Cairo in November of 1761, they settled in for a stay that would be their longest period together as a group in any one place. The results of the expedition—always excepting Niebuhr's single odyssey—are, therefore, in large part a product of their stay in Egypt. Just as Pietro della Valle had done in 1616, they landed at the port of Bulaq,[1] before making their way over "a very beautiful plain from which the flood waters had retreated a little"[2] to the Fatimid core of the city. In the intervening 145 years the city had expanded westward towards the Nile, so in Niebuhr's map it appears to be a journey of a little over 5,000 Danish feet, or just under an English mile, to the outskirts of the settled area. It had been double that distance in the seventeenth century. But Cairo in the middle of the eighteenth century was still a densely-populated core, a rough quadrangle laid out parallel to the river, on a roughly northeast by southwest axis.

---

1. Perhaps from the French *beau lac,* or "beautiful lake." But such extended etymologies are suspect. A more likely derivation is Redhouse's Turkish *bulaq,* "A spring, a fountain," 408.
2. Della Valle, *The Pilgrim,* 48.

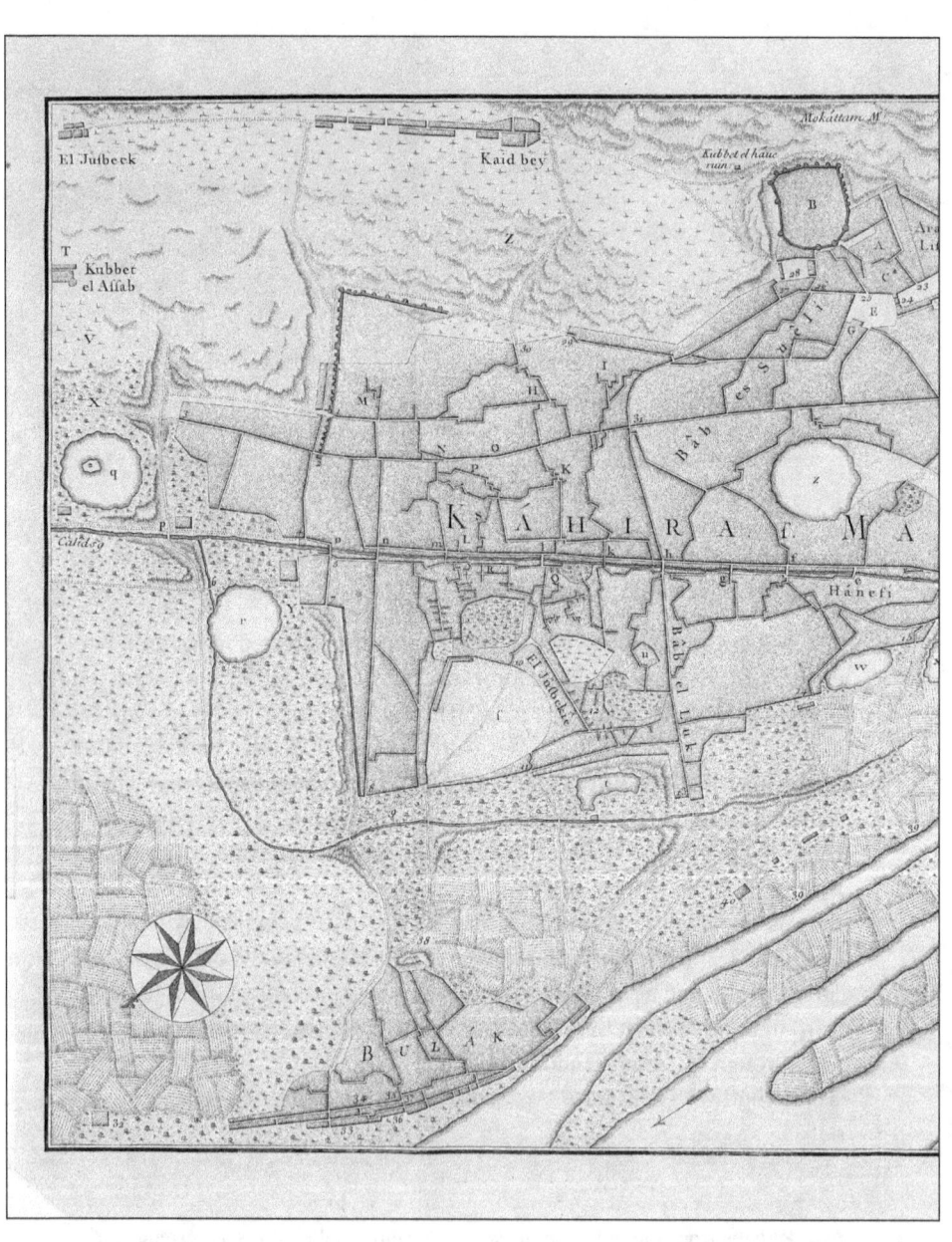

Tab. XII.

URBIS KÁHIRA
nec non
OPPIDORUM BULÁK
MASR EL ÁTIK ET DSJÍSE
ICHNOGRAPHIA
Auctore C. Niebuhr.

Niebuhr in Egypt

Oddly, for a capital city located near one of the major rivers of the world, it was not located on the Nile, but at some distance from it, and potable water had to be brought from the river on a daily basis. To the east of the quadrangle lay the cemeteries and the desert, and to the west was cultivation, the port, and then the Nile. To the south lay Old Cairo and the Roman (read Greek) fortress of Babylon, and to the north more cemeteries and the road leading to Syria. The view of the Fatimid city itself cannot have changed dramatically since della Valle's time, with the skyline dominated by the monuments rising in the distance above the piles of accumulated rubbish on the outskirts of the settled area. There was a reason why Cairo was called the city of a thousand minarets.

As they left the river and began the overland journey to the city proper, probably following the track leading to the "Bab el Hadid"[3] in its northwestern corner (see No. 8 in the map), Niebuhr and his companions would have first passed through a sown area before reaching the dregs of the flood in the Birket al-Azbakiyya or the lake that appeared seasonally with the rise of the Nile, so called after the Amir Azbak bin Tutuh, commander in chief to Qaytbey, who built a complex there in the late fifteenth century.[4] Above the minarets of the city, and further in the distance, they would have seen the Citadel with its Crusader-built walls. And, above the Citadel, overlooking all, lay the looming presence of the Muqattam, surmounted by the little mosque of al-Guyushi that Niebuhr would identify as "Kubbet el haue," or the Dome of the Wind.

They rented a house in the Haret al-Ifrang, or Frankish Quarter, where the few Europeans living in the city congregated. Niebuhr shows it (Q in the map)[5] to be just to the west of the *khalig* (Niebuhr's "calidsj"), or canal, and almost in the geometric center of the city, where today Shari' Muski crosses Shari' Bur Said. The *khalig* formed the western boundary of the medieval Fatimid city. Begun during the Late Period by the Pharaoh Necho II (610–595 BC) and completed by the Persian Darius (521–486 BC), the canal led to Kulzum near present day Suez and connected the Nile with the Red Sea. It had been repaired by the Emperor Trajan in the second century AD, and

---

3. I have preserved Niebuhr's transliterations in the text between quotation marks. The "Bab es sueli," for example, is the more familiar Bab Zuwayla, and notes are introduced only where the correspondence does not appear to be obvious.

4. It originally consisted of a palace and a congregational mosque, as well as commercial structures and buildings. They all fell into disrepair and the last of the structures, the mosque, was razed in 1869. All that remains of the original quarter is the name. See Behrens Abu Seif's *Azbakiyya and Its Environs* for an excellent general history of the area.

5. The letters and numbers are taken from Niebuhr's own key to the map. It is shown in full in Appendix B.

The Mother of the World

then by Amr ibn al-Aas after the Arab conquest in 641 AD. But shortly afterwards it fell again into disrepair, except in the vicinity of Cairo where in the eighteenth century it was still ceremonially opened in the autumn with the rise of the Nile. In 1761 it no longer marked the western boundary of Cairo but instead bisected the city, introducing a welcome, but seasonal, source of water into a city that was otherwise dry.

The area to the west of the medieval city had a storied history. Outside the walls of the Fatimid core and home to Copts and successive waves of immigrants, including Mongols and followers of the Great Khan of the Golden Horde on the Volga, it occupied an intermediate position between the city proper and Bulaq. When the Nile was high, much of the area temporarily became a pond or lake, the Birket al-Azbakiyya. Still later, with the Mamluk urban project, it was the home to the summer pavilions and residences of many prominent beys as they increasingly built near the pond. With the coming of the French in 1798, it would become the European quarter and Bonaparte's headquarters would be chosen at the former residence of one Alfi Bey. After the French occupation, the Azbakiyya area became the center of European Cairo, and European consulates would line the streets that bordered the gardens that replaced the pond: the Russian, Portuguese, Austrian, French, Prussian, Swedish, Italian, Dutch, and Greek consulates were all within a half-mile of one another to the north and east of the gardens. The British consulate was on the west, near the original Shepheard's Hotel. But even before the arrival of the French, the area had a cosmopolitan character, and in 1761 it was home to a polyglot population of Franks, Jews, Copts, and other minorities.

The members of the expedition were not a happy group, although there is no hint of these tensions in Niebuhr's account. After the acrimonious exchanges on the voyage across the Mediterranean and the superficial truce struck in Constantinople, there followed the report of the arsenic bought by von Haven in Rhodes. The two professors, Forsskal and von Haven, still bitter enemies, had now settled into what to the others seemed to be a deadly rivalry. Niebuhr, in his quiet way, tried to reconcile their irreconcilable differences. The other two members, Kramer and Baurenfeind, appeared to be little more than spectators to the feud. It was probably best for the equanimity of all concerned that Forsskal and Niebuhr alone stayed in the rented house. Shortly after their arrival in the city, von Haven was offered an apartment in the house of M. Bezoardin, "the most prominent Frenchman in the country,"[6] also in the Frankish Quarter. He accepted the offer with alacrity, finding the society, the table, and the wine cellar there more congenial than the workaday atmosphere prevailing in the other house. Herr Baurenfeind took a room with

6. Hansen, *Arabia Felix*, 99.

the Capuchin monks. Dr. Kramer stayed with the Dutch consul who, along with the representatives of France and Venice, was one of three permanent European consuls permitted in Cairo. Now separated from those who seemed to be more interested in leisure than work, Forsskal and Niebuhr set about making the most of what would become their enforced stay in Egypt.

Niebuhr spent the next ten months walking the length and breadth of the city and its suburbs; keeping thrice-daily records of the temperature on his Fahrenheit thermometer; polishing his language skills; and traveling on the Nile to Rashid and Damietta and back to Cairo, as he prepared his maps of the river and the Delta. There was much to see in the *umm ad-dunya*, or the "Mother of the World," still the first metropolis of the Arab world. In addition to 500 years of Mamluk and Ottoman monuments, there were the antiquities of ancient Egypt, the nearby Pharaonic ruins. To a man of his catholicity of interests, the time in Egypt could not but be profitably spent. Niebuhr would fill 211 pages[7] of his *Travels* with observations and impressions of Egypt and the Sinai. Seventy years later Edward William Lane would provide an essentially sociological study[8] of the Egyptians of his time. Niebuhr's view is more that of an engineer, but of an engineer with a wide range of interests. The chapters on Egypt include speculation on the location of the ancient cities of the country; careful measurements of the pyramids and copies of hieroglyphs; a survey of the current inhabitants, the form of government and commerce of the city of Cairo; practical matters such as descriptions of hydraulic mechanisms, plows, and ovens for hatching eggs; the dress of the Orientals; the "diversions" during their hours of leisure; and an account of the expedition's journey to Suez and then into the Sinai Peninsula. We will examine his contributions in each of these areas in more detail below. But first, we will look closely at his map of the city, annotated with his own commentary, to give us a snapshot of Cairo in 1761–62. We will note what the city had been, currently was, and in some cases would become in the years that followed.

Niebuhr's Map of Cairo

Niebuhr spent the time profitably during the winter and following spring learning the city, the language and the habits of the inhabitants, about which he had known little when he arrived. The study of classical Arabic would

---

7. Throughout the number of pages are stated as they appear in the original German edition.

8. Lane's scant twenty-one pages on the "industry" of the country (in a book of 570 pages) include the profession of begging and the use of intoxicating substances.

have prepared him for more formal exchanges, but not for the colloquial language used by the average Cairene. And, as we have seen, there was almost nothing in Michaelis's meticulous tutoring that would have prepared him for Egypt. Only with the onset of summer and the end of overcast and rain, both enemies to astronomical sightings, would he be able to take productive trips on the river. So he devoted his early attention to the city of Cairo and its immediate environs. We can imagine him, dressed *ala* Turk, wandering over the city and its suburbs, sometimes in the company of a *sarraj*,[9] more frequently alone. The temperature, records of which he attached as an appendix to vol. I of the *Travels*, shows that the weather cooperated in this effort. From November 1761 through February 1762 the high was only seventy-four degrees. The low of forty-two degrees on the morning of February 3rd would have been downright balmy to this northerner, although the Cairo cold could be penetrating.

In the summer, the weather would be an enemy to physical activity. From a high of ninety-six degrees Fahrenheit at noon on May 29th, the mercury would rise slowly over the next three months, achieving highs of one hundred-one in June and July before falling slightly in August. The nights gave little relief, with average temperatures at 11 o'clock in the low eighties throughout the summer. Even the breeze from the north would have little effect if not cooled by the presence of water from the nearby *khalig* and the Birket al-Azbekiyya. And, as will see, it was not until August 12th of the year that the dam on the river was broken and water introduced into the city. But, even in the heat, the distances were manageable. The map shows the confines of the city proper, from the "Bab en nascha" (4) in the north to the "Bab aijubbeh" (17) in the south, to be just over 2,500 geometric paces.[10] That would be approximately two English miles, easily walkable in an hour at a moderate pace. "Masr el Atik," or Old Cairo, was another 2,000 paces, or a mile and three-quarters to the south. From east to west, the city was just under 1,600 paces, or a mile and three-eighths in extent and Niebuhr, from the house on the *khalig*, was roughly equidistant between the eastern and western boundaries.

For journeys to the outlying areas, to the pyramids at Giza in the west or the monuments in the cemeteries to the east—al-Qarafatain[11] or the "two

---

9. A kind of assistant to the beys, literally "a saddler" who, as we will see, lorded it over the average Cairene.

10. Each pace was approximately 4.6 feet. See Chapter 10 for a discussion of Niebuhr's units of measure.

11. From *qarafa*, "cemetery, specif., that graveyard below the Mokattam Hills near Cairo" (Wehr, *A Dictionary of Modern Written Arabic*). The northern and southern parts together make up the dual form, *qarafatain*.

cemeteries"—he probably rode a donkey, accompanied by a janissary or a *sarraj*. But even the most distant of those outlying areas, the plateau at Giza on which the pyramid complex lay, was only a half-day's journey from the city, including the river crossing. Given the restrictions on Franks riding even a donkey, however, we suspect that he went about the city on foot, keeping a low profile and avoiding where possible the grandees whose trajectory through the streets was eased by footmen who did not spare the staff to clear the way. Besides, his own two feet were the surest measure of the distances he would carefully lay out on his map. His disclaimer as to what the map was *not* is vintage Niebuhr: "one will not find a history of the city here; I have described its location and its size as I actually found them."[12] As with the map of the Delta, his intention was to fill a gap in European geographical knowledge. The quote at the head of the chapter shows the difficulty of the undertaking. But he provides enough detail for us to flesh out a portrait of the city as he saw it with his own eyes.

In the map produced as Plate XII of vol. I of the *Travels*, Niebuhr confines himself to the major landmarks and thoroughfares, noting the odd orientation of the streets and adding that since the small dead-end lanes were in quarters where a foreigner was unwelcome, he could only indicate a general heading. Nevertheless, considerable detail is indicated by means of the several keys to the map, showing mosques, residences, churches, bridges over the *khalig*, and the many gates into the city. They are reproduced in full in Appendix B. Niebuhr's Arabic is also preserved for those interested in the place names in that language. Both the detail and the Arabic were unprecedented in the European treatment of the city of Cairo.

Before looking at these details, we will first make a rough survey of the outlines of Niebuhr's map. At the top it is bound by the range of the "Mokattam," below which lies the "Karafel" or cemetery, with the mausoleum of "Kaid bey " and the sepulchre of the Imam "Schafei" prominently featured. The central part of the plate is occupied by the shaded contours of the city itself. Within this rough quadrangle lay the original medieval city, which constituted only the northeastern quarter of the plan as Niebuhr shows it, as well as the later extensions to the south towards Ibn Tulun (Niebuhr's "Teilun") and to the west into the area of al-Azbakiyya (Niebuhr's "El Jusbekie"). This rough quadrangle, bisected by the *khalig* which originates near "Fostat" on the Nile, is approximately 2,500 by 1,200 geometric paces, or just two and one-quarter square miles in extent. To the right and the south lie "Fostat," and "Masr el Atik" or Old Cairo. Still to the right, but now west of the river, appear "Dsjise," "Dukki," and "Bulak Tacruri." "Dsjise," or Giza,

---

12. *Travels*, vol. 1, 109.

The Mother of the World

is shown by Niebuhr as a small settlement of roughly 200 by 600 geometric paces with a sal ammoniac oven and several pottery factories. Niebuhr speculates that the settlement is old, judging from the height of its environs which appears to consist of refuse carried outside and dumped beyond the built-up area. "Dukki," approximately 1,100 geometric paces, or just under an English mile to the north, is shown as a small collection of structures some 50 by 150 paces in extent. "Bulak Tacruri," another 150 paces to the north, is approximately the same size. Individual structures are not identified in either of the last two villages.

Through this area flows the Nile, moving from right to left in its northward progress towards the Delta. On the right, in midstream, lies the island of "Rodda," with its "Mikkias," or nilometer, followed by what is today the single island of Gezira, shown as two separate, parallel sausage-shaped and unnamed islands.[13] Swinging back east and across the river, now in the bottom left-hand part of the plate, we see the shaded cluster of streets and commercial buildings that represent the port of Bulak. Moving up the plate, or eastward from Bulak, we see first the cultivated area before reaching "El Jusbeckie," or the pond. Finally, to the north of the city, leading to the left and off the page, there are the two ponds, the "Birket es scech kammer" (q) and "Birket er roteli" (r), and then the road leading north to Syria.

We will now examine more carefully the detail in Plate XII. The irregular quadrangle that describes the city, penetrated by many gates, suggests that it was surrounded by walls, although only the Citadel, the massive structure to the southeast with its crenelations and half-round towers and a fragment to the northeast, are indicated as such in the map. Niebuhr distinguishes between this interior, or more massively built wall, and an outer wall that encompasses the remainder of the city. The interior wall, beginning with Niebuhr's "Bab el futuch" (2) and including the "Bab el nasr" (1), "Bab el ghreiib" (30), "Bab el machruk" (29) and the "Bab es sueli" (31), enclosed an area of the city that was only a fraction of its original size. But the limits of the Fatimid city continued from the "Bab es sueli" to the west and the *khalig*, which had originally served as a kind of moat, separating the city from the varied populations who lived in the outlying areas to the west.

---

13. According to Niebuhr's calculation, the overall extent of the two is approximately 2,500 geometric paces, or just over two English miles in length. Without wanting to put too fine a point on it, the insert in the modern *Lehnert & Landrock* map of the city shows Gezira to be 3,500 meters, or 11,480 English feet in length. Niebuhr's 2,500 geometric paces would be 11,475 English feet. What seems clear is that Niebuhr's distances are very accurate.

Tab: XIII.

Bâb el fitûch, ein Thor zu Káhira.

The Mother of the World

From its western terminus at the "Kantaret bab el chark" (h), the city boundaries then ran to the north and the "Bab en nascha" (4), before turning east again and returning to the "Bab el futuch" where they had begun. Outlined by the walls to the north, south, and east with the *khalig* to the west, the Fatimid city admitted access through massive gates on the north and south, the Bab al-Futuh and the Bab Zuwayla, both built in the eleventh century. These gates were connected by Shari' Mu'izz li-Din Allah, so called after the second of the Fatimid Caliphs, which served as the main north-south artery and roughly bisected the city. It is shown but not identified as such in Niebuhr's map. There had also been an outer wall, built by the Fatimid general Jawhar bin Abdullah in the tenth century. But it was constructed of brick and at some distance from of the *khalig*, far enough to allow pleasure palaces to have been built in the intervening space. Of this wall, only fragments remained even in Maqrizi's time.[14] In 1761–62, these fragments consisted of two sections, the first from the north-western gate on the *khalig*, or the "Bab es scharie" (5), west to the "Bab el Hadid" (8); and the second from the "Bab el machruk" (29) south to the "Bab el karafe" (21), west of the citadel.

The intention of Salah ad-Din al-Ayyubi (our Saladin) in the mid-twelfth century had been to enclose the whole Fatimid complex in a strong line of fortifications, commanded by the fort, or Citadel. The plan was for the northern wall to be extended to both the east and west. To the east, it was to terminate in a new tower, the Burg al-Zafar and, from there a new wall would run south along the hills to the Citadel. In the west it was to be extended to the Nile, after which it was to run south to a point due west of the Citadel, whence it would turn east and join with the system of fortifications of the Citadel itself. But the grand design was never completed. The eastward and westward extensions of the northern wall did take place, the former to the abovementioned Burg al-Zafar, traces of which remain today. However, the southward extension of the towers and crenelations towards the Citadel ended after some 400 paces. These massive walls are shown as the heavy, L-shaped line in the upper left portion of Niebuhr's map. They still exist today, and have been incorporated into buildings at the eastern limit of the city. To the north, the wall has recently been unearthed and restored.

Although not in its massive form, the northern wall was extended to al-Maqs[15] on the Nile, the course of the river being considerably to the east of where it was in Niebuhr's time. This is the section Niebuhr describes as a portion of the outer wall, between the "Bab el Hadid" (8) and the "Bab es

---

14. Al-Maqrizi was a historian born in Cairo in 1364. His history of Egypt was mainly topographical. He died in the city in 1442.

15. It is near the "Bab el Hadid" (No. 8) in Niebuhr's map.

scherie" (5). The wall running south along the Nile was never begun, nor was the eastward extension to the Citadel. However, the greater security afforded by the wall to the north allowed the area to be developed, which eventually culminated in the neighborhood of the Azbakiyya. The other section of Niebuhr's outer wall, to the east of the city between the "Bab el machruk" (29), the Citadel (A, B, C), and the "Bab el karafe" (21), appears from the "Historical Plan of Cairo" in Brill's *First Encyclopedia of Islam*[16] to have also been built, albeit on a smaller scale, by Salah ad-Din. Niebuhr's supposition that only the massive, interior wall was built by Salah ad-Din is, therefore, probably incorrect.

The Gates of Cairo

Whether surrounded by the monumental Ayyubid walls or lesser ramparts, the expanded core of the city could be accessed, according to Niebuhr, by no less than thirty-one gates. Beginning with "Bab el nasr" (1), Niebuhr's list moves westward to the familiar "Bab el futuch" (2), then north to the "Bab el medbach" (3) before turning west again to the "Bab en nascha" (4). This takes him to the *khalig* and the western limit of the Fatimid city. Gates numbered (5) through (16) are all to the west of the *khalig* and in the area that developed in the late Mamluk or Ottoman periods. The gates in these outlying areas were not monumental structures comparable with the Bab al-Nasr, Bab al-Futuh, or Bab Zuwayla. But as the urbanization of Cairo continued to the south and west—especially since the internal walls were now pierced by streets and bridges over the *khalig*—security required that even these outlying areas be protected by walls and gates.

Gates numbers (17) to (30) in Niebuhr's list swing east and then north, encompassing the areas of Sayyida Zaynab, the mosque of Ibn Tulun, and the Citadel, before turning west again and returning to the Fatimid city walls at the Bab Zuwayla. These gates provided security to the south and east of the city, as those above did to the west. Of these southern and eastern gates, only the "Bab el 'assab" (25), a monumental structure built in 1754 by Radwan Katkhuda al Gulfi to enclose the *azab* (or "bachelor") quarters in the Citadel and the mosque of Ahmad Katkhuda al-Azab, remains today. The gate led into the lower enclosure of the Citadel from Niebuhr's "Romele" (E), or the old hippodrome. As might be expected, the names of the gates were indicative of their location: those outside Sayyida Zaynab (18) and Ibn Tulun (19) were named after the mosques that served as prominent landmarks in the area, and the "Bab al karafe" (21) led to the northern cemetery.

16. Vol. II, 831.

The Mother of the World

Also known as the Gate of Qaytbey, it was built in 1494 and only a fragment remains today. To the north, Niebuhr's "Arab Lissar" (23) is still marked by a mosque of that name. The "Bab kara meidan" (24) and the above mentioned "Bab el assab" led, respectively, into the city and the Citadel. The Citadel, the center of government in the mid-eighteenth century, was itself divided into strongholds of the several ruling factions. As we will see below, it became an elevated residential quarter where members of these factions lived. There were gates within the complex itself as well as its own external gates that made it independent from the rest of the city. Both the interior and exterior walls boasted of fortified towers that allowed the inmates to bring artillery to bear on potentially troublesome elements, whether in another of the enclosures, or in the city below. Niebuhr mentions[17] that the doors of "Sultan Hassan" (G) were walled up in his time since, at one point, it had served as the site of a battery directed against the Citadel.

Within this rough system of walls and gates lay the eighteenth-century city itself. The *khalig*, once the western limit of the medieval city, increasingly came to bisect the metropolis as it spread to the west. For much of the year it was dry and particularly unwholesome for those living in the vicinity, since it was filled with refuse and often functioned as an open sewer. But with the rise of the Nile it was opened and briefly provided the pleasant prospect of running water and cooling breezes through the center of the city. Lane would devote ten pages in his *Manners and Customs of the Modern Egyptians* to the festival of the opening of the canal, or the "Mosim el-khaleeg." Niebuhr describes the same event from his engineer's perspective. In the first place, he doubted that anyone had ever scientifically measured the width of the Nile at the *fum*, or head of the canal. So, "using a base of 233 feet, and two adjacent angles of 83° 10' and 92° 10', I found that near Djise the Nile was 2,946 feet wide."[18] As for the height, his surveyor's eye finds a stratagem to check the official figures: on the Giza side of the river there was a wall made of stone. Counting the number of stones above the water provided a ready reference for the height of the river. Comparing the level in June 1762 to the same in January of the same year, he found that it had fallen twenty-four feet. More particularly, he found that on a day in August 1762 just before the opening of the canal, the actual rise was only one-third of what the government had announced.

The rise and fall of the Nile was a geographical problem of the first order to European science, and it would be more than another hundred years before

---

17. *Travels*, vol. I, 119.

18. Ibid., 126. Niebuhr probably means Danish feet, which were about 3% shorter than an English foot.

the source of the river would be established with any certainty. As we might expect, Niebuhr was interested in the problem and speculated on the reasons for the changes in the level of the river. The river began its annual rise at Cairo in the middle of June, a period that, according to Niebuhr's informants, lasted forty to fifty days. Niebuhr attributes the rise to the rains in Abyssinia, which, like the Yemen, experienced heavy precipitation in the summer months. Leaning heavily on Lobo,[19] he would seek the source of the Nile in the highlands of Abyssinia. We now know that the river has two major sources: the Blue Nile flowing from the highlands of Ethiopia, and the White Nile, proceeding from the lakes of Central Africa. But with the waters of the White Nile being largely lost in the *sudd*, or marshes, of the southern Sudan, some 90 percent of the flow that reaches Egypt comes from the Blue Nile and Abyssinia.

The flow of the river was of a more practical concern to the authorities in Egypt: the actual rise was a closely guarded secret, allowing the government to select the day when the *khalig* would be opened. That day would determine the collection of the land tax, which traditionally could not begin until the Nile had reached sixteen cubits at the nilometer. On August 8, 1762, it was announced that the river had reached the necessary height and that the canal would be opened. However, the dry bed had not been properly cleaned and it was not until the 12th that water flowed into the city. There were rules that specified the series of subsidiary canals that could then be opened. Generally, none of these was opened until three days after the opening of the main canal to allow the level of the water to rise. Niebuhr tells us that water was not let into the Birket al-Azbakiyya that year until the 18th of August.

Bridges over the Khalig

Over the *khalig* Niebuhr identifies fourteen bridges, from the "Kantaret fum el chalidsj" (a), or the "mouth" of the canal on the Nile, in the south to the "Kantaret ed dahher Bebers" (p) in the north. Several were called after gates or principal routes of the city, such as the "Kantaret bab el chark" (h),

---

19. Jerome Lobo was a Portuguese missionary who was posted to Goa in 1622. Realizing an ambition to reach Abyssinia and the celebrated Christian Kingdom of Prester John, he arrived in Tigre in 1625 at the head of a mission of eight priests. He found the Emperor receptive and was made vicar general. While there, he traveled to the highlands and found springs that he believed to be the source of the Nile. After an extraordinary series of misadventures, he eventually returned to Europe where he published his *History of Abyssinia*, in Portuguese, in which he describes the sources of the Nile as these springs in the Ethiopian highlands. The book appeared in French in 1728 as part of *Relation historique d'Abyssinie*. The French was translated into English by Dr. Samuel Johnson, who added his considerable weight to a defense of Lobo against his detractors, principally the Scot James Bruce. Niebuhr probably read the French translation.

The Mother of the World

perhaps a misprint for Bab al-Luq since it is a the terminus of that thoroughfare, "Kantaret el muski" (l), after the principal commercial street of the medieval city, and "Kantaret bab es scherie" (n), connecting with the principal route west to the port of Bulaq. Others were named after prominent personages or mosques, such as "Kantaret Sunqur" (f), surely after the fourteenth-century Mamluk whose mausoleum-madrasa was nearby, "Kantaret Abdrachman Kichja" (g), certainly the same Abdel Rahman Katkhuda who was the governor of Cairo in 1762, and "Kantaret el Emir Hossein" (k) that led to the celebrated mosque of that name. For reasons of security, the gates of the Fatimid city originally were not allowed to open onto bridges over the *khalig*. However, it appears that by 1762 this stricture had been allowed to lapse and, from Niebuhr's map, it appears that many of the bridges provided direct access between the eastern and western portions of the city. Of the fourteen bridges Niebuhr identifies as passing over the *khalig*, most[20] appear to be late Mamluk and then Ottoman additions, as the development of the area to the west of the canal required greater facility to cross back and forth between the two halves of the city.

Major Landmarks of Cairo

The major landmarks of the city were also identified: the citadel, with its several quarters (A,B,C); the major mosques including the "el Ashar" (H), "Teilun," and "Sultan Hassan" (G); the Khan al-Khalili (O); Coptic (L), Greek Orthodox (M) and Armenian (L) Churches as well as the residences of several Christian patriarchs (I, M). Niebuhr provides a brief commentary about each, although most were known from the accounts of previous travelers:

> Other accounts of Egypt already contain descriptions of quarters and the principal edifices of Cairo that are so detailed that it is sufficient for me only to indicate their location on the map. However, I will still say a word about each.[21]

The outlines of the Citadel are indicated with its separate enclosures, each harboring rival contenders for power in the city. Niebuhr calls them "the residence of the reigning Pasha" (A), known today as the southern enclosure; "the quarter of the Janissaries" (B), today the northern enclosure; and "the quarter of the Assabs" (C), the lower enclosure. The southern enclosure was largely a ruin, Niebuhr commenting that the reigning pasha spent little

20. All but "k" through "n" in the key to the map.
21. *Travels*, vol. I, 113.

time there, preferring to spend long periods of time in the countryside. It contained the mint where the coinage was debased, the silver and copper being of a baser alloy than that struck in Istanbul. Also here was the "Palace of Joseph" where the *kiswa*, or covering of the Kaaba in Mecca, was made each year. Surprisingly, this Joseph was not the Joseph of Genesis but, Niebuhr tell us, Joseph, "father of Modafar and son of Ayub" otherwise known as Salah ad-Din. But there was another purported connection between the citadel and Scripture. According to another religious—but not scientific—traveler, Benjamin of Tudela, the Jews in Egypt in the twelfth century referred to the Citadel as Soan, apparently believing it to be the site of the city of Zoan of the Old Testament.[22] Niebuhr was given a tour of the remains of the palace by the superintendent of the artisans and afterwards entertained with coffee in his home. There was a secret room with a spectacular view of the city below and the pyramids of Giza in the distance. Nearby, were thirty columns of red granite, all that remained of the Ablaq Palace where the Mamluk Sultans once held public audiences. For the most part, however, in 1761 there remained only mean hovels scattered among the monuments of the southern enclosure.

The quarter of the janissaries, or the northern enclosure, had traditionally been the fortified residence of the imperial army in the city. However, the janissary corps was no longer a disciplined military force, as it was increasingly bent to the will of the fractious beys. The officer corps was made up of former Mamluks who now owed their allegiance to the beys in the city, not to the Pasha or the Sultan in Istanbul. It was from the janissary quarter that the expulsion of an unpopular pasha was announced with a cannon shot. Niebuhr remarks that the quarter was full of houses, and by the middle of the eighteenth century the whole of the Citadel had become an elevated residential district.[23] The consequent decline in martial qualities was noted by Niebuhr who says that the people no longer feared the janissaries and had the temerity to commit thefts in the neighborhood of the Citadel. And this particular German tourist was not particularly impressed with the nearby Joseph's well, one of the greatest "curiosities" of Cairo, commenting that the excavation through soft limestone was comparatively easy when compared with other ancient monuments, such as the pagodas he would later see in India. They had been carved from much harder material.

22. Others would make it the city of Tanis in the Delta. It is mentioned in Numbers, Psalms, Isaiah, and Ezekiel as a city in Egypt, thought to be the northern capital of the Pharaoh.

23. See Lyster, *The Citadel of Cairo*.

The Mother of the World

Below the southern enclosure was the lower enclosure, or the quarter of the "assabs." By the beginning of the eighteenth century the *azban*—or "bachelors" in Arabic because of their unmarried state—had come to rival the janissaries as a military force in the country. Unlike the janissaries, who were made up of Christian-born levies from Rumelia, or European Turkey, the assabs were freeborn Muslims. As we have seen, a new gate into the Citadel, Niebuhr's "Bab el assab" (25), had been recently built, in 1754, to cement their rise to power. However, they shared in the increasing domestication of the military, and theirs had also become became a residential quarter, honeycombed with shops and houses.

West of the "Bab el assab" and outside the Citadel lay the "Kara meidan" (D) and "Romele" (E) where parades and military exercises were held. Further to the southwest, near the mosque of Ibn Tulun, Niebuhr shows a ruined citadel called "Kalla el Kabsch" (F) or the Fortress of the Ram, which, as we have seen, was associated in popular belief with the ram sacrificed by Abraham in place of Isaac. Basing his comments on Jean Leon and Marai, Niebuhr states that the fortress had been built by Ahmad Ibn Tulun in the year 1298 during his rebellion against the Caliph al-Muwaffaq in Baghdad. However, the expected invasion of the caliphal forces was repelled well before they reached Egypt, and the fort did not come into use. By 1761 it was a ruin. Niebuhr spent several days copying the hieroglyphs on a nearby Ptolemaic sarcophagus being used as a watering trough. From Niebuhr's description, the sarcophagus would have been just outside the madrasa of Qaytbey, not to be confused with his mosque on the island of Rhoda or his mausoleum in the northern cemetery.[24]

The mosques "Dsjamea al ashar" (H) and "Sultan Hassan" (G) are shown in their familiar places, and Niebuhr includes a short note on the academy of the "el ashar" with its four muftis, one for each of the orthodox Sunni schools. But the number of mosques in Cairo was so numerous that it would be tedious to list them all or try to place them on his map:

> I remember hearing ... that a poor sheikh could spend an entire year here visiting each day a different mosque, where he would be given drink and lodging without payment ...[25]

Niebuhr notices the function of the minaret, the Muslim prejudice against the bell (they were attached to beasts of burden) and the generally Spartan furnishing of mosques. Even today, the general absence of decoration in the mosques is noticeable. The little splash of color provided by subsequently-applied *iznik*

---

24. For a description of its ultimate destination, see below in the chapter on the "Antiquities of Egypt."

25. *Travels*, vol. I, 117.

tiles in the mosque of Aqsunqur make it stand out in a generally dun-colored field and, for that reason it is now referred to as the "Blue Mosque." Cairo is no Samarkand. When Niebuhr notes that "the administrators soon enrich themselves, while the mosques become gradually impoverished"[26] he referred to the common practice of the diversion of *waqf*, or endowment, revenues to personal use. It would be an increasing problem as competition for the wealth of the country accelerated in the eighteenth century.

Surprisingly, three of the four churches Niebuhr shows as "The Patriarchal Church of the Copts" (I), "St. Nicholas, a Greek Church" (K), "A Coptic Church, under which there is an Armenian Church" (L), and "The residence and Church of the Greek Bishop of Mt. Sinai" (M) still exist today in these same locations, three of them in the old Fatimid city. Niebuhr's (I), still advertised as the Coptic "Batriarchat," lies in the warren of the Harat ar-Rum, off the Darb al-Ahmar just before it turns west towards the Bab Zuwayla. It was built in the sixth century and, between the years 1660–1788 served as the patriarchate.[27] The second church identified by Niebuhr, (K) the Greek Orthodox church of St. Nicholas, was built in the tenth to the eleventh century and was enlarged in 1850 to its present size. Niebuhr saw the previous and much smaller structure. Located in the Atfet al-Kanissa al-Gedida, just off al-Azhar Street and several hundred feet west of the mosque of al-Azhar, it is still today the seat of the Greek Orthodox patriarchate in Cairo.

The church of the Holy Virgin (L) now sits perhaps forty feet below the surface of the nearby street. Today a sign outside the entrance states that it was built in the fourth century. Maqrizi says that it was "formerly known by the name of the physician Kabilun, who lived about 270 years before the appearance of Islam,[28] which gives color to the statement. The associated church of St. Mercurius once served as a side chapel of the main church and was built in 1773 by Ibrahim al-Gawhari[29] on the site of the Armenian church of St. John the Baptist, Niebuhr's "Armenian church." Finally, the residence and church of the bishop of Mt. Sinai (M) is today a ruin, although the shell of the walls is still identifiable where it appears in Niebuhr's map, a hundred yards inside the Bab al-Nasr. As we will see, the members of the expedition would probably have paid several visits to the bishop in an effort to secure permission to enter the monastery of St. Catherine in Sinai.

26. Ibid., 120

27. Later in the century the papal residence was transferred to Azbakkiya when Ibrahim al-Gawhari received permission of the sultan to build a church in that area.

28. That would confirm the date but Meinardus doubts this and dates it to the tenth century. See his *The Historic Coptic Churches of Cairo*.

29. The Coptic secretary to Ali Bey. See below, p. 135.

The Mother of the World

Other areas of the map remain obscure. The "Jewish Quarter" (S), or Haret al-Yahud, to the east of the *khalig* and within the old Fatimid city, no longer exists. As we will see below, it was an area of narrow streets and mean buildings, where every effort was made to avoid attracting attention to such wealth as may have existed within. "El Muristan" (P), or the quarantine station, appears to be the same as the thirteenth century hospital of Sultan al-Mansur Qalawun. It was a wonder of the age when it was built in 1285. It is not to be confused with the maristan of Muayyad Sheikh off Shari' al-Maghar, one of the most striking and beautiful of the Mamluk monuments still standing in Cairo. Niebuhr says that the hospital provided music for the patients through the generosity of Abdel Rahman Katkhuda. However, the number of patients was "very small, considering the grandeur of this establishment."[30]

Development of the City

By the beginning of the Ottoman period in 1517, Cairo had already outgrown the rough quadrangle formed by the original, one-square-kilometer Fatimid core, although the Ottomans tended to fill the interstices of earlier Mamluk expansions rather than enlarge the boundaries of the city. The city-center soon acquired a commercial character, while residences increasingly were located in what became the suburbs. Urbanization to the south had taken place over the three centuries of the Mamluk period—roughly, the early thirteenth to the sixteenth centuries—before reaching the vicinity of the mosque of Ibn Tulun. That mosque dated from the late ninth century and had originally stood in splendid isolation from the city. To the west, development of the area of al-Luq had already begun with the a belvedere and series of hippodromes built by the Sultan al-Salih Nagm ad-Din Ayyub (1240–50), one of the Ayyubid intermediaries between the Fatimid founders and Mamluk successors. The area of Bab al-Luq was, some maintain, so called after the fecund *ardd al-luq*, the ground being so fertile after the subsiding of the flood that the seed had only to be "tossed"—*yulaqa* in Arabic—on the receptive ground for it spring to life. A decade later, it was settled by Mongol refugees, originally 2,000 horsemen and their families, befriended by the Sultan Baybars. Others of their nation came and the area took on a decidedly Mongol look. Over the next several centuries it maintained its different, often questionable, reputation, as a quarter of jugglers, acrobats, "drinkers" of hashish, and prostitutes. But it wasn't until the fifteenth-century Mamluk urbanization project that the area came into its own.

30. *Travels*, vol. I, 120.

Niebuhr in Egypt

To the north of al-Luq was the garden known as Bustan al-Dikka, where the Fatimid caliphs celebrated the opening of the canal. In the thirteenth century, however, a shift in the course of the Nile to the west opened up more land and this coincided with a desire on the part of the Mamluk Sultan Qalawun (1279–90) to expand into the western suburbs of the city. The main impetus for development was the foundation of the Nasiri Canal in 1325, "dug by order of the Sultan to facilitate the transport of grain to the Delta village of Siryaqus, where the Sultan had established a great *khanqah*, or monastery for the Sufis . . . "[31] A portion of this canal, then dry, appears in the Niebuhr map, beginning just west of the "Birket en nassarie" (w) and running in a roughly parallel direction to the *khalig*, before looping to the east two hundred paces past the "Bab el Hadid" (8) and the road to *Bulaq*. It rejoins the *khalig* just past the "Birket er roteli" (r). It did not appear to be in use in Niebuhr's time.

Development of the area later known as al-Azbakiyya began with this initiative on the part of Qalawun. Building was at first concentrated on the western bank of the *khalig*. Development was not uniform and there were periods of contraction as well as expansion. Niebuhr suggests as much when he suggests that "it is impossible to know for certain whether the city has grown or shrunken in size in recent centuries." A series of catastrophes—famine, plague, and Timur's invasion of Syria—brought the first period of expansion in the late fourteenth century to a close, and much evidence of that activity had disappeared by the early fifteenth century. But the areas of economic importance suffered less in the downturns. The pleasure domes and belvederes may have fallen into disuse, but the small trades—textiles, food processing, and other commodities-related activities, as well as slaughterhouses, dairies, mills, and oil presses—remained. Copts and other minorities practicing these trades continued to cling to the margins of the *khalig*.

With the establishment of the *atabak* Azbak's foundation in the fifteenth century, the area received newfound attention. Between 1476 and 1485 Azbak dug a huge pond of fifty feddans on the site of the old Batn al-Baqara, connected it with the Nile via the Nasiri Canal, laid out a paved walk around the pond and built an elegant palace at its southeastern corner. Other amirs followed suit and soon a congregational mosque and a commercial center grew up among the residences. The pond became a place for the launching of pleasure boats and gathering place for people, not always of the most reputable kind. In periods of low water, horses grazed in the dry bed. Although it had acquired an urban character of its own, many amirs still preferred to build to the south of the Bab Zuwayla, and the Azbakiyya was still not the preferred

---

31. Behrens Abu Seif, *Azbakiyya and its Environs*, 9.

The Mother of the World

residential area. It suffered from the vicissitudes of Mamluk rule, including periodic pitched battles between competing factions among the buildings in the neighborhood. After the Ottoman conquest, most of the houses were plundered and moveable decoration carried away by the victorious Turks. Later, however, movement to the south and west resumed and even accelerated, many of the structures built in a "neo–Mamluk" style.[32]

In 1761–62, Niebuhr's map shows a pattern of irregular streets laid out to the west of the *khalig* from the "Kantaret es Sabba" (c) in the south to the "Kantaret bab es scherie" (n) in the north. The area still maintained its Coptic and *Farang* (or "Frankish") flavor. Copts lived near the *khalig* to the south of the pond (f), where Niebuhr shows north-south and east-west arteries and many side streets. In general, he informs us that the houses in Cairo were of unbaked brick, of only one story in height and, therefore, not as large as the houses in Europe. In addition, there were large expanses of gardens and water when the Nile was high. So, Cairo was less densely populated than a European city of similar size. We have seen that the residences of the consuls of France and Venice were just to the west of the *khalig*. The Harat al-Ifrang was located in this area, between "Kantaret el muski" (l) and al-Maqs (near the "Bab el Hadid," 8). There were thirty-five of these *hara* or quarters when the French arrived, each of them often with only one means of ingress and egress. Niebuhr gives a latitude (31° 2' 58") of "the street where the French reside." He was able to set up his instruments and take celestial sightings from the roof, we suspect, at night or early in the morning when his activities would not be noticed by passersby.

The Ponds of Cairo

Of the lakes or ponds of Cairo, many were dug as the upper classes began to build in the area to the west of the *khalig*. Niebuhr lists nine ponds, which, he notes, still contained water in February and March of 1762. They appeared and disappeared with changes in the water level and the building activity in the area. The names often changed as well. Maqrizi also list nine ponds in the early fifteenth century, but only his "Birket er roteli" and "Birket en nassarie" are included in Niebuhr's list. Certainly, the most prominent in 1762 was "Birket el Jusbekie" (f), described by Savary some 10 years after Niebuhr

---

32. See Raymond, "Architecture and Urban Development: Cairo during the Ottoman Period, 1517–1798." In the west, in particular, the building of fifteen mosques, twelve fountains, and four baths in the years between 1726 and 1798 contrasted with the nine mosques and two fountains built during the preceding two centuries.

as "an immense basin surrounded by palaces of the beys."[33] In addition, there was "Birket el fil" (z)—or "pond of the Elephant," so called from the long, trunk-like projection that extended to the southwest—in the center of what was still the favored residential district to the south of the city. Its development during the Ottoman period, beginning early in the seventeenth century, was facilitated by the relocation of the odoriferous tanneries whose effluents previously dominated the area.[34] It was here that the inhabitants of Cairo would have witnessed the social event of the season in 1760-61, the marriage festivities of a Mamluk of Ali Bey. Jabarti provides the details:

> The nuptials were celebrated fittingly and with great pomp at Birket el Fil. This took place at the time of the flood of the Nile in 1174. A large part of the Birket was covered with planks, which floated on the water and bore the large number of visitors who came to see the decorations, witness the marvels created for the wedding, and to be amused by the jugglers and magicians. Lights hung from the neighboring houses, inhabited for the most part by emirs and notables, cast their reflections on the waters of the lake and created a beautiful impression. From each of these houses charming concerts could be heard. Sumptuous tables, covered with the most diverse and delicious dishes, were provided for the guests.
>
> The celebrations lasted an entire month, during which period the gates of the city of Cairo were left open throughout the night, so as to allow everyone to attend the great festivities and experience the pleasures, which were a kind of enchantment. The emirs, notables, the old, commanders of ojaks, merchants, Copts, Europeans, Jews, etc. all offered Ali Bey presents.[35]

Had the members of the Danish expedition been in Cairo at the time they surely would have been witness to the celebration, as they were to later similar events on the Birket al-Azbakiyya. But in late 1760, Forsskal had just arrived in Denmark, Niebuhr was on his way to Copenhagen from Göttingen, and Baurenfeind and Kramer had just been selected for the expedition. Von Haven seemed to be out of touch.

Finally, there was "Birket el hadsj" or "pond of the Hajj." It was several miles to the east and does not appear in the map of the city, although Niebuhr shows it separately in Plate XIV entitled "Camp of the Pilgrims

---

33. Behrens Abu Seif, *Azbakiyya*, 21.

34. See Raymond, *Problems of the Middle East in Historical Perspective*.

35. Jabarti, vol. 11, 218. My translation from the French. Jabarti also says (Vol VII, 168) that the arcades in the area were later closed because of the presence there of people of loose morals and the pervasive odor of hashish when they assembled.

before their Departure for Mecca." In 1762 the pilgrimage caravan departed on 22nd of May and Niebuhr hurried out to witness the departure. As we have seen, normal commercial traffic was disrupted due to the unsettled state of relations between the authorities in Cairo and the Bedouins of the Hejaz. But there was no stemming the great tide of the faithful undertaking the yearly pilgrimage to the holy places of Islam, and the escort of Mamluks and Maghribi irregulars led by the redoubtable Hussein Bey would ensure safe passage down the west coast of the Arabian Peninsula. Niebuhr remarks that "I doubt I could show the disorder which prevails among the travelers who have pitched their camp there."[36] The scene was one of tents pitched helter-skelter over the plain, the roaring or braying thousands of animals, distraught relatives taking their last-minute leave, and the businesslike work of the victualers and camel-men who would attend to these thousands. Like an island in the midst of it all, the Amir el-Hajj reclined in his sumptuous tents with the three little brass cannons deployed in front, we suspect primarily for effect. But after a few days of travel, when loads had been adjusted, weak animals had fallen by the wayside, and a rough order-of-march developed, the massive flow of human beings and animals would take on some semblance of order.

Outlying Areas

Beyond areas described above, Niebuhr also shows settlements that would grow in time into the neighborhoods of greater Cairo. To the south, there was "Fostat" and "Masr el Atik," or Old Cairo. He provides a short history of the successive cities of "Babylon, Masr, Fostat, and Kahira," drawing on the accounts of Patriarch Eutychius, Benjamin of Tudela, Sherif Idrisi, Abulfeda, and Marai. The island of "Rodda" is shown with the "Mikkias," or nilometer, at the southern tip. This structure, dating from 861 AD, consisted of a stone-lined pit into which water entered when the river was high and its level could be measured on a column graduated in cubits. The island itself was so like the descriptions of previous travelers that Niebuhr could add little that was new. Sherif Eddris would have noticed scant difference between the Rhoda of the twelfth century and the island seen by Niebuhr in 1762. Oddly, a separate "Dsjesiret el Mikkias" is shown to the south in the Niebuhr map. Similarly, what is today the island of Gezira, is shown by Niebuhr as two, long unidentified islands. We can only assume that the always-changing course of the river—what the *Encyclopedia of Islam* calls

---

36. *Travels*, vol. I, 122.

the everlasting "struggle with the Nile"[37]—accounted for these differences from the islands of today. The hatched areas on the map appear to be cultivation, and indicate that the islands were devoted primarily to agriculture.

Back on the east bank of the river and at the northern extremity of "Fostat," Niebuhr shows the intake to the aqueduct, "a very high structure with five wheels where water is raised from the Nile and sent beneath the walls to a reservoir near the Citadel." Nearby, is a "large mosque" (46) and 450 paces to east is a structure identified by Niebuhr as "the Mosque of Abu Saki who built the aqueduct" (47). The mosque is actually that of Abu Sa'oud Mohammed al-Garahi (founded in 933 AH, or 1584 AD), the error probably due to Niebuhr's mistaking "al-Garahi" for the "al-Ghuri" who *did* build the acqueduct. Originally built by Sultan al-Nasr Mohammed in about 1311, the aqueduct was extended to its full length in 1505 by Khansowa al–Ghuri. On the other side of the aqueduct, 400 paces north on the riverfront, was the impressive edifice of "Kasr el ain" (42), a dervish monastery, topped by a superb dome. The monastery was founded by Abu Mohammed Mahmud Badr ad-Din al-Aini, the Syrian-born Sufi who settled in Cairo in the late fourteenth century. He was appointed inspector of pious institutions in 1401 and his fortunes waxed and waned as several Mamluk Sultans succeeded to power. He was *qadi*, or judge, of the Hanafites from 1426 to 1438 and an influential courtier throughout most of his life. His written works on history, poetry, and Sufi doctrines, were numerous. He died in 1451 at the ripe old age of ninety-one and was buried in the monastery that he had founded.[38] Niebuhr's brief description of the Dervish "curiosities," includes a shoe twenty-two Danish inches in length. The name Qasr al-Aini lives on as a major thoroughfare in Cairo, but the structure with its "superb dome" was pulled down in the nineteenth century.

The "Coptic Church" (48) shown to the northwest of the "Mosque of Amru" (49) is probably the complex comprising the churches of St. Mercurius (Abi Seifein), St. Shenute (Amba Shenudah) and the Holy Virgin of Damshiriya (al-Adra ad-Damshiriya), all of which date from at least the eighth century. The large convent that sits on the site today was built in 1913, although there was probably a convent associated with the churches from their beginning. Further south, the oldest place of Muslim worship in Cairo, the "Amru Mosque" was very poorly maintained in 1762, as it has been intermittently between restorations later in the eighteenth and twentieth centuries.

---

37. *First Encyclopedia of Islam*, vol. II, 820.
38. Ibid., vol. I, 213.

The Mother of the World

To the south of the mosque of Amr was "a spacious area surrounded by a wall, like an old castle, occupied today only by Christians" (50).[39] The complex included several churches and cemeteries, both Greek and Coptic, and a monastery for Coptic women. A grotto under one of the Coptic churches (Abu Serga) was particularly venerated as having been occupied by the Holy Family during their sojourn in Egypt. This is the area occupied today by the Roman fortress, the Coptic museum, the Hanging Church, the Greek Orthodox church of St. George, and the Ben Ezra synagogue. Finally, completing the southern structures on the east bank of the Nile, there was a grain storehouse (51)—Niebuhr remarks that it is one of the "supposed corn storehouse of Joseph" (Gen, 41:56)"—a bazaar (52), and the customhouse (53) where duty on merchandise on its way to Upper Egypt was paid.

To the north, "Bulak" is shown as a substantial settlement separated from Azbakiyya by approximately 600 geometric paces, or 900 yards. Niebuhr speculates that it is the Litopolis of the Greek authors. He shows the building (32) where the new Pasha was received when arriving by the river, and the old arsenal (34) from the days when a fleet was maintained at Suez. There were also warehouses for rice, saltpeter, wood, and saffron (33), the customhouse (36), a brickyard (38), and a large covered market (37). In 1762 Bulaq was now the port of Cairo, having long-since replaced Fustat in commercial importance. All merchandise passing on the Nile between Cairo and Rashid or Damietta went through Bulaq, although in Fustat there was still a warehouse from which grain was shipped to the holy cities of Mecca and Medina.

To complete our survey of Niebuhr's map of the environs of Cairo, we move from Bulaq across the northern outskirts of the city, past the "birket er roteli" (r) and the "birket es schech kammer" (q), with a short detour to the north and the district of "Matare." There, in the village of Matariya, was shown the sycamore venerated by Oriental Christians as having offered shade to the Holy Family during their flight from the murderous intentions of Herod (Matt 2:13). Nearby was a spring that had reportedly given them refreshment. Niebuhr, a good Lutheran, is skeptical of both claims. The mention in the Bible of the Egyptian sojourn is slender, but among Egyptian Christians the story had blossomed into a comprehensive itinerary in the country, leading the Holy Family eventually to Assiut in Upper Egypt before returning to Bethlehem on the death of Herod. Both the tree—with names carved in its gnarled and now lifeless trunk—and the spring are still shown today. To the east of the city were the "Mokattam" and the "Karafel" or cemetery. The ruinous "Kubbet el haue" shown on the Muqattam opposite the

---

39. *Travels*, vol. I, 113.

Niebuhr in Egypt

Citadel is, as mentioned, the mosque the Fatimid Wazir al-Guyushi, who ruled Egypt from 1074 to 1094.[40]

In the cemetery itself, Niebuhr shows the sepulcher of the Imam Shafei with the laconic note that "Mohammed women throng to this place, especially on Friday, some out of devotion, others for a promenade."[41] As we will see below, Egyptian popular religion had become extraordinarily important to a population effectively deprived of influence in the conduct of the affairs of the country and they were fond of public celebrations. Some commentators—Jabarti among them—were disgusted by the spectacle of the lower orders who were inevitably attracted to these festivals. Seven hundred paces north of the "Kubbet el haue" is the large mausoleum complex of "Kaid bey" which, Niebuhr notes, was in fact a small village. The mausoleum, one of the most splendid of the Mamluk monuments, would later be memorialized by David Roberts in the print "The Tombs of the Califs" and it now appears on the face of the Egyptian one-pound note. The series of buildings Niebuhr shows another 500 paces to the north of Qaytbey are surely the *rab* of Qaytbey, the *takiya* of Ahmad abu Seif, the complex of Sultan al-Ashraf Barsbey, the tombs of Qurqumas and Gani Bak, and the complex of Barquq.[42] A further 750 paces to the north and at the edge of his map he shows "El Jusbeck," a complex of 5 buildings, with the note that "El Jusbeck, who built a large mosque in Cairo in the district which bears his name, rests in the Mosque of Kaid bey to the northeast, and his tomb is also surrounded by a number of houses."[43] This is puzzling[44] since the only remaining monument of Azbak today is the mosque and *madrasa* in the vicinity of the mosque of Ibn Tulun. We have already seen the development in the area to the west of the *khalig* that took his name, as well as the mosque to which Niebuhr refers above. In 1761–62, this extensive necropolis was—as it remains today—a place for the living as well as the dead, with rude hovels scattered among the stately Mamluk monuments. It would have been peopled by residents as well as the family members coming on feast days or anniversaries to remember the departed. Finally, the "Kubbet

---

40. Now open to the public after serving for years as a military installation. The name "Kubbet el haue," or the "Dome of the wind," is also applied to the heights above Aswan where the Agha Khan's mausoleum, a replica of the Guyushi mosque, is located.

41. *Travels*, vol. I., 116.

42. See Williams, *Islamic Monuments in Cairo*.

43. *Travels*, vol. I, 118.

44. No source that I have seen mentions a tomb of Azbak in this area. The scale of the buildings shown by Niebuhr is very rough and this complex is shown to be much smaller than that of Qaytbey. But, given its orientation to the west of the road between the monuments, it appears to be the complex of Qurqumas and the mausoleum of Sultan Inal. See Williams, *Islamic Monuments in Cairo*.

The Mother of the World

el Assab" (T) was a former fortress of the "bachelors "before they moved into the citadel. It was a ruin in 1761–1762, although it was still the place where the beys met the Pasha if he was coming overland from Istanbul.

It is remarkable how a solitary European, in the midst of a hostile populace, was able to make a map of such detail and accuracy. Niebuhr mentions that there were areas he knew well, and it is clear that he spent a great deal of the ten months in Egypt moving around the city and familiarizing himself with its monuments, large and small. His only tools were a small pocket compass, which he must have used surreptitiously to check the general bearings, his two legs, and his ten fingers, probably supplemented by a *subha*, or Muslim rosary, to keep track of the hundreds. It is also seems clear that, given the accuracy of the map, that he walked the same streets over and over again, checking distances and verifying the orientation of the structures and geographical features. We can form a picture of this obvious Frank appearing in the streets of the city, first coming under the gaze of a suspicious and hostile public—shopkeepers, *bawwabs, sarrajs,* and commoners—before becoming a familiar and less-threatening presence as the months passed by. Even then, however, extreme caution was necessary. In the Ottoman Empire of the mid-eighteenth century, not to mention the Europe of the period of the Seven Years War, anyone suspected of making a map of a city would have been immediately apprehended and taken to the authorities. Even so apparently benign an activity as copying the hieroglyphs on the sarcophagus near the "Kalla el Kabsch" (reproduced in Plate XXX) was suspect. On several occasions, his drawings were taken by a *sarraj* and not returned. As we will see below, a little humor and a judicious use of *baksheesh* often eased the way. To the end, however, the Egyptians were puzzled by the curiosity of this odd European.

Collection of the raw data—distances, compass bearings, names, details, and anecdotes—would then have to be followed by the careful task of verification and plotting on the first tentative outlines of the map. This Niebuhr probably did in the mornings when he could work undisturbed in the house in the Frankish Quarter. The raw material for the map that eventually appeared as Plate XII of vol. I of the Travels, as with the other detailed maps of Oriental cities, was subject to the constant vicissitudes of travel. There could be no question of back-up copies, and such material as was not periodically shipped to von Moltke would have been carried in his baggage for another five years before he returned to Copenhagen. We will see below the unfortunate fate of some of that baggage in Mocha.

When Niebuhr says he doubted that any European before had attempted to make a detailed map of Cairo he was certainly correct, and a search for previous or contemporary maps of the city yields only a few crude representations, with little but the most obvious features depicted. His effort was, of course, overtaken by the French when they issued their own, far more detailed map of the city in 1812. But the French ruled the country, spared no effort in their research, and brooked no interference from the populace in the preparation of their *Description de l'Egypte*. For Niebuhr, without official support and working entirely on his own, the difficulties were great and the dangers considerable. But he was not a man to let such matters stand in his way. The map of Cairo was his first sustained effort at mapmaking in the Orient, and although it would not be replicated again in terms of detail, it was an important harbinger of things to come.

7

Government

> ... the people of Egypt are not now allowed to indulge that fanatical rudeness with which they formerly treated unbelievers; and hence European travelers have one great cause for gratitude to Mohammed 'Alee. (Lane, *Manners and Customs of the Modern Egyptians*, 562 n. 10)

WHEN LANE WROTE THE above lines in 1836 a great deal had happened in Egypt since the arrival of the Danish expedition 75 years before. Most importantly, there had been the French invasion and the subsequent rise to power of Mohammed Ali. After a brief period of unrest, the Albanian pasha instituted changes that, among other things, made living in Egypt safer for Europeans and accelerated the diffusion of European science and methods throughout the country. In 1761, as we have seen, Egyptians were still actively hostile to outsiders. The Europeans were clearly in Egypt on sufferance and they suffered, with other non-Muslims from certain of the Caliph Omar's "conditions:" they could not ride horses[1] and were required to dismount from an ass in the presence of a Turk, they couldn't ring bells in their churches or read their scriptures in a loud voice, a Christian man could not marry a Muslim woman, and they couldn't drink wine in pubic. Other conditions imposed on local Christians and Jews included the wearing of distinctive clothing, the keeping of their crosses and swine from the public view, building their houses no higher than those of the Muslims, and

---

1. Butler says that these were only the "contingent" conditions whose enforcement depended on the terms of the treaty binding the subject people and the Muslims. The more stringent conditions, which were binding in all cases, included prohibitions against acts or words that would bring Islam or its founder into disrepute. Niebuhr says that certain European consuls, however, did ride horses on the days when they had an audience with the Pasha. But Christians were not permitted to pass, even on foot, in the vicinity of Sayyida Zaynab, a mosque in the vicinity of the Bab Al–Nasr (probably that of al-Hakim), or in the *qarafa*, the cemetery. Butler, *The Arab Conquest of Egypt*, 448.

the burying of their dead in silence. These conditions were enforced with greater or lesser severity, depending on the general state of relations between the communities. Relations were often dependent on outside factors. Periodic outbreaks of anti-Christian feeling and violence, often based on some often remote, real or imagined, insult to the Prophet or his religion, brought about a scrupulous enforcement of the strictures. At more peaceful times, certain of them fell into abeyance. For Europeans in Cairo in the middle of the eighteenth century, the capitulations were decades in the future and those who lived in the city were very careful residents indeed. The recent events in Damietta[2] suggested that they should conduct their everyday lives with the utmost caution.

How different it was to be later can be judged from the remarks of Richard Lepsius on another European scientific effort, the Prussian expedition to Egypt of 1842–45:

> The whole journey, of which this is a very hasty sketch, was one of the most fortunate expeditions which has ever been undertaken for a similar purpose. None who participated in it suffered from the climate or the accidental casualties of a journey. We travelled under the very powerful and, in every way efficient protection of the Viceroy. We had an explicit and written permission to make excavations, wherever we should consider it desirable, and we employed it, to acquire a number of interesting monuments for the Royal Museum in Berlin . . . The scientific results of the expedition have, in almost all respects, surpassed our own expectations.[3]

In contrast, in the middle of the eighteenth century there had been little protection from the beys, and the members of the Danish expedition carried away little from Egypt except a few manuscripts bought under conditions of the utmost circumspection. Such scientific results as ensued were entirely fortuitous. So far from suffering no insult from the climate or "accidental casualties" of the journey, only a single member of the Danish expedition would live to tell the tale.

The Cairo of 1761–62 was a flourishing metropolis that had long-since adjusted to its status as a provincial capital rather than the head of an empire. With the coming of the Ottomans in 1517 and the end of independent Mamluk rule, Egypt had become only another remote province of the Ottoman Empire, albeit the most wealthy one. Much of the decoration of the city's Mamluk monuments and residences, delicate dadoes of multicolored

---

2. See chapter on the Delta.
3. Lepsius, *Letters from Egypt, Ethiopia, and the Peninsula of Sinai*, 24.

Government

marble, had been stripped and shipped to Istanbul, along with almost everything else that was of value and portable. In true Central Asian tradition, the Turks had also shipped the artisans and craftsmen who had been the authors of this flowering of Mamluk art and architecture to the capital along with the artifacts. The same was true of the fisc. The wealth of the country, always extracted by coercion or the bastinado from the reluctant *fellaheen*, was now sent to the Ottoman capital rather than retained in Cairo. The change in destination probably meant little change in the life of the average Egyptian.

Mamluk Rule

The Mamluk[4] dynasty, a byword for rapacity and violence, had begun with the importation into Egypt of a caste of Central-Asian military slaves in the thirteenth century. They had later turned on their masters and seized power. But the phenomenon of slave rule had begun long before the nominal Mamluk era. The first of the usurpers was Ahmed Ibn Tulun, the son of a Turkish slave who supplanted his Abbasid overlord in 868 AD and succeeded in combining Egypt and Syria under his virtually independent rule. On his death, the Abbasids in Baghdad briefly resumed their rule, at least in name, until another Turk, Ibn Toghej, an Ikshidite[5] from Fergana in Central Asia, seized power. He ruled for just under 40 years before he was, in his turn, supplanted by a Berber eruption from the west that finally realized its ultimate goal, the conquest of Egypt, in 969. The Berber, or Fatimid, dynasty lasted for the next two hundred years, incidentally making Egypt nominally Shia, the dynasts claiming descent from Ali and Fatima. Their building program gave to Islamic Cairo the appellation *Fatimia* that it retains to this day. Then, in 1170 another people, this time Kurds of the Ayyubid family, seized power in the person of the famous Salah ad-Din. They ruled for 80 years before the last Ayyubid, Turan, was assassinated by Mamluks in 1250. This ushered in a formal period of rule by slave dynasties, first largely Turkish or Central Asian and then largely Circassian, with assassination often the favored instrument of dynastic change.

Circassian, or *Cherkess*, was the general name for the people who occupied the northwestern Caucasus as far south as Abkhazia, and they would play a prominent role in Egypt nearly to the end of the nineteenth century. Divided into ten tribes whose names appear as difficult to pronounce as

---

4. From *malaka*, to possess. The term *Mamluk*, or "owned, in possession of" or "belonging to," expressly referred to a white slave. Black slaves, at the opposite end of the political and social spectrum, were called *abid*.

5. From an old Persian princely title.

they would be unprofitable to list, they were nominally Muslim, although they included Orthodox Christians who appear to have been their primary export. Neither religion appeared to be deeply rooted among them. The custom of *atalik*, or handing over of children to the upbringing of strangers, may account for the ease with which they sold their offspring, both male and female, to slave dealers. They were of a very warlike disposition and exhibited racial characteristics that, however unscientific they may seem today, lay at the heart of the special place of these people in the Muslim East. They were white, the men "more light than dark-eyed," although "some observers say that beauty for which the women were renowned is over-rated."[6] Nonetheless, the export of Circassian women as domestic slaves to good Turkish homes continued well into the twentieth century and, as we saw above, was a major factor in the Turkish resistance to Western pressure for an end to the slave trade. There were many middlemen involved in the traffic in Caucasian slaves, and the establishment of a Genoese trading station on the Black Sea coast of the Cacusus in the fourteenth century facilitated their transport to Italy and the Near East.

Slave rule lasted until the coming of the Turks in 1517. However, the Mamluks were nothing if not tenacious and they found ways to prevail under their new suzerains in Istanbul. Bought and brought to Egypt from their native lands, the later Mamluks were primarily Christian in origin. They were selected for their martial qualities and, in the absence of any check, had waxed arrogant and then powerful:

> From the first, insolent and overbearing, the memluks began, as time passed on, to feel their power, and grew more and more riotous and turbulent, oppressing the land by oft-repeated pillage and outrage. Broken up into parties, each with the name of some Sultan or leader, their normal state was one of internal combat and antagonism; while, pampered and indulged, they often turned upon their masters. Some of the more powerful Sultans were able to hold them in order, and there were not wanting occasional intervals of quiet; but trouble and uproar were ever liable to occur.[7]

Cut off from family ties, they owed such allegiance as they possessed only to their masters. More frequently, they owed none at all and the most ruthless of these rootless men rose to the top. The first, or Bahri Mamluk dynasty, so-called since they were quartered on the island of Rhoda on the *bahr*, or river, set the tone for the next half-millennium. No less than twenty-four

---

6. See the *First Encyclopedia of Islam*, vol. II, 836.
7. Muir, *The Mameluke or Slavre dynasty of Egypt*, 4.

Bahri Mamluks, mainly Central Asian in origin, ruled in the following 132 years, an average of just over five years per reign. They took the title Sultan, and attached to their names the honorifics *al-Muizz, al-Mansur, al-Zahir, al-Ashraf, al-Adil*—the Mighty, the Victorious, the Conspicuous, the Noble, the Just—among many others. Mighty, victorious, and conspicuous they may have been, but only infrequently were they noble and just.

The Bahri Mamluks were succeeded by the Burgi Mamluks, so-called since they were quartered in the *burg*, or Citadel. Instead of Central Asia this next wave of slaves came primarily from the Caucasus, but they represented a continuation of the pattern. In the remaining 135 years before the coming of the Ottomans, an equal number of Burgi Mamluk Sultans reigned. Some reigns were long and al-Ashraf Qaytbay ruled for thirty years during which period, as we have seen, he became the greatest Mamluk builder of them all. But others were short, and in the single year 1421 three men occupied the throne following the death of Al-Muayyad Sheikh. But whatever else they did, the Mamluks made Egypt the first principality of the Muslim World, and they are responsible for the bulk of its Islamic monuments.

Nominal Ottoman Rule

With the coming of the Ottomn Turks in 1517, the back of Mamluk power had theoretically been broken. The country was now under the rule of the Ottomans, and the violence and rapacity of Mamluks presumably a thing of the past. But the reality was more complex. On the arrival of the Ottomans the Mamluks had not gone quietly. In fact, they had really not gone at all. At first they were forbidden to wear Turkish dress or shave their beards, measures intended to maintain the distinction between them and the new rulers, although the stricture was reversed shortly afterwards. But Mamluk revolts in 1523 and 1524 convinced the pragmatic Turks that some accommodation with the existing powers in the country was necessary, and in the years to come, representatives of the two settled into a state of mutual tolerance. As long as the sovereignty of the Ottoman Sultan was recognized, regular payments were made to the treasury in Istanbul and Egyptian contributions to the military forces—a kind of standing army on demand—continued, the authorities in the capital were content and the beys were given relatively free reign in Egypt.[8] The seven regiments (*ojaqs*) of the Ottoman garrison[9]

8. The details of this review of the history of Ottoman Egypt can be found in Winter.
9. They were:
   1. Janissaries, the largest, richest and most powerful. They were the *Mustahfizan Qal'a-i-Misr* (Guardians of Citadel), were infantry, and were entirely Turkish.

were originally heavily Turkish. But the incorporation of Mamluks into the formal ruling structure and the increasing demand for soldiers for service in the expanded sphere that Ottoman Egypt now included—the Red Sea, Abyssinia, the Hejaz, and the Yemen—led to infiltration of the Arabic-speaking "sons of Circassians" (that is to say, *fellaheen*) into the army.[10] By the middle of the eighteenth century the Egyptianization of the institution had profoundly changed its original Ottoman character.

At the same time, the "Beylicate"—or rule by the beys, an institution peculiar to Egypt in the Ottoman Empire—had reasserted itself and the Sheikh al-Balad, or senior bey, was the most powerful man in Cairo. The use of the titles "pasha" and "bey," the former corrupted into "basha," are among the most lasting Ottoman legacies in Egypt, and an understanding of their difference is a key to understanding the role each played. Both are Turkish words, although pasha may come originally from the Persian "pad-shah," or "emperor, sovereign, monarch, king." Nominally a position of three tails[11]—the horse tail having long been the symbol of authority among the Turks of Central Asia—the Pasha was the official Ottoman representative in Cairo, and the administration of Egypt was theoretically in the hands of the following officers of the Ottoman *diwan*:

1. The Pasha or Vali;
2. the Qadi Asker, or chief judicial officer;
3. the Qapudan[12] of Alexandria;
4. the Qapudan of Damietta;
5. the Qapudan of Suez;

---

2. *Azabs*, also infantry and also entirely Turkish.
3. *Muteferriqa*—elite Sipahi or cavalry, they often were provincial governors or *kashifs*.
4. *Chawishes*—also elite Sipahi, also often provincial governors or *kashifs*.
5. *Cherakise ojagi*—Circassian, mounted soldiers or *gundis*, they represented an effort to integrate the Mamluks into the Ottoman system, but the *agha* or *kahya* (or *katkhuda*, deputy commander) had to be Turkish.
6. *Gonulluyan*—mounted volunteers.
7. *Tufenkjiyan*—mounted musketeers.

See Winter, *Egyptian Society under Ottoman Rule*, 39.

10. In spite of Artin Pasha's contention that the Mamluks did not mix with the local population, it is clear that many did. See the "Memorandum" attached as an appendix to Muir.

11. The Sultan in Istanbul had a rank of six tails and the governors of the more important provinces, of which Egypt was one, carried the rank of three tails.

12. A captain or commander in the Imperial Ottoman Navy, from the Italian *capitano*.

6. the Defterdar, or executive of the Imperial Treasury;
7. the Amir al-Hajj, or director of the pilgrim caravan;
8. the Serdar al-Hajj, or commander of troops accompanying the pilgrim caravan.

However, the decline in the importance of the position of pasha had been inexorable, if gradual. In 1623, for the first time, the army had refused to accept a pasha sent from Istanbul, and by the time the Danish expedition arrived in the city this assertion of independence from Turkish rule was a regular phenomenon. In just over a year in 1761–62 three pashas came and went. Niebuhr noted the changes:

> While I was in Alexandria, the Cairenes drove out one Pasha. *Mustafa Pasha*, who had been Grand Wazir twice, and afterwards again attained that powerful post, received the order to proceed to *Jidda* . . . but he remained in Egypt. The Cairenes chose him as their Pasha, and so managed things that the Sultan, although he was not pleased with either the new Pasha or the Egyptians, nonetheless named him as governor of Egypt. But he occupied the post for only about seven months . . . and had to give way to another Pasha . . . and the new Pasha died quite suddenly the following night.[13]

We saw above that forty-eight sultans exercised power during the 267 years of Mamluk rule. In the 281 years of the Ottoman rule in Egypt, 110 pashas served as governors, an average tenure less than half that of the unruly Mamluks. Part of the reason is that service in Egypt was considered a vast *muqata'a*, a source of revenue to be exploited,[14] and the pasha was given a relatively short time to make his fortune. If there was anything that characterized Ottoman rule in the middle of the eighteenth century it was this institution, and the baleful effects of rule by the pashas will follow Niebuhr through the lands under Turkish control, even extending into Rumelia, or European Turkey, where the tyrants were typically Christians. But there was a frequent disregard of the nominal ruling council, or *divan* and, as we will see, beys sometimes occupied these offices themselves or simply established bases of power independent of them and behaved as they wished.[15] So far from being a formal rank, a "bey" (written *bik* in Arabic) was instead a title given to "the sons of Pashas, and of a few of the highest civil functionaries

---

13. *Travels*, vol. I, 133.
14. See Winter, 250.
15. See Cezzar, *Ottoman Egypt in the Eighteenth Century*.

... and popularly, to any persons of wealth, or supposed distinction."[16] It was largely an honorific, but the informal nature of the title was a mirror of the informal nature of the rule. For the superficial Turkification or Ottomanization merely papered over the reality that the Mamluks—Circassians, Georgians, and Mingrelians[17]—were soon again in effective control of Egypt. In the middle of the eighteenth century the Pasha, or formal ruler of the country, was still sent by Istanbul. But he was a tool in the hands of the beys and was dismissed at their whim.

There was another factor in the survival of the Mamluks and that had more to do with religion than power. Oddly, given their foreign origins and reputation for violence and rapacity, the Mamluks were seen by Egyptians as better Muslims than the notoriously irreligious Turks. The Sultan in Istanbul may have taken the title of *Khadim al-Haramain al-Sharifain*, or Keeper of the two Holy Places, but in Egypt the Mamluks were considered the more legitimate custodians. Mamluk reverence for local shrines, particularly the sepulchre of Imam al-Shafei, founder of one of the four orthodox schools of Sunni Islam, compared favorably in local eyes with laxity of the Turkish soldiery. While the cultural and linguistic differences between the Turks and the Egyptians grew, the same differences between the Egyptians and the Mamluks narrowed. In spite of the fact that the Mamluks did not constitute a hereditary caste and their numbers were replenished by regular, and often very substantial, purchases in the Caucasus, they did produce offspring with Egyptian women and these further clouded the racial picture and diluted Ottoman exclusivity. Winter suggests that as prospective marriage partners, Egyptian women often despised the Turks and preferred the Mamluks. By the middle of the eighteenth century, the numbers of Mamluks had increased greatly, with prominent beys owning hundreds or even thousands of these men. At the same time, the coin had been debased. No longer was ownership of slaves the exclusive province of the beys, and it was not uncommon for even tradesmen to own Mamluks.

---

16. Redhouse, *A Turkish and English Lexicon*, 375.

17. Another Transcaucsian principality on the Black Sea coast that exported its sons. The Mingrelians were closely related to the Georgians, with whom they lived in close proximity and shared membership in the Greek Orthodox Church. Lavrenti Beria, Stalin's feared and hated head of the NKVD, was Mingrelian and when he and Stalin discussed sensitive matters they often spoke in the Mingrelian dialect which only they and the Georgians understood. Along with the Imeretians, Gurians, Lazis, Svanetians, Masoks, Kevsurs, and Georgians proper, they were inhabitants of the tiny area that for centuries was periodically overrun by Mongols, Persians, Turks, and Russians. It is no wonder that they were tough.

The Government in 1761-62

In 1761-62 the government of Egypt was again in the hands of the strongest of the competing factions of Mamluk beys. Niebuhr says that they were named by Istanbul, but the Egyptians proposed the candidates. No longer sultans, they would nonetheless continue to rule the country until the coming of the French expedition in 1798, although the influence of a foreign mercenary caste would continue long after that. The French squares may have decimated the Mamluks at the battle of the Pyramids in 1798, and Mohammed Ali may have massacred all the remaining Mamluks he could lay his hands on in 1811. But the new Pasha replenished the coffers with levies of new slaves for his European-style army. That army was officered largely by Turks and Circassians and it was not until Urabi's revolt in 1882 that a native Egyptian officer class could make its views felt—albeit briefly—in the affairs of the country.

To say that the beys ruled Egypt, however, was not to say that this rule was uniform or predictable, and factions competed for power with often-deadly results. That made Mamluk rule even more lethal for the populace, who did what they could to stay out of the crossfire. Control of the various districts was exercised by different beys and they moved about the city with "a great display of magnificence,"[18] although the real powers behind the scenes were often content with less show:

> I knew an old and rich merchant who had only one servant, and mounted an ass when he went out to tend to his affairs; but he had placed several of his slaves in positions of importance . . . and they were always ready to defend their benefactor.[19]

There were other contending factions as well, complicated by the fact that there was frequent overlapping among them. The janissaries in the Imperial Ottoman regiments were the largest, richest, and most powerful of the military factions, and they had traditionally allied themselves with the artisanal class in a symbiotic commercial relationship. Many janissaries were themselves merchants, involved in the profitable coffee trade, and Niebuhr and his companions would travel from Suez to Jidda in the company of several of these merchant-warriors. Like travelers before and after them in this part of the world, they would arrange themselves among the crates of merchandise that made the "Cairo" ship little more than a floating bazaar. Then, there were the merchants (*tujjar*) themselves, who represented the other half of this powerful janissary-merchant alliance.

18. Travels, vol. I, 134.
19. Ibid., 134.

Niebuhr in Egypt

The tax farmers (*multazimeen*) were another contending faction in the country. They collected land as part of their fee, acquired substantial land holdings, and were often prominent men in their own right, not infrequently being Jews or Copts. The *ulama*, or clergy, were often themselves large landholders and *multazimeen*, and this further clouded the picture. And finally, there were the village sheikhs (*shuyukh al-balad*), the representatives of the *multazimeen* in the countryside. They were often Bedouins whom the Ottoman authorities had placated by conferring power over the settled population. But however cloudy the relationships among the factions, there was one constant in the country: at the bottom of the heap was the farmer, or *fellah*. He may have been the ultimate source of the wealth of Egypt, but for over two millennia the Egyptian farmer had been exploited by the foreign masters of the country. Enjoying only usufructory rights, he worked but did not own the land and waged a constant struggle to avoid the levies of the taxman, by whom he was tyrannized and abused. In the words of Jabarti, himself a *multazim*, the *fellah* was more degraded than the slaves. However, in spite of his degradation, the *fellah* was a free man and he maintained a semblance of self-respect by his sullen resistance to the *multazim*, and by his stratagems to avoid taxes. The number of blows of the bastinado he could absorb before revealing his wealth became a source of perverse pride to the long-suffering *fellah*.

Niebuhr leaves us with a snapshot of the situation as it existed in Cairo, from November of 1761 through August of 1762. As we have seen, the representatives of the Sultan had already become playthings in the hands of the beys:

> Now, as their way of thinking usually differs from that of the Pasha, whom they regard as their tyrant, it often happens that the Egyptians depose this governor of the Sultan if he hasn't the astuteness to incite the different parties against one another, and at an opportune time lend his support to one or the other of them.[20]

Niebuhr says that most of the beys in his time were purchased from Georgia or Mingrelia, brought to Cairo and sold for "60 to 100 piasters" apiece. Chosen for their spirit and martial qualities, they were educated by their owners as they would their own children.

> When a master notices in one of these slaves extraordinary capacity or fidelity, he often spares no expense to raise his protégé to a place even higher than he occupies himself....[21]

---

20. Ibid., 133
21. Ibid., 134.

Government

Abd al-Rahman Katkhuda was such a man. Niebuhr says that he had raised many protégés to positions of prominence and this, as much as any formal position, was responsible for his power. He was the effective governor of Egypt in 1761 and had won the "affection" of the populace through pious works,[22] affection being remarkable enough a characteristic among the Mamluks that its mention is notable.

An understanding of the titles taken by the rulers is another key to understanding power in eighteenth-century Egypt, although, like pasha and bey, many were used nominally and often did not refer to a specific rank. The word *katkhuda* appears again and again in Niebuhr's travels through Ottoman lands. Along with its vulgar form of *kikhya*, it was also Turkish in origin[23] and could mean a great many things, from "a steward in a great man's household" to "the title of the colonel of the first Bulak regiment of the old Janissaries." According to Niebuhr, the katkhuda or kikhya was formally next in rank after the Aga of the janissaries, although like many other Turkish words, it was widely adopted in Egypt as an honorific. As we will see below, there was even a Jewish kikhya appointed in Istanbul to look after the affairs of the Jewish subjects of the Empire. The *aga*, too, was a Turkish rank, used in the military for the position below the commander of a battalion. But it frequently appeared in its adjectival form, *agassi*, and was also used informally by the chief eunuchs of the Empire, both black and white, positions of great power.

The Beys

Niebuhr says there were officially twenty-four beys in Egypt but that their number was never complete. He speculates that the diminution was a function of declining revenues: the pie simply wasn't large enough to divide twenty-four ways. He lists eighteen beys, including several unfortunates whose heads would later be paraded through the Bab al-Nasr, and the list is itself an interesting study in the relationships that prevailed among them. They were:

1. *Khalil Bey*, had been bought, raised and furthered by Ibrahim Kikhya; and 1762 he occupied the position of *Defterdar* or treasurer of Egypt.

---

22. Most of his many works in Cairo were restorations of previous buildings, although in 1754 he did build a small mosque in the Nubian Quarter. See Behrens-Abouseif, *Azbakiyya*, 55. Among other monuments, his *sabil* (public fountain) in Shari' Muizz li Din Ullah was recently restored by the German Archaeological Institute.

23. See Redhouse, *A Turkish and English Lexicon*, 1524. Even today, the eighteenth-century mosque of 'Uthman Katkhuda bordering on the old Opera Square is known to every Cairo taxi driver as *masgid al-kikhya*.

2. *Hossein Bey* was in the same year *Emir el Haj* or leader of the Egyptian caravan. He had also been a slave of Ibrahim Kikhya.

3. *Ali Bey*,[24] another slave of Ibrahim Kikhya, was *Sheikh el Belled*, or Governor of the city of Cairo. He was called at the same time *sogair* or the *small Ali Bey*. But the following year he, with his party, forced the reigning Pasha to forbid Abd el Rahman Kikhya, who was at the *Birket al-Hajj* in the company of friends going to Mekka, to return to Cairo, but to depart with the caravan. In this way Ali Bey became as powerful as Abdel Rahman Kikhya had ever been. His power was not to last long; for he was also forced to leave Egypt and was sent to Gaza. But in 1768 he returned to Cairo, killed in one night four Beys, and forced the Pasha to forbid four other Beys, who had taken flight, to return to Cairo. From that time, he was the head of the party that remained, and everyone submitted to his orders. It was known from the gazettes that he even banished the Pasha, and he openly declared himself against the Sultan. But he was driven out in his turn by another Bey named *Mohammed abu Dahab* and had joined *Daher Omar*, Sheikh of Acca. I could not find out exactly what positions the other Beys held. The following is the order after the above-mentioned three:

4. *Othman Bey*, formerly a slave of Ibrahim Kikhya.

5. *Hassan Bey*, a slave of a certain Soliman Aga, Kikhya of the Tsjaus.[25]

6. *Hassan Bey*. He had been a slave of Omar Bey, the elector. He was called *Hassan Bey Reduan* to distinguish him from the previous man.

7. *Khalil Bey*, surnamed *Belsie*. He was the son of a certain Ibrahim Bey and a man of distinction with a Mohammedan father.

8. *Hassan Bey*, surnamed *Damud*, a slave of Soliman Aga, Kikhya of the Tsjaus.

9. *Saleh Bey*, a former slave of a certain Mustafa Bey el Kerd.

10. *Othman Bey*, surnamed *Abu Seif*, i.e. he who handled the sword skillfully. He was a Turk by birth, from Constantinople, and had never been a slave. Ibrahim Kikhya, in whose service he was, raised him to the highest positions among

---

24. The *Ali Bey* called "*Bulut Kapan*."
25. Heralds and messengers of the corps of the janissaries (Redhouse, 711).

his household troops, and finally obtained for him the title of one of the Beys of Egypt.

11. *Khalil Bey*, surnamed *Es sekran*, that is, the drunkard. He had been a slave of the current Emir el Hajj, Hossein Bey.

12. *Ahmed Bey es sukari* was the son of a Mohammedan sugar merchant of Cairo, and thus was not a Christian by birth. He was raised up in the house of Ibrahim Kikhya. In my time he was governor of Suez and so, as it were, was in exile.

13. *Ismail Bey*, a slave of Ibrahim Kikhya.

14. *Mahhmud Bey*, a slave of Othman Kikhya.

15. *Hamsa Bey*, was the son of one Hassan Bey Abassa, and so a Mohammedan by birth.

16. *Mohammed Bey*, surnamed *Hanefi*. He had been a slave of Soliman, Aga of the Tsjaus.

17. *Mohammed Bey Dali*. He was the son of Ismail Bey ed Dali, and so a Mohammedan by birth.

18. *Ali Bey*, called *el kebir*, i.e. Ali Bey the Great, as much to distinguish him from the *Sheikh el belled* as because for several years he had been very powerful in Cairo. During the time I was in Egypt he was banished to Gaza; but a little before our departure, he returned to Cairo and died very soon thereafter. His death was attributed to a poisoned pelisse, which a false friend among the Beys presented him as a mark of esteem. He had been bought and raised by Ibrahim Kikhya. I believe he was the same man who, it was said, was the son of a Georgian priest; that his parents, brothers, and sisters had come to visit him in Cairo; that the aged father had returned, but one sister and two brothers had stayed; and that the latter, after becoming Mohammedans, had been made by *Ali Bey* the governors (*Kashefs*) of small districts.[26]

So, of the eighteen, thirteen had been slaves and five had been freeborn Muslims, the latter including one Turk. But the former slaves were clearly in control, and there was even a certain perverse honor that attached to their servile origin. Here, Niebuhr clarifies the often-confusing references to the two Ali Beys. Ali Bey *al-Saghir* (or "the small") is the Ali Bey *Bulut Kapan* of Jabarti and, so, the Ali Bey of greater fame. It is he who impressed his stamp on Egypt for the period of over twenty years, from his rise in the early 1760s to his death in 1783. Niebuhr informs us that he was the Sheikh al-Balad, or

26. *Travels*, vol. I, 134–36.

governor of the city of Cairo, during the period the Danish expedition was in Egypt. However, several years of intrigue lay ahead, including his own ultimately successful action against Abd al-Rahman Katkhuda detailed below. Niebuhr tells us that in the process he also suffered exile, also to Gaza, before he returned in 1768 and siezed control of the city.

The Diwan and Public Order

After the beys, Niebuhr lists the following in order of importance in the *diwan*, or governing council, of Cairo:

1. the *Aga* or commander of the *Metaffaraka*[27] regiment;
2. the *Katkhuda* of the *Tsjauschan*;[28]
3. the *Aga* of the *DsJumlan*;[29] regiment,
4. the *Aga* of the *Teffekshun*;[30] regiment,
5. the *Aga* of the *Tsjaraksa*;[31] regiment,
6. the *Aga* of the Janissaries;
7. the *Aga* of the *Assab*[32] regiment.[33]

As we have seen, the *ojaq* of these "bachelors" was originally subordinate to the janissaries and a bitter enmity between the two corps developed. The assabs were originally charged with guarding the gates of the Citadel, and the monumental gate at the lower enclosure is named after them. In addition to the heads of these formal regiments, there were several other kikhyas of regiments, men of the law, and ecclesiastics about whom Niebuhr was uncertain. These, he observes with characteristic self-deprecation, were beyond the ken of a "simple traveler who only passed through the country."

Under Mamluk rule, according to Niebuhr, security in Cairo was superior to that in most cities in Europe, in spite of the chronic state of civil strife among the beys. Murder and other crimes of violence were rare among the populace, although there is the suggestion elsewhere that petty crime was

---

27. According to Redhouse (p. 1714), originally holders of fiefs and attached to the Grand Vezir.
28. Probably a herald or messenger (Redhouse, 740).
29. Probably a camel-mounted regiment.
30. Vulgar for *tufengi* or "musketeer," (Redhouse, 574)
31. Circassian.
32. Or "bachelors." We saw them earlier in discussing the *Bab al-Assab*.
33. *Travels*, vol. I, 137.

Government

widespread at this time. To this eighteenth-century German visitor who made it his business to know all the districts, the city was comparatively safe in 1761:

> one hears less of theft and murder here than in the greatest cities of Europe . . . The perpetual fear of being surprised by the magistrates keeps the criminal population in check. I saw the fear and dread among the Egyptian populace whenever I met an officer in the street . . .[34]

The dread was understandable, as punishment could be cruel, by hanging, impalement, or flaying alive. Often, petty criminals were sent to the Ottoman fleet as rowers. The instruments of public order appeared to be functioning well. Judges sat in their chambers and janissaries patrolled the streets. The *serrajs* lorded it over the populace and, as we have seen, lent their own measure of caprice to governance. The masters of guilds, of which there were some 240 in thirty major categories, knew their members and monitored their activities.[35] Even footpads had their provost and it appears that there was honor among thieves: on application to the provost and payment of a small fee, a stolen item could often be recovered.

The accuracy of weights and measures was maintained in each quarter, whose gates were often closed at night. Offenders, particularly those caught fleecing the public, were very summarily dealt with. Floggings were administered freely and it was not uncommon for a cheat to be hanged on the spot. Those abroad after dark were required to carry lanterns and woe betide the man who was not so equipped. We have already seen that the quarters were often closed communities unto themselves and that a stranger, especially a Frank, was immediately noticed. *Bawwabs* monitored the comings and goings by day and night. Janissaries also manned a small room near the gates and slept at their posts by night. In the event of trouble among the beys, the gates were closed and the ensuing riot and upheaval kept out of the quarter. Guilds were often a substitute for the municipal institutions that typically did not exist in Islamic countries and often had social, political, and religious functions. But there was another side of the coin: as we will see below, along with the Sufi brotherhoods they could be a source of resistance, not to say rebellion, against the established political order.

---

34. *Travels*, vol. I, 137–38.
35. See Lewis, "The Islamic Guilds."

Portents of the Future

Occasionally, a strong bey would assert his control and the country would settle into a period of centralized and more organized rapacity. These periods were probably more tolerable than the anarchy that prevailed at other times. In 1761, when the Danish expedition arrived in Cairo, such a period of consolidation was just beginning and the political and economic system of the country would undergo a profound change over the next ten years. Suspended, as it was, between a violent and turbulent Mamluk past and a firm, if oppressive, future under Mohammed Ali, the country shared characteristics of both periods. In fact, there is the suggestion[36] that this period represented not the last gasp of Mamluk rule but rather, under the relatively enlightened tyranny of Ali Bey Bulut Kapan, or "the Cloud Catcher,"[37] a harbinger of the future under Mohammed Ali. But if Ali Bey had his head in the clouds, he dreamed with his eyes open to make the dream happen.

In 1761 he was in the early stages of establishing a strong centralized rule. By 1768, he would have replaced the janissaries with a mercenary corps and dispersed, exiled, or exterminated his opponents, not excepting the early confederates in his rise to power. As we will see in the chapter on commerce, these moves were not made in isolation from forces that were integrating Egypt into the European economic system. Many representatives of the old order were replaced since they were less susceptible to exploitation under the new. In the process, Ali Bey would become not just the most ruthless and successful of the beys but, in his centralizing tendencies, a glimpse of the future. As with the Albanian pasha after him, the stakes were high. Some merchants were enormously wealthy, with fleets of ships, buildings, and other holdings worth millions of *paras*, and sheikhs controlled vast tracts of land and the products they produced. A certain Sheikh Humman in Upper Egypt owned 12,000 bullocks working his cane fields alone.[38] None of this escaped the attention and, ultimately, the grasp of Ali Bey.

What now changed was the effective end of the collegial system in which the conflicting interests of the rootless and violent contenders for power had been reconciled by a kind of balance of power. In its place, came centralized rule by a single man. As a first step Ali Bey ousted the regiments that had traditionally exploited the urban (*iqtaat*) and rural (*iltizimat*) tax

---

36. See Marsot, *Egypt in the Reign of Mohammed Ali*. Much of the information on the commercial developments of the period come from this source and her sources, Jabarti, and Raymond.

37. The derivation is from Winter, 25. See above Niebuhr's discussion of the two Ali Beys.

38. Marsot, 11.

farms and brought the control of the entire fiscal machinery under his own hand. He targeted both the head and the body of the military, starting with the former. He "massacred his opponents and exiled his notables . . . had them followed, strangled, and killed; he exterminated them root and branch . . ."[39] He then changed structure of the army, set aside the Ottoman regiments and hired mercenaries. This led to an increased need for cash for the purchase of the mercenaries and the modern arms to equip them and, so, to even more extortionate demands on the population. But along with the newly centralized extortion, Ali Bey firmly established the rule, if not of law, at least of a kind of clearly understood order. That meant Ali Bey himself and he ruled absolutely.

Along with changes in the military structure and the tax system, Ali Bey also brought the customs officials and the merchants, or *tujjar*, under his control. In his time Niebuhr reported that Jews controlled the customhouses of Bulaq, Masr al-Atik, Alexandria, and Damietta. So great was their influence with the beys that all the customhouses were closed on Saturday, the Jewish Sabbath, and goods could not pass on that day, even if owned by Muslims or Christians. But because of their close links with the janissaries, Ali Bey dismissed the Jewish customs officials and in their place appointed Syrian Christians, Melkites who had fled Syria in the first quarter of the century, it is thought, to escape Greek Orthodox persecution. Not only were the Syrians more tractable, but they brought with them even more inventive ways of extorting money from the populace. They were now encouraged to displace not only the Jews, but also the Muslim traders, and having settled largely in Damietta and Rashid, they became Ali Bey's men in these cities. Ali Bey's two secretaries were Egyptian Christians, Muallim Rizq and Muallim Ibrahim al-Gawhari and the impoverishment of the Muslim *tujjar* continued apace. The rise of the Christians was spectacular and their control of customs allowed them to tighten their grip on the trade of the country. Their alliance with Ali Bey extended to his foreign adventures, and after he had conquered the Hejaz, these transplanted Syrians gained control of the lucrative Red Sea trade as well. Christians may have gained great influence in key sectors of the economy, including the rice trade and exports to France, but control by Ali Bey, not religion, was the operative principle and their tenure was always tenuous.

A short epilogue will tell the ultimate fate of Ali Bey and his reforms. Jabarti tells the story. By various means, including extortion and persuasion, he amassed a great fortune and in the middle of the 1760s his ascent was rapid. The rise was not without the aid of Abd al-Rahman Katkhuda,

---

39. Jabarti, quoted in Marsot, 11.

the patron of many of the boys. Ali Bey had actively courted his friendship and this was reciprocated, also for reasons of policy. However, beneath the surface friendship, Ali Bey was biding his time. Early in 1766 he made his move. Abd al-Rahman Katkhuda, too important to assassinate, was sent to Suez, whence he was later escorted to exile in the Hejaz. The move provoked agitation in Cairo, as Abdel Rahman was well loved by the people.

> For these reasons, terrible riots were feared on the day of his exile, but everything passed peacefully and the inhabitants of Cairo were in a kind of stupefaction by the blow leveled at Abdel Rahman.[40]

Simultaneously, moves against other prominent beys took place. Several took flight and Ali Bey's forces followed several of the fleeing men to Lower Egypt, where they had taken refuge in the mosque of Ahmad al-Badawi in Tanta. Many were killed and Ali Bey's army returned to Cairo with their grisly trophies:

> They entered the Bab al–Nasr amidst an immense cortege, preceded by servants carrying the heads of the victims on silver plates, and calling on the crowd to invoke the blessing of God on the prophet Mohammed. These heads were six in number, namely: those of Hussein Bey, Khalil Bey al-Sakran, Hassan Bey Shabaka, Hamza Bey, Ismail Bey Abou Madfa, and Soliman Agha al-Wali ... The entrance took place on the 17th of Muharram.[41]

But with the death by poison of Ali Bey in 1783,[42] centrifugal forces again asserted themselves and the traditional Mamluk contest for power resumed. There were uprisings and riots and impositions and extortion again became commonplace as the firm hand at the center was removed. The state of "quasi–permanent civil war" again characterized the country. Egypt was also visited by plague and famine in 1785, 1791, and 1792 and economic hardship and depopulation reigned. By the time the French arrived in 1798, Egypt had descended into the customary chaos. It was a descent from which Mohammed Ali, following in the steps of Ali Bey, would raise the country with his own, more organized and systematic regime of exploitation.

---

40. Jabarti, *Merveilles Biographiques et Historiques ou Chroniques*, vol. II, 222.

41. Ibid., vol. III, 6.

42. Jabarti says that Ali Bey was buried in the lesser *qarafa*, near the mausoleum of the Imam Shafei. The grave, next to that of Abd al-Rahman Katkhuda, his patron, is still there today but lies in what is now the courtyard of a private home. It can be viewed only in the presence of the owner.

8

Inhabitants

> Most of the inhabitants of the city of Cairo are Arabs, Turks, and other Mohammedans from all the provinces of the empire of Turkey . . . After the Mohammedans, the Coptic Christians are the most numerous. They are the descendants of the ancient Egyptians . . . After the Mohammedans and the Copts the community of the Jews is probably the greatest. (*Travels*, vol. I, 131.)

The majority of the residents of Cairo in 1761–62 were native-born Sunni Muslims, most of them of the Shafei school. Niebuhr gives no figures for the population of Egypt, although he does mention that the city of Cairo was probably not as populous as it once had been. Many factors, of which plague was the most important, periodically reduced the numbers and it was estimated that one visitation in 1734 had killed a sixth of the population of the city. We have seen from Niebuhr's map that the quarantine station was the hospital of al-Mansur Qalawun in the center of the Fatimid city, and quarantine may have reduced the most virulent outbreaks of the disease. But Cairo was a thriving commercial hub, and it was impossible to stem the flow of goods and travelers from Africa, the Red Sea, the Maghrib, Turkey, and the Mediterranean who passed in and out of the city on a daily basis. When infection was abroad in the population, it spread rapidly, due to a lack of understanding of contagion, compounded by the understandable habits of a grieving people. In its most contagious stage, the pneumonic strain of the disease passed from individual to individual by respiration—simply breathing on another person—and one can imagine funerals in the early onset of the epidemic, where friends and family members would gather to console one another, all the while, unknowingly, spreading the disease. As the virulence of the epidemic increased funerals were probably a luxury few could afford as the normal procedures for dealing with death were overwhelmed. Bodies would remain unburied in the streets until the epidemic

Niebuhr in Egypt

abated and the authorities could deal with the accumulated burden of the dead, only to await the next onset. The favorite place to pray for deliverance from the plague was the little mosque of al-Guyushi on the Muqattam, and it was a sanctuary to which Cairenes were unfortunately driven throughout most decades of the eighteenth century.

Precise numbers are difficult to come by, but the population at the time the Danish expedition was in Cairo cannot have been too different from the figure on the coming of the French in 1798. In 1800, according to the *Description de l'Egypte*, the population of the city was estimated to be 263,700. It had declined somewhat in the last decades of eighteenth century due to plague, compounded by famine, economic exploitation, and factional strife. But the figure gives a rough guide to the magnitudes at the time Danish expedition was in the city. According to the French, the ethnic breakdown was as follows: indigenous Muslim Egyptians, some 200,000; foreign Muslims, 25,000, of whom Turks were 10,000, Maghribis 10,000, and Syrians 5,000; religious minorities, 25,000—Copts 10,000, Greeks 5,000, Syrian Catholics 5,000, Jews 3,000, and Armenians 2,000; the ruling class of Mamluks and Turks, 12,000; plus assorted European merchants and members of religious orders. The percentages, if not the absolute numbers, were probably broadly representative. We will examine each of the categories in what follows, fleshng out the rough numbers with what we know from other sources to give a clearer picture of the Cairo Niebuhr would have seen during his year in the city.

As might be expected, these ethnic groups tended to live in separate neighborhoods. Niebuhr tells us that the gates to most neighborhoods were closed and locked at night and the comings and goings of strangers were closely watched. The Arabic word *hara*—quarter or section—divided the Fatimid city into religious or ethnic ghettos. There was Harat ar-Rum, to the north and east of the Bab Zuweyla, where the Greeks once lived. As we have seen, the Greeks had two religious centers in the Fatimid city itself, one the church and residence of the patriarch of Alexandria and the other of the bishop of Mount Sinai. But although the word *Rum*, or Romaean, was traditionally used to refer to the Byzantine Greeks, there were other European influences as well and the quarter still contains the shell of what appears to be a Spanish or Italian baroque church whose name and provenance are unknown. Outside the Fatimid city, the Greeks also had the church of St. George to the south in Old Cairo. The Armenians had a "pretty little church ... near the Kantaret jedid"[1] as mentioned above.

The foreign consuls and merchants—and the members of the Danish expedition—lived outside the walls of the Fatimid city and to the west of the

---

1. *Travels* vol. I, 132.

khalig in the Frankish Quarter that would become the center of European Cairo in the nineteenth century. European consulates would later line the streets that bordered the Azbakiyya Gardens, so-called after the park on the site of the old pond planted by Mohammed Ali in 1837, after deciding that the Europeans needed a promenade.² In 1761–62, only the French, Venetians, and Dutch were permitted to maintain permanent consulates in the city and, as we have seen, the members of the expedition took up residence in two of those establishments. There were European merchants in the city, probably numbering in the low hundreds, and it was only with the nineteenth century and the reforms of Mohammed Ali, that foreigners would come in large numbers. They would ultimately populate the cemeteries that grew up in the area between Old Cairo and the Fatimid city.³ According to Niebuhr, there was no shortage of European monks in the city—Jesuits, Capuchins, Franciscans, and Fathers of the Propagation of the Faith—all zealously proselytizing. The Muslims permitted their activities, as long as they confined their attentions to Oriental Christians. If there was another constant in Niebuhr's travels it would be the presence of these representatives of the Roman Catholic Church in lands with large numbers of Oriental Christians. The Counter-Reformation was alive and well in the East, the Pope recognizing the challenge to his authority not only by the Protestants but also by the ancient communities of Copts, Nestorians, Jacobites, and other believers not in communion with Rome. Here, as elsewhere, the Latin missionary activities provoked frequent disputes that represented opportunities for the beys to levy fines on the contending parties.

The Turks congregated in the district of the Khan al-Khalili in the confines of the old Fatimid city, although they also had centers of influence in areas to the south towards Ibn Tulun. The Muayyad mosque, incorporated into the Bab Zuwayla in Burgi Mamluk times, was a Turkish center, as was the Maridani mosque on the Shari Darb al-Ahmar that connected the Tulunid area with the southern gate of the Fatimid city. Maghribis, or North Africans, tended to live around the mosque of Ibn Tulun. The Harat al-Yahud, or Jewish quarter, was to the east of the *khalig* and, so, within the irregular quadrangle of the Fatimid city. As they had for centuries, the Copts lived in the Harat an-Nassarah, or the Christian quarter, to the west of the *khalig* in the area of al-Maqs. It lay both north and south of the pond, or Birket al-Azbakiyya. As we saw above, the ruling beys increasingly built

2. See Behrens-Abouseif, *Azbakiyya*, 84.

3. Each dispensation or nationality had its own cemetery and there would eventually be Armenian Catholic, Roman Catholic, German, American, English, Greek Orthodox, Maronite, and the "British Protestant Burial Ground."

their homes and belvederes on the shores of the pond, the earlier expansion to the south towards Ibn Tulun having come to close.

Egyptian Muslims

Native Egyptian Muslims were the vast majority of the population and they lived in the residential districts represented by the small, often one-way streets that crisscross Niebuhr's map. Although they did not constitute the ruling class, it would be a mistake to suggest that they had no influence in the affairs of the country. Egyptians were deeply religious, and religious sentiment lay very close to the surface in the country. It periodically made itself felt, to the discomfort of the minorities and the authorities alike. The Turkish occupation at first deprived the *ulama*, or Islamic scholars, of many of their privileges. But by the middle of the eighteenth century, they had reasserted themselves and the last several decades of the century represented the height of their influence in Ottoman Egypt. They played a role as intermediaries between the rulers and the ruled, and often served as "the last recourse of the oppressed subjects"[4] against the oppressors. A kind of local religious aristocracy grew up to rival the military and civic aristocracy filled by the Mamluks and Ottomans. The *ashraf*—descendants of the prophet through Ali and Fatima—increasingly came to play a role in Egypt through the Ottoman times. Most had their origin in the Hejaz, although they were later Egyptianized, and they provided a religious counterpoint to the role played by the secular ruling class. By the second half of the eighteenth century the Naqib al-Ashraf, or chief of the descendants of the prophet, was an Ottoman functionary and the only administrative position in the country held by a native Egyptian.

A parallel development was the growth of the great mosque and college of al-Azhar, founded in Fatimid times. The importance of the position of the Sheikh al-Azhar correspondingly grew with that of the institution itself. The Azhar became the virtual nerve center of Cairo, and demonstrations by students and teachers were frequent occurrences, most often having to do with perceived threats to the ascendancy of Islam. They often forced the authorities to take actions against non-Muslims they probably would have preferred to avoid. Real or suspected insults to Islam abroad, for which local minorities often took the brunt of the blame, were common causes for popular unrest. The Ottoman exercise of *Realpolitik*, with alliances that did not always conform to religious biases, was a constant source of reproach against the irreligious Turks. But even in the absence of outside influences,

---

4. Winter, 109.

there were ample opportunities for religious contention in Egypt itself, and the passions found an outlet in the streets. As we have seen, Niebuhr had a healthy fear of the religious passions of the Cairenes. Most Azharites were poor, Arabic-speaking Egyptians from the villages, steeped in little but traditional *fiqh*, or Islamic jurisprudence.

There was also a growth in the importance of Sufism, the popular expression of Islam with a belief in saints and miracles that gave the religion a more human face. Ironically, given the reputation of the Turks, it was the traditional Turkish attachment to Sufism that led to its increased importance in Egypt, although the following of the strictly Turkish orders was small. The orders—the Qadiriyya, Shadhiliyya, Rifaiyya, Saadiyya, Khalawitiyya, and Ahmadiyya—provided a religious framework that reached into every aspect of everyday life. The Ahmadiyya was the people's cult and had the largest number of followers. There were no special Sufi dress requirements, and the orders did not impose cloisters, there being no "monkery" in Islam. The *zawiya*, or small neighborhood mosque or prayer room, was the Sufi center and hospice. Many prominent men were Sufis and Jabarti was a Khalawati Sufi. The Mulid an-Nabi—or the Prophet's birthday—was the greatest Sufi event in Cairo, celebrated near the sepulcher of Imam Shaf'i. It attracted hordes of common people, small traders, jugglers, acrobats, animal tamers, snake charmers, and prostitutes, to the disgust of more respectable Egyptians such as Jabarti.[5] Cairenes were fond of celebrations, and favorite picnic grounds on public holidays were Matariyya north of Cairo, the pyramids at Giza, al-Basatin, Rodha, Qasr al-Ayni, and the great ponds al-Fil, al-Azbekiyya and ar-Roteli. The ruling beys themselves were often members of the Sufi orders, although as practical military men they were less superstitious than the general public. No matter how ruthless and exploitative they may have been, the Mamluks of all periods aspired to be remembered as builders of *zawiyas*, *sabils*, and *tekkes*—small mosques, public fountains, and hospices for the poor—and many of the monuments of Mamluk architecture that survive to this day are a product of this pious impulse.

Niebuhr tells us that the city also contained a liberal sprinkling of people from "other provinces in the Empire of Turkey:" in addition to the Turks and Maghribis from Barbary, there were Syrians, Arabs from the peninsula, Africans from the Sudan, Tatars, and *Agam*, or Persians. To the south, as we will see, regular caravans still traded in the traditional goods of

---

5. A visit on a Friday morning to the same area today is to one of the liveliest outdoor markets in the city, with everything on display—*bikiyya* (junk, from the Italian *vecchia*, or "old"), clothes, fabric, ceramic tiles, fruits and vegetables, sweets, meat, and animals of every description—ruminants, fowls, and rabbits. In short, everything but the prostitutes, although one can't be certain.

Africa, including human beings. But there was also the yearly caravan west to Timbuktu through the Fezzan,[6] the largest oasis in the Sahara south of Tripoli, where the object would have been gold as well as slaves. In 1761–62, black slaves were at the bottom of the economic and social order, although Jabarti would have reserved that dubious honor to the *fellaheen*.

The *Dimmis*

The two most numerous *dimmi* peoples were the Copts and the Jews. The word appears to come from the Arabic root *damma*—"to blame, find blameworthy, dispraise, criticize, find fault . . ."[7] While the derivation may be accidental, it is probably appropriate in that, as we have seen, Christians and Jews were often blamed for natural disasters in the city. They were rarely mentioned except negatively, in terms that have traditionally been used to describe minorities. They were distinct communities, a small number of whose members had attained great wealth, often occupying positions that gave them virtual financial control of the country and exercising influence disproportionate to their numbers in the general population. It was a position that made them simultaneously favored and vulnerable to extortion by the authorities and to periodic outbreaks of violence by the populace. The wealthy, of course—Muslims included—had always been targets of the beys. But with the superficial Ottomanization of Egypt after 1517, the natural religious predilections of the Egyptians came to the fore in a kind of national resistance to foreign influences. This complicated the lives of the *dimmis*, even though they were as Egyptian as the Muslims. Some would argue that they were more so. Oddly, as we have seen, the Mamluks—originally the sons of Balkan Christians—were seen as natural allies of the people against the conspicuous irreligion of the Turks in Egypt.

The Copts were the most numerous natives after the Muslims and they shared with the Jews the general opprobrium reserved for unbelievers, although they were, in general, more aggressive in seeking redress for alleged wrongs. As *dimmis* they paid the *jizya*, or head tax, in return for freedom in personal matters. As mentioned above, the conditions of Omar

---

6. Joining this caravan was the professed object of Burckhardt on his arrival in Cairo in 1812. Ultimately, he was unsuccessful, instead going south to the Sudan and Suakin, and thence to Jidda before returning to Cairo. He died of dysentery in Cairo in 1817 and is buried in the cemetery to the east of the Bab al-Nasr. A plaque recently placed there by the Swiss Embassy marks his very Ottoman tomb.

7. According to Hans Wehr (p. 360), *dimmi*s are "free non-Muslim subjects living in a Muslim country who, in return for paying the head tax, enjoyed protection and safety."

were enforced with greater or lesser severity, often depending on the state of relations between the Muslim World and Christendom, or sometimes in response to a local incident that provoked a popular outcry. These incidents had been regular occurrences during the Mamluk period and the story of an ambassador from the Court of Aragon, who came to Cairo in 1305 to recover a Christian prisoner, was probably typical. The prisoner was released but, before the Aragonese had left Alexandria, the Sultan had a change of heart and sent a messenger demanding a ransom. The Spaniards not only refused the ransom, but took the Sultan's messenger with them when they departed. The incident led to violent public demonstrations, the conditions of Omar were enforced with particular severity, Christians and Jews were dismissed from the government, and attacks against both were organized with official complicity and approval. An attack against Alexandria by the French in the next century had the same effect.

The beys, in general, were motivated less by religious considerations than the prospect of gain, but they often used the former to extract the latter: mediation between Muslims and Copts, Copts and Jews, or Latins and the Orthodox Copts was a source of revenue for many a bey. The Copts shared with the Jews some of the same positions of influence with the authorities such as tax farming, the Jews in general milking the city-dwellers and the Copts farming the rural areas. The fiscal year was still fixed according to the Coptic calendar, a holdover from Pharaonic times, which was more appropriate to the agricultural year than the Muslim lunar calendar. The two communities shared as well the chronic resentment of the average Muslim for what was perceived as their overweening influence with the authorities. In general, the Jews were despised more than the Christians although it was a relative benefit that was probably of scant consolation to the Copts. And the two communities had their own frictions as they often competed for the same privileged positions, often traded in the same commodities—alcohol, jewelry, gold, and silver—and shared the occupations of money-changing and lending. Conversion of Christians to Islam was fairly common, conversion of Jews extremely rare.

We will see below that not all Copts were wealthy. Niebuhr would find the Copts of the Delta village of Zifta to be welcoming but very poor, their very receptiveness to this particular European a harbinger of difficulties to come. The precarious position of the Copts in Egypt would not be helped when Europeans later began to arrive in large numbers, not as merchants and travelers, but as rulers. The two sides—European and Copt—cautiously tested one another, some Europeans feeling that they would find natural allies in these local Christians, and some Copts believing that the Europeans should translate this natural alliance into favoritism. It didn't work out that way, and one side of the mutual disenchantment was articulated by Lord Cromer. Variously reported in Egypt to this day as ranging from merely hostile to downright inflammatory, Cromer's disparagement of both communities—Muslim and Christian—was actually more nuanced.[8] The expectation of some Copts that they had finally found a champion after centuries of persecution did not endear them to their Muslim neighbors and suggested that perhaps, after all, they had divided loyalties. In 1761 there seemed to be little question of divided loyalties, if only because the Europeans were represented by merchants who competed for their business and Roman Catholic religious orders who competed for their souls.

According to Niebuhr, the Copts in Cairo lived in large districts "in the vicinity of Birket el Jusbekie, Kantaret el charq,[9] Bab Sheikh Rihan and other places . . ."[10] A glance at his map shows these to be where we would expect them to be: the Azbakiyya area, which had been the home to minorities long before it became the favored area of the beys in the urban expansion of the eighteenth century. It was originally al-Maqs, a port when the Nile flowed much farther to the east than it did in the eighteenth century. Throughout its history it had a heavily Christian flavor, effectively a suburb outside the walls of the Fatimid city, but from which the Copts who had important administrative and financial functions still had reasonably easy access to the centers of power. The first urban expansion in the twelfth century had been southwards towards Ibn Tulun, and this left the area to the west to the Copts and other minorities. The area eventually became the main Coptic quarter. As we have seen, the economic activities in the area had traditionally been those of light industry—textile weaving, food processing, slaughterhouses, mills, oil presses, vinegar presses, and money changing. In Niebuhr's map of the city, the "Kantaret Bab el chark" appears in the area where the road from "Bab el Luq" crosses the *khalig* and enters

---

8. "My own experience leads me to the following conclusions: first, that owing to circumstances unconnected with the difference of religion, the Egyptian Copt has developed certain moral attributes which also belong to the Egyptian Muslim; secondly, that owing to circumstances which are accidentally connected with, but which are not the consequences of his religion, the Copt has developed certain intellectual qualities in which, mainly from want of exercise, the Egyptian Muslim seems to be deficient; thirdly, that for all purposes of broad generalization, the only difference between the Copt and the Muslim is that the former is an Egyptian who worships in a Christian church, while the latter is an Egyptian who worships in a Mohammedan mosque" (Cromer, *Modern Egypt*, vol. II, 205–6).

9. Elsewhere Niebuhr calls it "Kantaret bab el charq." This appears to be both a bridge over the *khalig* and a lesser gate into the Fatimid city. "Charq" could mean "crossing, transit." Alternately, as we saw above, it may be a misprint for "Bab el Luq."

10. *Travels*, vol. I, 131.

the Fatimid city; and the "Bab es shech rihan" appears some 400 paces to the west of the *khalig* in the vicinity of the "Birket en nasserie."

Niebuhr mentions that the Copts had only two churches in Cairo, by which he means the confines of the Fatimid city, although there were others to the south in Old Cairo. Because the Copts were traditionally forbidden to build new churches, the two were outside of the areas where the Coptic population increasingly came to be concentrated. As we have seen, the patriarchal church of the Copts was located within the original walls of the Fatimid city, northeast of the Bab Zuwayla. The church dates from the fourth century and still exists today, although the patriarchate has moved north to Abassiyya. The second was in the Harat Zuwayla, to the east of the *khalig*, and so within the city itself. It was, according to Niebuhr, built over an Armenian Church. As we saw in chapter 6 above, it still exists today as the Church of the Holy Virgin.[11] From the fourteenth to the seventeenth century it served as the Coptic patriarchate. The present-day church of St. Mercurius, part of the same complex, was built on the site of the Armenian Church of St. John the Baptist. That church was demolished in 1773, eleven years after Niebuhr saw it, and the present church was built in its place.[12]

The Copts had a history of persecution and martyrology in Egypt that began long before the arrival of the Arabs. As we saw above, the Coptic calendar is dated from the persecutions of the Roman Emperor Diocletian in the year 284, and they endured successively as overlords pagan Romans, Byzantine Christians, and a succession of Muslim Arabs, Berbers, Kurds, and Turks. With the coming of the Muslims, to quote Butler referring to an earlier change of suzerains, "the Copts settled down in a measure of tranquility under one more of those changes of masters which had constituted their political history from time immemorial."[13] Forsskal left a commentary on the Copts, which was found among his papers by Niebuhr and printed in his *Description of Arabia*. The little excerpt is interesting not so much for the information it contains as for the confirmation of attitudes—religious and otherwise—of the Copts in the eighteenth century. We see there manifestations of a deep suspicion of Rome and a fear of Latin proselytizing, as well as the mildness of temper so characteristic, at least until recently, of the Copts. Most interesting, in view of the fact that their own language was the ultimate successor to that of the ancient Egyptians, was their own tendency to banish knowledge of the hieroglyphs to the arcane world of Hermes, the Thoth of the Pharaonic pantheon. The numbers in the *Description de*

---

11. Ibid., 134.
12. See Meinardus, *The Historic Coptic Churches of Cairo*.
13. Butler, *The Arab Conquest of Egypt*, 89.

l'Egypte—10,000 Copts and 200,000 indigenous Egyptian Muslims—would seem to understate the traditional claim of the Copts that they constitute 10% of the population of the country. But it should be remembered that these numbers described only Cairo and that many Copts, reacting to the twin pressures of persecution and conversion after the arrival of the Muslims, had decamped in great numbers for Upper Egypt and constituted a larger percentage of the population there.

The Jews

After the Copts, the Jews were the most numerous of the minorities. Given the explicitly biblical focus of the Danish expedition we would expect Niebuhr to pay Egypt's Jews particular attention. He does but, absent Michaelis's meticulous guidance, his description is a factual recital of their current situation with only passing speculation as to their supposed Mosaic past in Egypt. The Jews had significant influence in the country:

> There are not only Pharisees or Talmudites, but also Karaites, who have their own synagogue; although they are a very small number. The Talmudites are on a very good footing in Egypt. For many years they have controlled all the customhouses, namely those of Bulak, Masr el Atik, Alexandria, and Damietta. They have been able through gifts and other similar means to gain more protection from this republican government, than in other provinces of the empire of Turkey, where the customs officials are under the control of the Pashas, or the Director General of Customs in Constantinople. One proof of the fact that the Jews have great influence with the authorities in Cairo is that the customhouse is closed on Saturday, and no merchandise is allowed to pass on that day, even if it belongs to Mohammedans or Christians.[14]

As we have seen, one of the early actions of Ali Bey would be to replace Jews in these positions of influence. In 1768, he had Yusuf Levi, the director of the Alexandria customs house, beaten to death and his property confiscated. The next year, it was the turn of Ishaq al-Yahudi, the *multazim* of the customs station at Bulaq.[15]

The Harat al-Yahud, or Jewish quarter, lay on the eastern border of the *khalig*, and so within the boundaries of the Fatimid city, and covered an

---

14. *Travels*, vol. I, 131–32. Niebuhr appears to mean by "Talmudites or Pharisees" those who held to normative Judaism and accepted the law contained in the Torah and Mishnah. They were called Rabbinites by the Karaites.

15. See Winter, 210.

area of 6 hectares near the goldsmith's suq. It was a ghetto of narrow, exceedingly unclean streets and shops that remained largely a city unto itself. The unimposing external aspect often concealed wealthier interiors, the intention being to draw as little attention to itself as possible. We can, however, imagine the residents quietly following the faith of their fathers behind the walls, and the walls had a storied past. When Moses ben Maimon, or the famous Maimonides, arrived in Cairo in 1166 after being expelled successively from Cordoba and Fez, he took up residence within the ghetto.[16] It is there that he wrote, in Arabic, his most famous works including the *Siraj* (or "Light") in 1168, a commentary on the codification of the Oral Law; the *Mishnah Torah*, a religious code, in 1180; and the *Dalalat el-Heinin*, or "Guide to the Lost," in 1190. He died in December of 1204 and his body was carried to Tiberias where he was entombed. The Jewish cemetery lay to the south of the city and the path taken by mourners was always a matter of contention. The direct route would have taken them through the *qarafa*, or Muslim cemetery to the east of the city. Periodically, when relations between the Muslims and the *dimmi*s were strained, the Jews were denied passage through that hallowed ground and were forced, instead, to make a wide detour, carrying their dead westward to the river before turning south.

Although, from the Jewish point of view the Ottoman period was an improvement over the rule of the Mamluks, the Harat al-Yahud was still subject to pillage in periods of public unrest. Under the Mamluks, the Jews had led a comparatively tranquil existence,[17] but one that was subject always to the caprice of the authorities who exacted tribute from their wealthy subjects, often through hideous cruelties. The Jews and Copts were favored victims. Periods of official favor alternated with outbreaks of popular discontent and persecution, the spark for the latter often coming, as we have seen, from abroad. But there were ample opportunities for discontent closer to home and the blame for plague and famine was often laid at the door of the minorities. As we have seen, the plague regularly visited the country, carrying off tens or even hundreds of thousands, and the minorities, particularly Jews and Christians, were held responsible. The same was true of famine. In the eighteenth century, popular disturbances occurred during the famines of 1714, 1715, 1722, 1723, 1731, and 1733 and the minorities suffered disproportionately. New strictures were enacted or old enforced

---

16. There is some dispute as to whether Maimonides lived in the city itself, or to the south in Fostat. The most commonly accepted version is that he lived in the Haret al-Yahud.

17. See Fargeon, *Les Juifs en Egypte*.

with increased severity. The period between 1736 and 1770, however, was one of relative prosperity in Egypt.[18]

The number of Jews in Egypt had increased greatly after their expulsion from Spain in 1492 and by the sixteenth century Spanish Jews constituted the majority in Egypt. As Sephardim, their first language would have been Ladino, a dialect whose basis was medieval Castilian, although it incorporated words and expressions from Greek and Turkish and was written in Hebrew characters. Oriental Jews increasingly filled prominent positions in the government, generally involving the finances of the country. This earned them the same suspicion and hostility with which the Muslim population regarded the Copts. A Jew was often sent as a *chelebi*, or envoy extraordinaire by the Sultan in Istanbul, and a Jew commonly served as *Sarraf-Bashi*, or Controller of the Mint.[19] But the fall was often proportionate to the rise, the wealth of the minorities being too tempting a target for the local authorities. We have already seen Qaytbay as the greatest builder among the Mamluks. But to the minorities, he was "a cruel tyrant, the perfect prototype of the Mamluk."[20] He financed his construction programs by means of systematic extortion, although he seems to have been an equal-opportunity extortionist, targeting not only Jews and Christians but wealthy Muslim notables as well.

The coming of the Ottomans in 1517 marked a turning point in the fortunes of the Jews of Egypt. In general, the Jews were relatively well treated in the Ottoman dominions, often through the influence of a prominent Jewish physician at court, and Suleiman the Magnificent (1521–66) even invested a Jew with the title of kikhya, in this case a kind of defender of the nation. The Kikhya was not slow to bring to the attention of the sovereign a report of mistreatment of Jews anywhere in the empire. Another Jew, Joseph Nassy—incidentally the Duke of Naxos, Andros, Paros, Antiparos, Milos, and the twelve islands of the Cyclades—became the special protector of the Jews of Egypt.[21] But this improvement in their situation came at a price: the Jews were considered by the Egyptian populace to be pro-Ottoman, and riots against them were organized in 1516 for their alleged role in the betrayal of the Mamluk forces at the battle of Merj Dabik that brought the Ottomans to power in Egypt.

Jews were now seen as a potential fifth column, having assisted in the subjugation of the country by outsiders. As the Ottoman period progressed,

---

18. See Winter.
19. See Fargeon, *Les Juifs en Egypte*, 129.
20. Ibid., 135.
21. Ibid., 140.

control of the fractious Egyptians lessened and the Ottomans interfered less in the affairs of the country as long as the tribute was paid and the levies were raised. This left ample scope for tyrants to resume their exactions. There was some measure of protection against the beys, however, since the Jews farmed taxes for the janissaries and this led to a kind of informal alliance. But again, just as under the Mamluks, periods of power and influence for individual Jews were followed by harsh repression of the community.[22] The rise to prominence of Jews in positions such as controller of the mint left them subject to equally precipitous falls when, for example, the coinage was debased or unpopular measures were taken to introduce new money. While certain of their co-religionists rose to positions of great influence, the vast majority of the Jews of Egypt labored quietly as craftsmen, money-changers, goldsmiths, tax-collectors, and bankers. The latter occupations required craft and guile, traits for which the minorities—not only Jews—became famous, or more often, infamous. They often used their own languages as a kind of code in the arcane world of finance.

As we have seen, most of the Jews in Egypt were "Pharisees or Talmudites." The Karaites,[23] a sect rejecting the Oral Law, were a small minority and were subject to periodic pressure to convert by other Jews themselves.[24] As with Christians of various sects, the quarrels between Karaites and Rabbinites were often brought before Muslim courts and many a bey found it profitable to mediate between the contending factions. There were two Karaite synagogues in Cairo, and one still exits in the Abassiyya area today.

However, Niebuhr's mention of some Jewish settlements had more to do with ancient rather than recent history:

> About two German miles to the northeast of Heliopolis is a great heap of ruins of some ancient city that the Arabs call *Tel el Ihud*, i.e. mound of the Jews, or *Turbet el Ihud*, i.e. graves of the Jews. It is beyond question that the Land of *Gosen* was in this part of Egypt. Perhaps here lay the famous Temple of the Jews

22. While the demolition of churches and synagogues in Egypt had been frequent and well documented under the Mamluks, the same was not the case in Ottoman Egypt. See Winter, 223.

23. The Karaites originated in the eighth century in Persia and Karaite communities sprang up throughout the Byzantine Empire and as far as Muslim Spain. The Palestinian and Egyptian centers soon surpassed those of Persia, although events such as the Crusades and internal Jewish persecution itself conspired to reduce their number. The tiny Karaite community in Egypt emigrated en masse to Israel in 1948. See *The New Standard Jewish Encyclopedia*.

24. Many Egyptian Karaites were converted to Rabbinism by Abraham, the son of Maimonides, in 1313, and by Joseph Del Medigo in 1616. See Fargeon, *Les Juifs en Egypte*, 235.

> built by *Onias*, and not in Heliopolis, as is commonly thought. So it should still be possible to find Jewish monuments here. I saw it only on my departure from Cairo and, to be sure, at a distance of two leagues. I was told that there were two villages nearby, named *Shebin* and *Miniet Demata*.[25]

Niebuhr was right in his conjecture about Onias, although the ruins were of a settlement in the Ptolemaic period and had nothing to do with the "Land of Gosen," the part of Egypt allegedly allotted by Pharaoh to Joseph and his family. It is commonly accepted today that the site covers the ruins of Leontopolis, built by the High Priest Onias in the second century BC with the permission of Ptolemy VI Philometor.[26]

There were other reputed sites where Jews had once lived:

> Mr. Forsskal heard from the Arabs of *Kaidbey*, a village near Cairo, of various other names of places in this part of Egypt that they said were also known to have been inhabited by Jews. As some of these places may merit visits by Europeans, I will therefore insert here the accounts Mr. Forsskal received:
>
> 1. *Liblab, Aejn Sajidna Musa* is 2 1/2 leagues from Kaidbey. There was supposed to have been a sweet water spring here.
> 2. *Maerqab Sajidna Musa*, on the summit of a mountain 1 1/2 leagues from Liblab, on the side of Old Cairo or Masr el atik.
> 3. *Tartur l'Yehudiae*, i.e. *tiara judicae feminae*, four, or according to another account, six leagues from Kaidbey. Found there are the ruins of an old fort.
> 4. *Faesqita bataqiae*, two leagues to the northeast of Kaidbey. There are springs of sweet water here. The earth and surrounding hills are reddish. There is nothing else notable.
> 5. *Tanur Pharaun* or *Gebel Pharaun*, is the name of a mountain one league to the east of Kaidbey.
> 6. *Qabur l'Ihud bemderuthe*, six leagues from Kaidbey. There was once a large city here inhabited by Jews (these are perhaps the same ruins as those above called *Turbet el Yahud*).

---

25. *Travels*, vol. I, 100.

26. It was excavated by Naville in 1887 and more thoroughly by Flinders Petrie in November of 1905. Petrie confirmed that the temple of Onias was patterned on the Temple in Jerusalem and, based on the design of the perimeter glacis and names on scarabs, that the main mound should be attributed to the Hyksos.

7. *Qalat rai*, seven or eight leagues from Kaidbey. The ruins of an ancient fort are found here, which is believed to have existed at the time of Moses. Mr. Forsskal also notes that he even heard from these same Arabs that the children of Israel passed over the Red Sea near *Aejn Sajidna Musa* to the south of Suez.[27]

Unfortunately, the credibility of the Arabs of Qaytbey with regard to these supposed ancient dwelling places of the Jews appears to be no greater than local opinions on the place where the Israelites crossed the Red Sea. We will see below Niebuhr's own investigations on the most likely place for the crossing, where the Arabs at every place on the littoral of the Red Sea would say that *there* was the place where the Israelites crossed. A few of these alleged dwelling places may have had some Jewish connection, although they were certainly of recent provenance and had nothing to do with the supposed period prior to the Exodus. The "Turbet el Yahud " is assuredly the Turab al-Yahud (*turab* being the plural form) that still exists today in the area of Bassatin to the south of Cairo. There, approximately three miles south of the Citadel and another nearly two miles from the Harat al-Yahud, in an extensive walled enclosure of some 15 acres,[28] is the Jewish cemetery to which the inhabitants of the ghetto carried the dead. Most of the tombs appear to date from before the eighteenth century.

It is curious that Niebuhr makes no reference to the Ben Ezra Synagogue, which, according to legend, was constructed by Judeans who came with Jeremiah in the sixteenth century BC, and so was also called *Kenisset El Shamiyin*, or the Palestinian Synagogue. Jeremiah supposedly used stone from the island of Rhoda, near the nilometer, at the place where Moses was found by the daughter of Pharaoh.[29] More to the purposes of the Danish

---

27. *Travels,* vol. I, 100. Possible meanings for some of the words in this catalogue are as follows, all taken from Wehr, *A Dictionary of Modern Written Arabic:* "Liblab"- a hyacinth bean (1005); "Marqab"—lookout, watchtower (409); "Tartur"—a conical cap of darwishes (652); "Tanur"—a kind of baking oven (118); "Gabur"—Giant? (133);"bemderuthe"—unknown; "Turbet"—graveyard, cemetery (112); "Qalat"—fortress (920).

28. It displays all the signs of neglect of other cemeteries in Cairo, Muslim not excepted. A highway bridge that connects the autostrade with the bridge over the Nile to the north of Ma'adi intersects the cemetery, with a few free standing tombs on one side and the expanse of low-lying tombs on the other. Most of the inscriptions are in Hebrew, but a few are in Arabic, English, or French. The latest date I saw was 1950. The area, still called Turab al-Yahud, is now an exceedingly poor Muslim residential neighborhood.

29. The little booklet published by the *Communauté Israélite du Caire* says that the present structure was rebuilt in 1115 AD on the site of the earlier synagogue by the Rabbi Abraham Ben Ezra of Jerusalem. It details the earlier history of the synagogue, specifying the date (c. 1392 BC) and reign (Merenptah, 19th Dynasty) during which

expedition, was the existence in the synagogue of the *Geniza*, or sacred repository, which allegedly contained thousands of documents which, Fargeon laconically remarks, "have been carried off by English and American scholars who were able to corrupt the guardians of the Temple."[30] As we will see below, von Haven bought a beautiful Sephardic Bible in Cairo, later to be called the Kennicott Bible. We don't know where he bought it, but Ben Ezra would not seem to be a candidate.

The search for evidence in Egypt of sites mentioned in the Bible would continue into the twentieth century, with equally disappointing results. Flinders Petrie would latter dig at Defna in the eastern Delta near the border of Sinai, in search for the reputed *Qasr Bint el Yahudi*, or "the Castle of the Jew's Daughter." He was hoping to find evidence of the Tahpanhes of the Book of Jeremiah, where it was mentioned that the Jews had fled after the capture of Jerusalem by Nebuchadnezzar. To his disappointment, he found nothing that indicated the presence of Jeremiah or the Jews.[31] The above references to the prophet Moses suggest that these sites were the stuff of fable, popular tales probably having to do with places whose provenance was unknown or had long-since been forgotten and had little historical value. The Muslim population shared with Christians and Jews many of the stories recorded in the Old Testament, and the Koran preserved in abbreviated form the story of the bondage of the children of Israel in Egypt and their deliverance by Moses:

> 49. And remember, We delivered you
> From the people of Pharaoh; they set you
> Hard tasks and punishments, slaughtered
> Your sons and let your women-folk live;
> Therein was a tremendous trial
> from your Lord.
> 50. And remember, We divided
> The sea for you and saved you
> And drowned Pharaoh's people
> Within your very sight.[32]

Moses allegedly lived and prayed at the site in the city of Giza, which it identifies with the Land of Goshen.

30. Fargeon, 225. He is probably referring to the visit of the American Professor Schichter in 1894, who took back with him a part of an old Torah which is now distributed among Columbia University, the Bodleian Library, the British Museum, and museums in Vienna and Turin.

31. See Drower, *Flinders Petrie*, 97.

32. *The Holy Qur'an*, translation and commentary A. Yusuf Ali.

Inhabitants

We have already seen that, to many Egyptians, the Qalaat al-Kabsch, or "the fortress of the ram" above the mosque of Ibn Tulun, was the place where Abraham had prepared to sacrifice Isaac. It was Enlightenment Europeans, however, with their passion for the reconciliation of Sacred Scripture with the new scientific method that bothered themselves very much about the details.

Cairo in 1761–62, then, was a teeming metropolis occupied by a small ruling class and a large mass of sullen subjects nursing—some quietly, some more nosily—their grievances at the action, or sometimes the inaction, of the ruling beys. The latter were interested largely in the exercise of power and the acquisition of wealth, although many were builders of public works that served as memorials of their rule. The distant suzerain in Istanbul allowed them a great deal of latitude as long as the coffers were regularly filled and the raw material for the army periodically raised. The minorities—whether native Christian, Jewish, or European—kept a low profile, as they had for centuries. The mass of native Muslim Egyptians were, in truth, little better off than the minorities. But scapegoats were ready at hand when, as we have seen, affairs were not to their liking. Ultimately, it was the lack of cohesion of the great mass of subjects—whether Muslim or non-Muslim, Bedouin or settled—that allowed a small ruling class of Ottomans and Mamluks to maintain their sway.

9

Commerce

> Although Egypt is not as populous as it once was, the productions of the country are still great; and since it is so advantageously located for commerce, one finds in the capital of Cairo a multitude of wealthy merchants who maintain great trade with Europe, Asia, and Africa. Nearly all the products of India, Persia, and Arabia come to Cairo through the Arab Gulf. The Nile facilitates trade with Nubia, and the Mediterranean Sea with Syria, Turkey, Barbary, and Europe, while from the landlocked regions great caravans come yearly to exchange the precious goods from these lands for those things they lack. (*Travels*, vol. I, 141)

BY THE MIDDLE OF the eighteenth century, Egypt as a commercial entrepôt had weathered another cataclysm, occurring about the same time as the coming of the Ottomans, and nearly as disruptive: the opening by the Portuguese of the Cape route to the East. This interrupted existing trade routes, by means of which the commodities of the East had traditionally passed through Egypt and the Levant on their way to Europe. This was especially true of a commodity that came from Happy Arabia, coffee. For the century since its introduction to Europe, coffee had come from the Yemen through the Red Sea and then Egypt. But by the early part of the eighteenth century, the Europeans found direct trade with the Yemen to be a less costly source of the commodity, and increasing amounts were shipped directly to Europe from Mocha. Similarly, Cairo no longer monopolized the trade in the spices and products of the Indies. With the discovery of the monsoons, the seasonal winds that made regular trade with India possible, the products of the subcontinent had traditionally flowed to emporia on the Arabian Peninsula, whence they had made their way to Suez and Cairo before their onward passage to Europe. Control of the trade routes was enormously profitable for the Egyptians, not unlike their monopoly of fiber optic routes to the East in the twenty-first century, and they did what they could to maintain the traditional flow. But the forces of change were inexorable and by the middle of the eighteenth century the Red Sea system had largely been supplanted by the route around the Cape.

If the flow from India had been diverted, however, there was still much regional trade and Cairo remained a thriving center of commerce in the narrower confines of the Near East. As we have seen, European merchant communities in Rashid, Damietta, and Cairo itself still found profit in traffic to the Levant and to the European ports on the Mediterranean. If Alexandria, once the "queen of the Mediterranean" was a desert in 1761, Cairo, the "mother of the world" emphatically was not. From ancient times when it had served as the granary of the Roman Empire, Egypt's strength had been in agriculture. But by the middle of the eighteenth century, the forces of the world economy were transforming the country. With the rise of the Ottoman Empire it had become part of an Ottoman world market and by the middle of century the demands of what might be called a Mediterranean world market were becoming increasingly clamorous. Southern Europe needed not only the raw materials that Egypt traditionally produced, but also a market for its finished products. The trade deficit with Europe, accelerated as century progressed and by the time of Mohammed Ali, fifty years after the Danish expedition arrived in the country, Egypt would have become firmly wedded to the European trading system. But even before Mohammed Ali, the beys bought European arms to equip the forces in their chronic civil wars. Exports, largely of agricultural goods, were required to pay for the arms.

The Commerce of Egypt

We have already seen how the coming of the Syrians would profoundly change the composition of the commercial elites in Egypt. But in spite of this change, the three traditional components of Egyptian commerce remained the same, even though they may have changed in the proportions of the total they represented. They were the transit trade, the production of raw materials, and the manufacture of finished products. The main commodity in the transit trade, where two-thirds of the volume from the Red Sea was re-exported to Europe, was still coffee. It had traditionally been the main source of wealth for the merchant, class. In the early part of the century, 200 million paras worth of coffee yearly was exported to the Ottoman Empire and Europe, the equivalent of 50,000 pounds sterling, a major portion of which was clear profit.[1] But direct trade by the French threatened this monopoly. Beginning in 1708, French ships had called at Mocha in an attempt to buy directly the coffee they had

1. Marsot, *Egypt in the Reign of Mohammed Ali*, 2.

previously bought from the Turks in the Levant, the Turks themselves having bought it from the merchants in Cairo. The French were successful and the voyage led to a treaty of trade and commerce.[2] It is also interesting for the glimpse it offers of the increasingly international scope of trade: Portuguese Jews, so Laroque claims, had so debased the value of Peruvian silver that the merchants in Mocha would accept payment only in "Mexican piasters." Much of the explosion in commercial activity after the discovery of America was financed by American silver, which, according to Fernand Braudel, was simply recycled by Europe from the New World to the Orient.[3]

Not only was the grip of the Muslim *tujjar* on the transit trade being loosened, but coffee from the Antilles now threatened not only the trade monopoly but the production of the commodity itself. The plantations of the Americas, worked by slaves from West Africa, now produced coffee that competed in the world market with that from the Yemen. The Dutch, who in the time of Laroque loaded 700 tons of coffee yearly in a ship destined for Batavia on the island of Java, had also begun to send seedlings to their possessions in the east and threaten the Yemeni monopoly from the other direction.[4] They had first brought coffee trees to Java in 1696, although there was not a significant harvest until 1708 and Java coffee did not have a significant impact on the market until 1721–22. But East Indies coffee always suffered from a general preference for the Mocha product and, for the moment, the Dutch contented themselves with attempts to manipulate the price at Mocha. Egyptian trade in coffee with the French had been a relatively small part of the total trade in that commodity, and the trade with the Ottomans, by far the largest buyers, continued uninterrupted. However, by the middle of the century even the Turks were buying Antillean coffee, and the volume of the trade with Egypt had shrunk dramatically.

As we might expect, Niebuhr was interested in trade, as he was in so much else about Egypt. His figures for exports in 1762 show that 22,000–25,000 *fardes*[5] of coffee were imported yearly from Jidda and Suez. At the latter port, heavy imposts were levied. The trade, however, found a way around the impost and after Ibrahim Kikhya imposed a heavy tax on

---

2. See Laroque, *A Voyage to Arabia the Happy*.

3. See Braudel, *The Mediterranean*, vol. I. Braudel's comprehensive look at the relations, commercial and otherwise, between the worlds of Christianity and Islam, as seen through the prism of their relations in the Mediterranean, is an invaluable guide to the subject.

4. Furber, *Rival Empires of Trade*, vol. II, 253.

5. A *farde*, according to Redhouse, is "a single bale or bag of merchandise," from the Arabic word for "unit." No indication is given of its weight and, as we will see, standard measures were hardly standard and weights differed widely.

Yemeni coffee at Suez during the 1760s, *saidis* (or Upper Egyptians) found it more profitable to import the commodity directly through Quseir on the Red Sea. There, according to Niebuhr, the best Yemeni coffee was available as cheaply as that of Martinique. It was still forbidden to import American coffee into Egypt or export Yemeni coffee to Europe, although by means of gifts to the customs officials 4,000–5,000 *fardes* made its way annually to Venice, Marseilles, and Livorno. Most of the remainder went to Turkey, Yemeni coffee still being considered superior to that from the Americas. A complicating factor in the persistence of the Egyptian transit trade in coffee to Turkey was the Seven Years War (1756–63). As we have seen, English cruisers in the Mediterranean stopped French vessels as well as ships of other nations if they were suspected of carrying contraband. The members of the Danish expedition had already experienced this activity firsthand.

A second major item of the transit trade, according to Niebuhr, was gum arabic although it couldn't compete in volume with coffee. The gum, coming from several African species of acacia, was probably used medicinally and in the manufacture of candy. In October of each year 6,000–7,000 *quintals*[6] was brought from Tor and Mount Sinai. It was exchanged for clothes, arms, and other necessities in markets outside the city, the Bedouins being unable to endure the crowds and odors of the metropolis. In their infrequent forays into the city, they would often be seen with bits of cotton stuffed in their nostrils. The gum was full of impurities, which, Niebuhr says, were accepted by both parties as part of the bargain, the price presumably being discounted. The Bedouins never accepted credit and refused to take back defective merchandise. After cleansing, most of the gum arabic went to Marseilles and Livorno. 4,000–5,000 *quintals* worth of gum also came from Africa, along with elephant tusks, tamarin, slaves (often eunuchs), parrots, ostrich feathers, and gold dust. These products were exchanged for cloth, false pearls, yellow amber, and coarse fabric made for the African taste. As with the gum arabic from Sinai, most of the African product was sent on to Europe with a small amount remaining in Egypt.

Other imports from the east and south included a small amount of spices from the Indies, and senna from Yemen and Abyssinia. But the best senna still came from Upper Egypt and brought 60 purses (the equivalent of 20,000 German marks, according to Niebuhr) to the government each year. Two to three thousand fardes yearly of myrrh and incense came from Yemen and Arabia, most going on to Turkey, with a very little finding its way to Marseilles and Livorno. The mention of Livorno as a destination is an interesting study in the development of Mediterranean trade in the years after the expulsion

---

6. A *quintal* was probably the *qintar*, or 100 *ratls*, in Egypt the equivalent of about 45 kg.

of the Jews from Spain. Known as Leghorn today, 100 miles south of Genoa on the Ligurian Sea, Livorno was a small fishing village when the Florentines took it in 1421, but it grew into a major trading entrepôt. The grand dukes of Tuscany declared it a free port in 1593 and Jews were invited to settle there. *Marraños*—"new Christians" of Jewish descent—were promised immunity and the ghetto system was never introduced. By the middle of the next century the Jewish community of Livorno was among the most important in Western Europe and it had become a lively rabbinic, intellectual, literary, and commercial center. It will appear over and over again in this review of Egyptian trade.

As for the export of raw materials, Egypt annually sent 70,000–80,000 uncured skins to Marseilles, Italy, and Syria, with buffalo hides in particular going to Syria. The best hides came from the period of January to April when the animals were in grass in Lower Egypt. At one time the trade had been farmed by Ibrahim Kikhya, but in Niebuhr's day it was free again. Still powerful at the time Niebuhr was in Cairo, Ibrahim had been the owner of no fewer than a third of the eighteen beys who formed the divan of Cairo, and his commercial reach was still substantial. 15,000–18,000 *quintals* of saffron were harvested at the end of May and beginning of June and brought to market in Cairo at the end of June and beginning of July. There were four grades, the best coming from around Cairo and the poorest from Upper Egypt. Most of it was exported to Marseilles, Livorno, Venice, Izmir, Syria, and Jidda. Flax was harvested in July and sent largely to Turkey and Livorno. Raw linen went to Barbary, Marseilles, Livorno, Turkey, Syria, Jidda, and the Yemen. Cotton was grown primarily in Lower Egypt and the harvest fell in July, although the best time to buy was in December and January when prices had fallen. It was sent to Marseilles and Livorno.

Rice was a luxury crop, grown only for export, and control of the rice trade was the subject of rivalry among the beys. The rice harvest fell in October, but new rice was sold in December when the price was highest. In 1762 Europeans could load rice only at Damietta, with other ports being closed. The export of rice to France was violently resisted by the indigenous merchants, although the Syrians introduced by Ali Bey attempted to divert it to Europe to their own benefit. Abd al-Rahman Katkhuda al-Qazdugli[7] had three rice-growing villages in the Region of Damietta. Sugar cane, abundant in Upper Egypt, was harvested in June, but was not transported to

---

7. We saw him above as the virtual governor of Egypt during Niebuhr's stay. The meaning of the name "Qazdughli" is difficult to establish with certainty, but the suffix "*ugli*" is clearly "the family of" in Turkish. "*Qaz*" is the name of an ancient chieftain in the Caucasus (Redhouse, 1414) and "the family of Qaz" is a possible meaning, in the same way that another well-known Egyptian name, *Lazogli*, is "the family of Laz," either a Georgian tribe or Ladislas, the King of Serbia celebrated for a clash with the Ottomans in 1365 (Redhouse, 1618).

Cairo until November and December when the price was lowest. It was not well refined in Egypt and didn't find an export market, as Europeans could buy better American sugar at the same price. Yellow wax was produced by Egyptian Christians for liturgical purposes, a little of which was bought in the villages and exported to Livorno.

Industry

Sal ammoniac was produced in the winter, two thirds of which went to Marseilles and Livorno, and the remainder to Turkey. It was one of the industries Niebuhr examined in some depth and he and Forsskal visited the facility in Giza. Extracted from "the soot of burnt manure,"[8] sal ammoniac—or ammonium chloride—was used in a wide range of industrial processes in Europe, including tanning, refining of precious metals, textile printing, and dying in addition to its use as a fertilizer.[9] Although its production from natural sources, such as that described by Niebuhr, was of limited commercial importance, it was an example of a small self-contained industry that utilized a kind of recycling for which Egypt is still known today. Niebuhr left a careful description of the process.

Manure from horses, asses, camels, sheep, and cattle was gathered in streets by small girls. Mixed with straw that was the residue after the threshing of wheat, it was then dried in the sun. Formed into cakes, it was the staple household fuel of the *fellaheen*. The soot that rose from the cooking and heating fires adhered to the vaulted roofs in the rude kitchens. It was this soot that was then gathered and sold to ammonia factories. The process of extraction of the sal ammoniac from the soot required thick glass flasks in which it was placed, as well as an oven in which the flasks were subjected to sustained heat. A constant high temperature, maintained over a period of three days, caused the ammonia to rise into the necks of the flasks. After cooling the flasks were broken, the salt was removed, and the residue discarded.

What made the process particularly interesting was that the owner of the factory also made his own flasks. A coating of Nile mud and the residue from flax processing allowed the glass to withstand the high temperatures in the oven. Presumably, the glass was then reused to make the next generation of flasks. Seven to twelve *ratl*, or just under the same in English pounds, was extracted from each flask. A *qantar*, or 100 *ratls*, sold for 600 *paras*, or 200 Danish marks. The process in the factory required extreme care, including

---

8. *Travels*, vol. I, 154.

9. It is still used today in Asia as a fertilizer for paddy and upland rice, wheat, and other crops.

a firm packing of mud around the flasks and the maintenance of the constant temperature. The amount of the salt that resulted was a function of the care with which the workers had prepared the flasks and maintained the fire in the oven. If negligence was discovered, they were punished. All the characteristics of efficient use of scarce raw materials and recycling at every step seem to have been present in the manufacturing process. There is even the suggestion of the existence of an urban proletariat, toiling in what must have been noxious conditions, poorly paid, and subject to severe penalties for substandard performance.

A second industry was that of hatching eggs, for which Egypt had been famous since Pharaonic times. There were egg hatcheries near the "Bab es sharie" and the "Birket er roteli," the ovens built into mounds so that the constant temperatures necessary to the success of the operation could be maintained. The process required a great deal of experience as the temperature was judged by feel, not by thermometer. Owners brought their eggs to be hatched, each having placed his name or a brand on the shell. The eggs were spread on straw, which was laid on matting on the floor of the oven. They were very gently shifted twice during the day and four times at night. After the eighth day the eggs were examined under a lamp, those that would produce chicks separated from those that would not and the latter were thrown out. On the twenty-first or twenty-second day the chicks broke through the shells and the temperature was immediately reduced. Eggs were hatched only in the six coldest months of year, the incidence of bad eggs being too high in the hotter months. The chicks were sold immediately for 20 paras, or about one silver mark of Lubeck, for 30 chicks. The hatcheries, Niebuhr was told, were only in Cairo. They were all owned by the Pasha.

The manufacture of cloth had traditionally been specialty of Egypt, but earlier in the century it had suffered a "baisse spectaculaire." The water-powered mill, the device that was the engine of the industrial revolution in Europe, does not appear to have been used in Egypt in Niebuhr's time:

> I did not see water mills or wind mills in Egypt . . . The common people of Egypt grind their wheat by means of hand mills of the utmost simplicity as shown in figure A of Plate XVII.[10]

Meanwhile, European manufacturers were the beneficiaries of a revolution in technology that, with its savings in labor, simultaneously ruined the export market for Egyptian textiles and began to drive locally made textiles out the domestic market. First, there was the invention in 1733 of the flying shuttle. It was followed shortly thereafter by the spinning jenny, the water

---

10. *Travels*, vol. I, 150.

frame, and finally the spinning mule. As with most revolutions there were conspicuous winners and losers. The impact in Egypt was mixed, with the French now importing vast amounts of raw Egyptian cotton to feed their domestic mills. This could not, however, offset the loss of jobs in the manufacturing sector, and the male population of Cairo was increasingly driven to day labor to make ends meet. Marsot estimates that in the second half of the eighteenth century some 15,000 men—or 10–15 percent of the male population of the city of Cairo—were dependent on part-time work.[11]

Egypt and the World Economy

As the century progressed Egypt would increasingly be integrated into the world economy, but in a quasi-colonial relationship with Europe. The jargon of the dismal science now began to apply to the Orient, with "agro-capitalist production" of export crops, including rice and cotton, and the "commoditization" of the land on which they were grown. The populace was increasingly marginalized, and increasingly resentful of an alien exploiting class made up of Mamluks and Syrian merchants. The culmination of these developments was probably the coming of the French in 1798 which, in addition to its political aim of threatening the British position in India, had a commercial end as well: the securing of a source of grain to feed the French market. But, as we have seen, these forces were already at play during the period when the Danish expedition was in Egypt. Protectionist pressures built up in the country, since it was largely in cloth that the impact of European efforts to open the trade of Egypt was felt. By the 1780s, 60 percent of the total of Egyptian imports were of French cloth and Egypt was running a trade deficit with Europe. French textiles had largely replaced local production as local producers were caught in the trap of rising raw material costs while the prices of the finished products remained relatively constant.

However, during the period the Danish expedition was in Egypt the Seven Years' War suspended temporarily the inexorable process of change. Only 700–800 bales per year of the cloth of Languedoc was imported in 1762, Niebuhr tells us, primarily before Ramadan to make new clothes for the *Bairam*, or feast celebrating end of the fasting month. Sixty to eighty bales went on to Mekka in the pilgrimage caravan, both as presents for the Mekkawis and a part of traditional tribute to the Bedouins along the pilgrimage route. There was little need for European silk in Egypt, the gold and silver thread fabrics made on the island of Scio being more to the Oriental taste than the European product. Other imports included cochineal, a red

11. Marsot, *Egypt in the Reign of Mohammed Ali*, 15.

dye of which sixty to eighty barrels were brought in each year; and pepper with 400 bales, each of 300 ratls, from Europe; and cloves and other spices, also from Europe. Clearly, the last had been brought around the Cape before being re-exported to the Near East. Of manufactured goods, the Egyptians also imported tin, iron filings, cinnabar, ginger, needles, colored beads, glass, mercury, lead, knives, and paper. Of the last, a thousand bales were imported from Venice and Livorno, some of which was sent on to Jidda. It was used in writing, wrapping sweets, and window coverings. Some of it eventually made its way to the Yemen and India.

In addition to European war and mercantile pressures, there were internal reasons as well for the change in Egyptian trade. According to Raymond and others, the economic system underwent a profound change in the decade after 1760, mirroring the political changes outlined above. Originally based on a janissary-artisan symbiosis, the janissaries were increasingly Egyptianized and marginalized by the Mamluk beys. They increasingly lost not only their own influence but also their ability to protect the interests of their erstwhile allies, the artisans. The Mamluks had traditionally controlled the rural iltizams, or tax farms, but they now seized the urban iltizams as well to finance their civil wars. A new, three-way symbiosis now grew up to replace the old janissary artisan relationship: the *tujjar* made money, a portion of which was seized by the beys in return for maintaining order, while the ulama, increasingly themselves *multazimin* or tax farmers, labored to keep the population quiescent and docile.

The arrangement looks neat on paper, but it was fraught with contention as the old and the new, winners and losers, scrambled for a share of a pie that was declining in size. Many of the new *multazimin* were absentee landlords, leading to troubles with the landed peasants; the sharing of former Mamluk prerogatives led to clashes at a time of inflation and declining trade; the increased contention between the parties led to greater demands for more—and more sophisticated—arms; and periodic civil outbreaks occurred as the population was plundered ever more rapaciously. This led to even more severe repression. In time–honored Egyptian tradition—a constant if ever there was one in the long history of the country—an army of tax collectors extorted more and yet more from the long-suffering *fellaheen*. They may have worse than slaves but, unbowed, they managed to somehow find ways around the exactions. In the end, it has been the cunning of the Egyptian farmer that allowed the country to survive centuries of the most shameless exploitation by Persians, Greeks, Romans, Byzantines, Arabs, Turks, Circassians, Armenians, Jews, French, Italians, British, and other Egyptians themselves.

Business Practices

As described by Niebuhr, business practices in mid-eighteenth century Egypt were those that had characterized Mediterranean and Near Eastern trade from time immemorial. The activity was highly personal and depended on networks of contacts, family and otherwise. In an age before the collection of economic statistics was made a national responsibility, practical men engaged in businesses were the source of such scanty information as existed. As we might expect, Niebuhr's information about exports and imports and the productions of the country came from a French trader. Brokers brought buyers and sellers together, where exchanges were then made. Merchants gathered in *oqal*s and *khan*s and stored their goods in *caravnserai*s or *wikala*s. There were once hundreds of these establishments in Cairo, with the courtyard for the stabling of animals, the rooms on the ground floor for the storage of goods, and the upper floors for the merchants themselves.[12] In a commercial arrangement that is still puzzling to Westerners today, similar products and services were concentrated in close proximity to one another: there would be the printers *suq*, or the carpenters *suq*, or the sweet-sellers' *suq*, or the clothing *suq*. The Western marketing device of differentiating oneself from the competition would seem difficult when the competition are close neighbors and have been so for generations. Personal contacts, fostered over generations, acted as the instrument of differentiation.

But even with these cozy arrangements exchanges still took place. Most export and import business was carried on by correspondence: that is, correspondents in foreign countries acted on behalf of traders in the same way that correspondent banks act for other banks today. It meant that traders could expand their radius of operations without frequent travel or, when they did travel, without carrying large amounts of specie. The use of letters of credit and bills of exchange was widespread within the trading area. It was not unusual for a man to have deposited money with an establishment in one country and redeem the paper for an equal sum, less a small fee, in the currency of the correspondent in another country. Later, in the Yemen, we will see the Danish expedition deposit a large sum with a Banyan in Mocha and redeem the sum with his co-religionists in Taizz and Sana. Commercial networks—webs of influence, often based on common ethnic background—were the preferred means of doing business. Only the Bedouins insisted on payment on the spot. Bills of exchange worked only when there was an acceptable degree of mutual trust. There is also the suggestion by Braudel that it was Europe that learned

---

12. The only one in good repair is that of *al-Ghuri*, near the al-Azhar mosque. Others, including the *wikala* of Sultan Qaytbey just inside the Bab al-Nasr and that of the Amir Qawsun further along the same street, still exist but are in a ruinous state.

the use of bills of exchange from the Muslim world, and not the other way around.[13] However, the suspicion and hostility that still existed between the worlds of Christianity and Islam often made these arrangements fraught. The evidence suggests that Christian merchants in the Levant were not extended the same trust as Muslims, and that they were forced to borrow locally, at often-extortionate rates, to finance their operations.[14] In fact, the entry of the Europeans into the Levant caused commercial turmoil. Inexperienced newcomers, they always paid cash, "upset traditional habits, put old Venetian houses at difficulties and sent prices soaring."[15]

Even today, the degree to which trust is reposed in a known customer or purveyor makes the commercial transaction a more personal, leisurely affair: a glass of tea and discreet inquiries as to the health of respective families begins the proceedings; the merchandise can be taken and tried out before a deal is consummated; money can be paid the next day, or the next week, if there is not enough on the day of the deal; a handshake seals the bargain and a receipt is optional, often being drafted to suit the needs of the buyer; and if payment is made in cash, it is not rung up in a cash register but placed in a desk drawer. This argues not a naïveté on the part of the trader but a sophistication: surely, he understands his market. These practices—which seem to fly in the face of Western notions of transparency, accountability, responsibility, and liquidity—are all classics in the land par excellence of merchant capitalism.[16] Much of it comes from the fear of the taxman—or the exactions of the bey—in an atmosphere where tax evasion is not only a common practice but regarded as a positive duty.

Technology

If Niebuhr was no sociologist, he was a practical man, and he devoted eight pages to what might be called the technology of Egypt. This technology was in many respects unchanged from the time when the Israelites purportedly sojourned in the country. From the Pharaonic times the country had been primarily an agricultural one. And there was another reason for Niebuhr's interest, related to his biblical assignment:

---

13. Braudel, *The Mediterranean*, 817. This may have been through the agency of Jewish traders as intermediaries between the worlds of Islam and Christianity.

14. Ibid., 465.

15. Ibid., 466. It is not unlike the effect of expatriates on local communities today.

16. Among the least contentious of Cromer's observations was that in the late nineteenth century, there was hardly a flat rock between Alexandria and Wadi Halfa where there wasn't a Greek to spread his wares, if he could buy cheap and sell dear.

> It is known that with Egypt lacking rain, the inhabitants must use hydraulic machines to irrigate their fields . . . If Herr Lieutenant Niebuhr finds these machines different from those we see in Norden and Shaw, it would be useful . . . to have sketches . . . I would ask that he remember the words of Moses, Deutron. XI, 10 . . .[17]

It was one of the few questions in the *Fragen* that dealt with Egypt. Methods of irrigation had not changed much over the millennia, and the first machines Niebuhr noticed were the devices for raising water from the Nile and distributing it to the fields. The first was the *sakki et tur*, or machine moved by bullocks, and in Plate XV Niebuhr leaves a detailed sketch of the device, with its axle, wheels, twenty-two pots, and trough out of which the water flows. The second was the *sakki t'dir erridsjal*, or machine moved by human foot-power. Niebuhr saw it near the Birket al-Azbakiyya and he *did* remember the words of Moses (Deuteronomy 11:10):

> For the land wither thou goest in to possess it, is not as the land of Egypt, from whence ye came out, where thou sowdest thy seed, and waterdst *it* with thy foot . . .

His sketch of irrigation machines is completed by the bullock-drawn device he saw in the vicinity of Damietta, where the water level was relatively constant. There, the simple *shaduf*, or bucket with the counterweight, was used by a single man to irrigate a small area near the Nile.

Of water mills or windmills, as mentioned above, Niebuhr saw none. There was a public mill to grind wheat, turned by means of an ox, but most people used a simple hand-mill to grind their own. The larger, animal-powered mills were also used to grind beans, saffron, plaster, limestone, and oil seeds. Of plows, the implements used by Egyptians seemed crude and a thick bullock-drawn plank was used in place of a harrow. They still used bullocks to thresh wheat, just as they had in the time of the Prophets (Deuteronomy 25:4): "Thou shalt not muzzle the ox when he treadeth out the corn." A bullock-drawn sled called a *nauredsj* was used to drive over a heap of stalks six to eight feet in diameter and two feet high. Shovels and forks were used to winnow the resulting grains, straw, and stems. There were no carts or wagons used for transport in Egypt. The ancient Egyptians do not appear to have adopted wheeled transport, perhaps since they were a sedentary people and did not need the wagons such as the Scythians used to transport their goods over the steppes. Of the nearly 1,400 hieroglyphic signs listed by Budge in *An Egyptian Hieroglyphic Dictionary* only the war chariot is shown as evidence of wheeled transport. The chariot was a relatively latecomer to

---

17. *Fragen*, 91.

Niebuhr in Egypt

Egypt, being introduced by the Hyksos in the sixteenth century BC.[18] The Egyptians, however, learned their lessons well and their charioteers became "the dread of the world."[19] Individual animal-power was, again, the main means of transporting bulk commodities. For the excavation of earth, baskets were filled by hand and loaded onto asses, and it is not uncommon to see the same combination of *fellaheen* and *himeer* excavating for high-rises in Cairo today. The wood for shipbuilding and other nautical supplies were carried to Suez in camel-caravans, with especially heavy items such as anchors being suspended between two animals.

Niebuhr's description of commerce in Egypt in 1762 would not be complete without with a list of the weights and measures in use in the country. The same lack of standards is apparent here as is the case with measures of distance (see chapter on the Delta below): the same measure often meant something different, depending on the merchandise:

> The larger weights are measured in Cantars. But in Egypt a Cantar often varies greatly with merchandise: with some it is 100 rottels, others 102, 105, 110 to 150. Certain merchandise is measured in Ockes, of which 44, even 78, or 82 to 86 are a Cantar.[20]

The "Cantar" is the Arabic *qantar*, of which the standard today is still 100 "rottels" or *ratl*. Even given uniformity, an understanding was possible only in terms of some familiar measure. The following is a brief cascade of the weights and measures in use in Egypt in 1762, expressed in metric equivalents:

Measure	Local Equivalent	Metric Equivalent
Qirat		0.195 grams
Dirham	16 *qirats*	3.120 gams
Metkal	1 ½ *dirhems*	4.680 grams
Waqiya	12 *dirhems*	37.440 grams
Ratl	12 *wiqiyas*	449.280 grams
Uqqa	400 *dirhams*	1.248 kilograms
Qantar	100 *ratls*	44.928 kilograms

To those more familiar with the English system, a nominal *wiqiya* was 1 1/2 ounces, a *ratl* was just under a pound, an *uqqa* was 2 3/4 pounds and a *qantar*, at 99 pounds, was just under a hundredweight.

---

18. See the *Atlas of Ancient Egypt*, where the Hyksos irruption is dated to the 2nd Intermediate Period, or 1640–1532 BC.

19. See Redford, *Egypt Canaan and Israel in Ancient Times*, 214.

20. *Travels*, vol. I, 147.

Commerce

Finally, Niebuhr concludes his survey by observing that prices were subject to change on a daily basis, and that information from 1762 would soon be out of date. Those interested in current prices should obtain them from their correspondents.

10

The Delta

> It would be expecting too much of a traveler, who spent only a short time in Egypt, to furnish a complete map of the country ... So I have only provided a map of my itinerary, and the basis on which it was prepared ... I have indicated in my map only those things I saw myself or heard from those who knew the country. If all travelers did the same, one could appreciate the service that each would render to the betterment of modern geography. (*Travels*, vol. I, 71)

THIS STATEMENT BY NIEBUHR of his operating method, along with his characteristic disclaimer, will become staples in his account of the Orient. Advertising no more, or less, than what he saw with his own eyes or could reasonably verify from hearsay, he is anxious that the reader clearly understand the limits of his observations. As with much else, however, he is rather too modest and his map of the Delta would be unsurpassed in detail until the coming of the French some forty years later.

Chapters 5 and 6 of vol. I of the *Travels* contain the account of Niebuhr's travels on the Nile between Cairo, Rashid, and Damietta. He brought several advantages to the effort that others, before and after him, lacked. In the first place, most Europeans traveled only to Rashid and detail of the Nile between Cairo and Damietta was missing in most works. Besides, others hadn't been able to verify their positions with the astronomical sightings that were the keys to his position finding. Since the flow of the river was in a general north-south direction, longitudes were less important and latitudes could be established with relative ease. Careful sightings allowed him to establish positions of cities and villages with unparalleled accuracy and, so, calculate the distances between them.

The Delta

In addition, he listed the geographical detail—the names of villages, bends of the river, and major cities—in Arabic, as directed by Michaelis in the *Fragen*. However, as a relative newcomer to the country, he was anxious not to appear overly curious and so excite suspicion. So, he was not able to follow his own recommendation to befriend the Copts, whose position as scribes and bookkeepers to the beys, particularly in the rural areas, gave them an unmatched knowledge of the countryside. As we have seen, the Jews generally served this function in the urban areas. But he had learned during the first journey from Rashid to Cairo to avoid officialdom, whose only interest seemed to be extortion. It was a lesson he would apply through his journey. When not suspicious, the official

> never passes up an opportunity to make a profit, or make himself indispensable to the one in his charge; and to prove his zeal, at every turn he creates dangers where none exist.[1]

On the subsequent trips he contracted with a prominent merchant to use one of his own boatmen. The merchants were practical men, interested in profit of a different sort, and they made it their business to travel safely on the river. Their boatmen knew the names of the villages they passed and they agreed to travel during the day, which allowed Niebuhr to use his instruments.

To Damietta and Back to Cairo

Even in Egypt the winter was not a propitious time to travel and Niebuhr delayed his trips down the Nile until middle of the spring, 1762. The frequent overcast and occasional rain made astronomical sights difficult and this was a key to his practice of position finding. As we have seen, the time could be profitably spent in Cairo before he set out to the north again. If nothing else, there was the Egyptian dialect to learn before he could make use of the information he gleaned from his informants. So, it was not until late April that he set out on the first of these trips on the river. Baurenfeind accompanied him, along with a janissary and a cook. They left the port of Bulaq just after midnight and Niebuhr awakened on the 1st of May to find the boat underway on the river. The river looked very different from what he remembered from the previous November. Islands appeared where none had existed before and the water was so low that the boat grounded more than once. It would begin to rise again in June, before reaching its greatest height in late August or early September. Near Tahnie he took his instruments ashore to take a sighting but discovered that the crosshairs in the glass were broken. He repaired the

1. *Travels*, vol. I, 61.

instrument, but on the 2nd of May the *khamaseen*[2] reduced visibility nearly to zero. The wind died in the early afternoon but returned with redoubled force in the early evening. In between, he took a reading and calculated that the latitude of a nearby village was 30° 26′.

On the 3rd of May, about eighty miles from Bulaq, they paid a visit to the welcoming but exceedingly unhygienic Coptic community of Zifta. Niebuhr remarks that the church was full of fleas and that "a large number found refuge on me." The Coptic service lasted for several hours and during the unfamiliar and seemingly interminable liturgy, people came and went freely, those who remained often reclining on the floor. Some even appeared to be sleeping. The paintings in the church were crudely executed to his European eye and he found this exposure to the relaxed Christianity of the Orient to be very different from the stern Protestantism of his native Saxony. They left Zifta in the evening, although their progress was hampered by a northerly wind. The next day they passed a raft of pots, brought for sale from Qena in Upper Egypt. Qena was noted for its pots of porous pottery—the *zir Qanawi*—that allowed natural cooling by means of evaporation. Placed behind a screen of *mashrabiyya* (from the Arabic "to drink") the natural action of the breeze and the pottery conspired together to cool the contents of the jar. Later in the day, they passed Mansura where, in the year 1249 Saint Louis IX was held prisoner during the abortive sixth Crusade. Famous for a visitation at a later date, in 1798–1801, Mansura remains today inextricably linked with the French who gave the inhabitants, some say, their characteristic green eyes.

The next day, the north wind blew up again and this again impeded their progress. But it allowed Niebuhr to go ashore, and he took a reading that established the latitude of Mansura at 31° 3′. The surrounding fields were planted in rice which, we as have seen, was a cash crop farmed by the beys, although the Syrian Christians were just then engaged in a contest with them for control of the trade. The Syrians would eventually win out. The cultivators borrowed from moneylenders to finance the planting and were required to pay their taxes in cash. This was different from Upper Egypt where they were still able to pay in kind. The traffic on this major branch of the Nile was heavy, the river serving, as it had since Pharaonic times, as the major north–south avenue of trade and transportation. Every kind of commodity moved on the water and just before they reached Damietta they passed a convoy of twenty boats carrying bees. The hives were laid horizontally, 200 to a boat, and Niebuhr calculated that this meant about 4,000 hives in all. The *Sanjaq*[3]

---

2. From the Arabic for "fifty," it is traditionally the number of days between Easter and Pentecost when wind-borne sand and dust from the western desert dramatically reduces visibility in Egypt.

3. The head of an Ottoman administrative district.

The Delta

of Mansura was encamped nearby with his entourage, and the purpose of the procession of boats was to levy and collect the tax on the hives.

The party arrived in Damietta on the 5th of May. The city was the entrepôt for merchandise coming from the Levant, as Rashid was for that coming from Europe. There was one important difference: where Rashid was full of European merchants, there was not a single foreigner in Damietta. Niebuhr tells us that when the inhabitants noticed that the Europeans became attached to Muslim women, they had risen up and massacred them all. The French were now forbidden to set foot in the city and their trade was conducted through agents, of whom one was a Greek renegade whose debts in his own country had condemned him to an exile abroad. He had hoped to make his fortune in Egypt, but the star of his patron—one of the beys in Cairo—was no longer in the ascendant, and the man was now reduced to work in the Damietta customhouse. Renegades were relatively common in the Mediterranean, that debatable land between the worlds of Christianity and Islam. But they were despised by adherents of their old faith and never really trusted by those of the new, so they were at home in neither world. Niebuhr and Baurenfeind were not disturbed during their visit, dressed as they were in the Turkish style with turbans and *benisches* instead of tight-fitting trousers. Niebuhr made a rough map of the city—a detailed map would have been too dangerous—and determined that the latitude of Damietta was 31° 25′, a little more northerly than that of Rashid. Both cities were about two German miles, or just over nine English miles,[4] from the sea. Baurenfeind was not idle and he sketched a *Prospect der Stadt Damiat*. The map and sketch appear as Plates VII and VIII of vol. I of the *Travels*.

There were other nearby areas of interest as well, including the ruins of Sisutanis, which Niebuhr speculates was the Tanis, the Greek name of the ancient Dja'net and the modern Sa al-Hagar, the burial place of the pharaohs of the 21st and 22nd dynasties. He was right. We have already seen Tanis as the purported site of the biblical city of Zoan. But there were dangers if they ventured too far from the beaten path. The inhabitants were accustomed to relieving travelers of their baggage, and Niebuhr was cautious:

> this journey was too dangerous to attempt, as I knew the risk of losing my instruments, especially since it was only a side trip and our arrival in Happy Arabia, our primary destination, was near at hand.[5]

It was also on this trip that Niebuhr made a short excursion to the mouth of the Nile at Boghas. He speculated on changes in the coastline made by the

---

4. See below for a fuller discussion of the measures used in the *Travels*.
5. *Travels*, vol. I, 70.

gradual deposit of river-borne silt and measured very carefully the distance between the ancient fortress and the Mediterranean ...," so that other travelers can determine whether Egypt is relentlessly expanding, as some authors maintain."[6] Without making extravagant claims for Niebuhr, we can surely say that his scientific predilections and curiosity set him apart from the average traveler. We will later see him speculating about the number of years it must have taken for fossilized rock to have formed in the Muqattam to the east of Cairo, or the reason for the presence of fossil crustaceans in areas no longer close to the Red Sea. It would be another seventy years before Lyell suggested in his *Elements of Geology* that the world was millions rather than thousands of years old. The ensuing controversy would rock the scientific world and turn the religious establishment on its head. But it is not too much to say that Niebuhr, in his own methodical way, was an intellectual precursor of the later scientific revolutionaries. The two men, Niebuhr and Baurenfeind, arrived back in Cairo on the 15th of May.

Correlated to the names in the Delta appearing in Idrisi's *Geographia Nubiuensis*,[7] this was Niebuhr's first attempt at detailed mapmaking in the Orient. It was different from his plan of Cairo, which consisted of establishing the broad outlines of a city in a single, relatively small geographical area, before filling in the details of streets, buildings, etc. Here, he was dealing with an area of several thousand square miles, with differences in topography and different geographical settings. It was a good preparation for his mapmaking duties in the Yemen, where the topographical differences would be far greater. Using a combination of astronomical observations, dead-reckoning, and judicious questioning of the boatmen, Niebuhr lists in table form the 174 towns and villages they passed on the journey between Cairo and Damietta, with the compass bearing of the bend of the river at each location. A typical entry is that for "Miniet Samanud," the forty-seventh village to the east of the Nile on the journey to Damietta, with the river tending SSW, and the note that it was "a small city, where there are 5 minarets, of which some appear to have once served as bell towers in Christian churches."[8] Listed in both the Latin and Arabic scripts, the 174 entries consist of 100 towns and villages to the east of the river and 74 to the west.

A second list includes the 135 he saw on the first journey from Rashid to Cairo although, for the reasons mentioned above, he was not as confident of their accuracy. Finally, he gives a list of the 123 towns, villages and bodies

6. Ibid., 68.

7. See above p. 83 for a discussion of Sherif Idrisi.

8. *Travels*, vol. I, 80. Given that the general trend of the river in the lists is to the south, and opposite to the flow of the Nile, it seems clear that Niebuhr either oriented himself in that direction or made his observations on the return journey to Cairo.

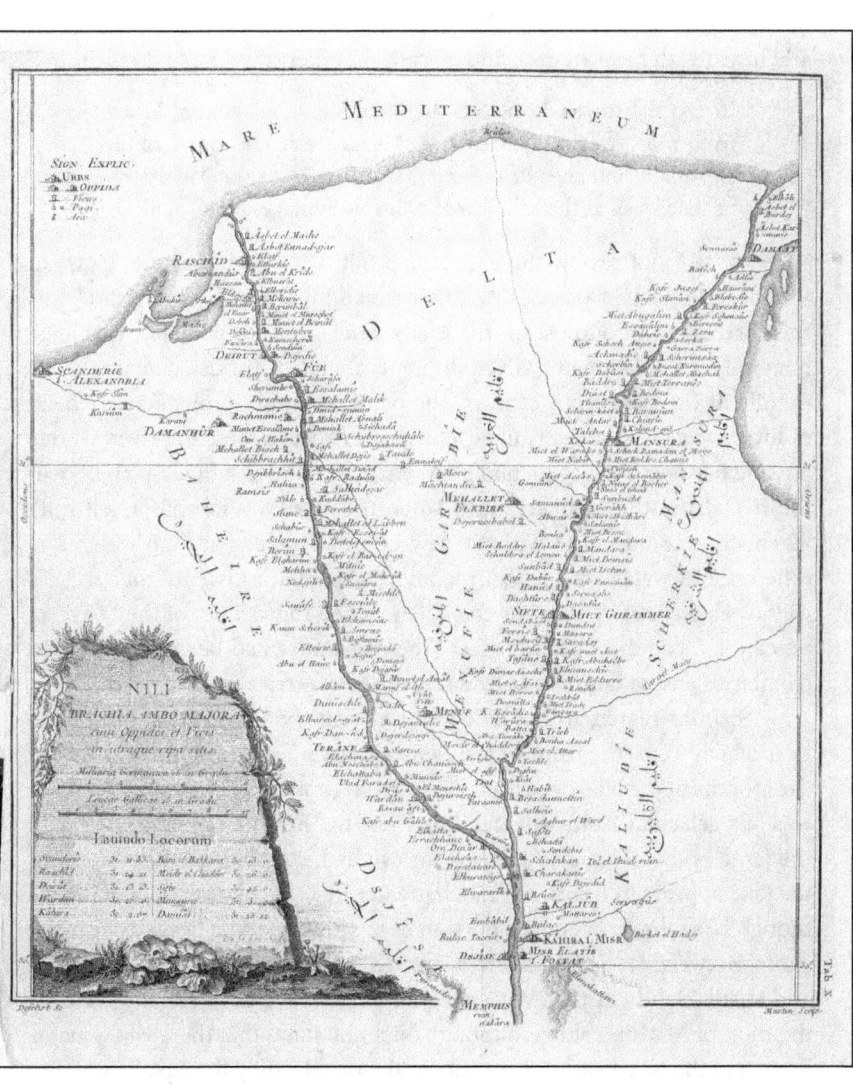

Niebuhr in Egypt

of water seen by Forsskal on his separate journey to Alexandria and return via Rashid. The results were plotted on a map that appears as Plate X of vol. I of the *Travels* as "The principal branches of the Nile from Cairo to the Mediterranean Sea." Niebuhr says that he has not attempted to correct the results of previous maps, since the course of the Nile was continuously changing in any case. Instead, he offers the product of what he saw with his own eyes and heard with his own ears, and he calls on other travelers to do the same:

> In Egypt I traveled only on the Nile; but I believe that I have fixed more accurately than any other traveler the course of the two main branches of the river, from Cairo to the Mediterranean, as well as the location of cities and villages on its banks.[9]

We will not review the entire map but, with Niebuhr, only pause at the more important places. "El koli, a small village with 4 batteries," was the northernmost outpost on the Mediterranean, approximately two German miles from Damietta. Given the general state of Turkish defenses in the middle of the eighteenth century, the batteries would probably have been of little use against a determined European effort to force the river. Goods passing between the Levant and Damietta had to be freighted up and down the river from the Mediterranean when the river was low, although fully laden ships could sail to within a few miles of the city at high water. The difference in water level at Cairo could be as great as twenty-four feet, as we saw above, but by the time the Nile reached the vicinity of the Mediterranean it was only about four feet. The remainder had been diverted into tributaries and canals in the Delta to be used for irrigation purposes.

Niebuhr regretted that he could not visit the little village of Matare, near which were "the ruins of *Sisutanis*, perhaps *Tanis*, whose name one of the mouths of the Nile once bore."[10] About five miles upriver from Damietta was "Es schaara," from which a canal ran to "Beheire" (Arabic for lake), the large body of water to the east known as Lake Manzala. The extent of the lake appears to have been much greater in 1761–62 than it is today, although Niebuhr shows only a portion of it with three small islands near the shore. Between "Es schaara" and Mansura there was little but a few villages and small islands in the ever-changing stream. We have already seen his mention of Mansura above, although Niebuhr states that there was a major branch of the Nile nearby. It appears in the map leading to Cape Brulos, the most northerly point of the Delta, and roughly midway between Rashid and Damietta. Another short branch ran from just north of Mansura to the lake.

9. Ibid., 72.
10. Ibid., 69.

The Delta

From Mansura, the boat would have beaten south against the current, although in season a northerly wind would have filled the lateen sail and assisted their progress. Villages and pigeon cotes would have appeared around the bends in the river, of which Niebuhr counted ten between Mansura and Samanud, the capital of the province of Gharbiya. At Samanud, there was another canal that, according to Niebuhr's informants, crossed the breadth of the Delta. After thirty-five miles upriver they reached Damalla, from which "Taraet Mues," another branch of the river, ran northeast to Beheire. *Taraet* is the Arabic for "canal," suggesting an artificial waterway; but a branch of the river had run to the lake from nearby Athribis since Pharaonic times. After another fifteen miles and the great bend of the river at Dighue where boats were held up to pay the toll, they reached "Kafr Karinejn," where another canal branched off to the northwest, to join the Rashid branch of the Nile forty-five miles north at "Mehallet el Labben." Five miles south of "Kafr Karinejn" was "Kafr Faraonie" where another canal branched off to the western arm of the Nile at Nadir, near "Manuf el Ale." From "Kafr Faraonie" to Bulaq and home was another thirty English miles.

To Rashid and Back

On the journey between Rashid and Cairo, Niebuhr was less confident of the accuracy of the information since it was gathered on the first trip upriver when he had neither the exposure to Egyptian Arabic nor the assistance of the boatmen of the later journeys. And he took compass bearings only on the portion between "Salhadsjar" (ancient Sais), some thirty-five miles upriver from Rashid, and Cairo. Nonetheless, the map of the delta is interesting if for nothing else than the picture it gives of the major branches of the Nile in the middle of the eighteenth century. Over time, the changing waterways in the Delta reflected a consolidation of the streams, probably man-assisted if not man-made, and the system of the two main branches and subsidiary canals was a product of thousands of years of irrigation engineering. The seven mouths of the river reported by Herodotus[11] had, by 1761, long since been reduced to the two that now bracketed the Delta: the Bolbitic emptying into the Mediterranean at Rashid, and the Phatnitic at Damietta. But others were suggested in Niebuhr's text, and we have seen that he knew of the Tanitic branch that had exited the Delta to the east of Damietta. Still further east, we know that the Pelusiac branch had reached

---

11. The Pelusiac, the Canopic, the Sebennytic, the Saitic, the Mendesian and the "Bolbotine . . . and the Bucolic . . . not natural branches, but channels made by excavation." See Herodotus, *The Histories*.

Pelusium on the Egyptian frontier. In the west, the Canopic branch had flowed northwest from "Kaum Scherik" (number 36 to the west of the Nile in Niebuhr's Rashid itinerary) before reaching the Mediterranean at Canopus, between Alexandria and Rashid. We have already seen his speculation that the twenty marble columns unearthed earlier that year near the little village of Abumundur marked the site of ancient Canopus. As for the Sebennytic branch, according to Butler it flowed north through the center of the Delta to "Paralus,"[12] surely Niebuhr's Cape Brulos. But, as the lists from Herodotus to those of the early twentieth century suggest, there was never complete agreement on the names of the branches, or which of them were natural and which artificial. There was, however, general agreement that the Nile had always been represented by multiple branches in the Delta and that the hand of man had played a large part in determining where they flowed.

Niebuhr's map of the Delta has shortcomings, the most obvious being that it largely confines itself to the two major branches of the river. With the exception of a few cities in the southern portion near the bifurcation, the area between these branches is a great empty space in the map. There are branches, or the beginning of branches, but they often disappear into the void. But if some of the detail is missing, what is provided is very accurate. And we should remember that Niebuhr was just passing through Egypt on his way to the Yemen. But his map of the Delta, an afterthought in the work of the expedition to Happy Arabia, was good enough to set the standard for detail and accuracy until the coming of the French some forty years later.

The Delta in History

The Delta has always suffered by comparison with Upper Egypt in terms of the attention given to it by Egyptologists. It is widely recognized that the Pharaonic civilization had been no less present here than in the south. But the presence of the water table only a few feet below the surface, the constantly-shifting bed of the Nile, and the greater population meant that what little building material survived inundation was reused repeatedly by succeeding generations, and little remained in its original state or in its original site. As for the less permanent remains—soft sandstone memorials, everyday objects, not to mention papyrus records—most were lost or destroyed under the joint influence of time and the pervasive humidity. But Lower Egypt was of interest, as well, for the practitioners of "biblical archeology," since it was thought that the Land of Goshen must have been somewhere in the eastern Delta. Unfortunately, and in spite of the efforts of Neville, Petrie,

---

12. Butler, *The Arab Conquest of Egypt*, 350.

The Delta

and other researchers, nothing remains—if indeed anything ever existed—to indicate the presence of the children of Israel in the area. Interestingly, Niebuhr's short review of the archeological sites in the Delta focuses entirely on the profane, and he resists the temptation to relate such evidence as existed to the explicit religious purpose of the Danish expedition.

A first site mentioned by Niebuhr was the ancient city of Canopus not far from Alexandria. The name was a misnomer, the product of a mixture of legends in this region of many myths, having been mistakenly called "Kanopus" by early Egyptologists after the pilot of Menelaus in the Trojan War.[13] In actuality, it was the ancient Egyptian city of Pwer-Gwati and the modern city of Abu Qir. But, however misnamed, it lent its name to the Egyptian funerary receptacles containing the viscera of the mummified bodies, and forever after they have been called "Canopic jars." Nearby, was the ancient city of Salhajar whence a Ptolemaic catafalque would be taken to Cairo and whose inscriptions Niebuhr would later copy. He mentions that in spite of its Arabic name it "must have been celebrated in the time of the ancient Egyptians."[14] And so it was, being in fact the city of Sais, the capital of the 5th Lower Egyptian nome during the 12th dynasty (1991–1783 BC). Later, during the 26th dynasty (664–525 BC), it was the capital of the country. Niebuhr made a special trip from Cairo to see the ruins, but was rewarded only with a view of considerable mounds, evidence that a large city had once been there, and a few columns that had been incorporated into the poor homes of the inhabitants. He did copy the hieroglyphs on a large stone set in front of an oil press, noting that some texts were carved in relief and others cut into the stone. The eastern Delta seemed a rich mine for antiquities, and Niebuhr mentions that Abusir[15] was similar to the name of the ancient Busiris; Samanud, which he notes "is probably located at the same place as the ancient Sebennytus";[16] and Trib, or Atrib, the site of the ancient city of Athribus and the capital of the 10th Lower Egyptian nome. Athribis was the Greek name of Hut-hery-ib, whose history probably goes back to the 4th dynasty (2575–2465 BC) although the earliest remains found have been those of the reign of Amasis of the 26th dynasty.

Closer to Cairo, at the village of "Matare," two leagues from the Fatimid city and near the old city of Heliopolis, there was a standing obelisk with legible hieroglyphs on all four sides. It was of Senwosret I (1971–1926 BC), one

---

13. *Atlas of Ancient Egypt*, 221.

14. *Travels*, vol. I, 97.

15. Not the Abusir near Memphis, of pyramid fame, but the Greek Taposiris Magna in the Delta at latitude 30° 57'.

16. *Travels*, vol. I., 98. As we have seen, this was the birthplace of Manetho.

of a pair (the other fell in 1158 AD) and the oldest surviving obelisk in the country. Heliopolis had once been once the site of many obelisks, including a pair of Tuthmosis III (1479–1425 BC), of which, as we have seen, one is now on the banks of the Thames in London and the other in Central Park in New York. On December 24th, 1761, however, there was still so much water around the monument that Niebuhr could only compare the hieroglyphs on the southern face with the sketch provided by Norden. He found the earlier man's copy to be exact. After setting up his astrolabe, he calculated the height above the ground to just over fifty-eight feet. After excavation, the accepted figure today is sixty-seven feet, an indication of the amount of sediment that had accumulated around the base over the centuries. The rise of the Nile in 1761 was five feet eight inches at the site, and Niebuhr suggests that the Temple of the Sun was surrounded by walls to protect it from inundation. The enclosed area is estimated today to have been 1,100 by 475 meters, although the site remains largely unworked. On Niebuhr's thoughts on the provenance of the ruins of Tel al-Yahudiya we have already seen his correct identification of the site with the Ptolemaic temple of Onias.

Niebuhr's Mapmaking

Niebuhr's review of the historical sites in the Delta is hardly exhaustive, and his speculation broke no new ground. But the journeys on the river give us a glimpse of his methods of mapmaking that would be put to good use in the years that followed. We have already seen that he was trained in the technical aspects of his duties by Prof. Johann Tobias Mayer, Professor at the University of Göttingen and the foremost German astronomer of his day. Mayer had made several improvements in instrument design, and Niebuhr carried with him the latest the age could offer in terms of technology. The first was a good Hadley's octant, or eighth of a circle, whose lesser arc and smaller size made it suitable for use at sea. For astronomical readings on land, he carried a quadrant, or quarter of a circle, two feet in diameter, and personally calibrated by Mayer. Here, the larger size of the instrument was manageable, and the quadrant remained useful for work ashore long after mariners had moved to the smaller sextant, or sixth of a circle. The problem with celestial sights on land was, of course, that there was no ready reference point. Unlike at sea, the horizon was not always immediately apparent, and geographical features often stood in the way. And correction for altitude above sea level was also necessary, although in the flat Egyptian Delta and near the river, the problem of altitude was lessened.

The Delta

Niebuhr took his responsibilities on land as seriously as he had at sea and, as we will see, his book is littered with the latitudes (and occasional longitudes) of Oriental cities. The latitudes are very accurate, in some cases more accurate than those taken by the French expedition. He also equipped himself for other tasks prior to his departure from Copenhagen. For surveying, as we have seen, he carried an astrolabe. The astrolabe was an ancient, multipurpose instrument, most commonly used for determining celestial altitudes. It came with varying degrees of accuracy depending on the scale. But by the eighteenth century, a compass was often attached and the instrument came in both mariner's and surveyor's versions. Niebuhr's astrolabe was a variation of the latter, stripped for surveying purposes to a circular plate with a centrally pivoted alidade. The circumference was divided into 360 degrees, with an inset compass and compass rose for orientation. Later known as theodolites, they could be mounted on a stand for greater stability. This Niebuhr did. His calculations of the heights of everything of interest he saw, from "Pompeii's Pillar" in Alexandria to the pyramids of Cheops and Khephren in Giza are examples of the uses to which he put his astrolabe. The device could also be used, by means of triangulation, to measure distances and Niebuhr was tireless in his calculations of, for example, the width of the Nile at Giza or of the Red Sea just south of Suez.

So much for Niebuhr's surveying equipment. But this was probably the easiest part of mapmaking. A far more difficult problem, clearly seen by Michaelis in the *Fragen*, was with the names of the places that appeared on the map. Names may have seemed relatively easy since they involved no complicated calculations or careful measurements. However, because of the difficulty of language, they were often the most troublesome. Niebuhr recognized it at once with this, his first detailed map, and his method is worth quoting in full:

> I have already mentioned in the Preface to my *Description of Arabia* that it is difficult enough to write correctly the names of strange villages and cities in their own language, much less in a foreign language, especially when they are pronounced by people with different dialects or, which is worse, who speak their native language poorly. For this reason, I have sometimes spelled completely differently the names of the same villages according to the pronunciation of different people. In order to arrive at the correct Arabic names of the villages and cities of Egypt, I noted on my journey all the names in European characters, and afterwards in Cairo I had them written by a master Arab scribe in the idiom of those who had accompanied me. By this means, I hope I have spelled most of the names correctly in Arabic characters,

and afterwards any European, of whatever country he might be, can in turn write them in the orthography of his own language, assuming that I have not always given correct European letters for the Arabic. Those who obtain lists of villages in Egypt should have the names read aloud, and then express them in European characters as their own ear tells them. The Arabs often write familiar *proper names* without diacritical points, and copyists may forget them or omit them altogether. This is why it is impossible for a foreigner to always read these names correctly, and for a translator to divine their correct spelling.[17]

He then goes on to list fifty names from the *Geographia Nubiensis* with the original Arabic and the versions as he heard and transcribed them. The differences range from the absence of the Arabic diacritics to uncertainty about short vowels, not to mention questionable transliterations, although that is also the matter of the ear of the intended listener. But the point was made: even with the greatest care, differences arose as unfamiliar sounds were heard, written in a system of symbols that often had no equivalents for those sounds, and then transcribed into the original language based on the imperfect version in the intermediate language.

The problem was complicated by the fact that the guide, not knowing the name of the location or village, often answered as if he knew. Any answer is better than no answer, as an interested traveler in Egypt even today will discover, often to his regret.[18] In many cases, official names were unknown to locals, or were themselves inaccurate.[19] There was also the matter of accent. Even today, the accent of a *sai'di*, or upper Egyptian, is immediately identifiable as different from that of a resident of Cairo. These differences would lead to differences in the written versions, both in the European and Arabic characters. Finally, the similarity of sounds could lead to confusion of one location for another: Niebuhr lists "Tucha" in the *Geographia Nubiensis*, and "Talcha" as he heard it. But this appears to be a simple mistake: "Toukh" and "Talha" are both towns in the Delta, one approximately thirty and the other about one hundred kilometers from Cairo, and it appears Niebuhr has mistaken one for the other. "Minia"—probably "port"—is often confused

---

17. *Travels*, vol. I, 72–73.

18. There is the probably apocryphal story of the Englishman who dotted his map with locations called *mish aarif*—"don't know." At least, his informant was being honest.

19. In what may seem a minor difference, the Mosque of *Aslam as-Silahdar* in Cairo is unknown today by that name to nearby shopkeepers, who know it only as *Aslan*. The street sign also spells it *Aslan*, with a "saad" and a "nun," although a purist would say it should be spelled with a "seen" and a "meem."

for "mit," lower Egyptian for village.[20] This also illustrates the problem with attempting to follow a map made several hundred years earlier: in many cases, names have changed and, in others, they have disappeared altogether.

So, the problem of language was considerable for Niebuhr. Even Michaelis, whose erudition in dead Semitic languages was enormous, undoubtedly underestimated the difficulty of sound and orthography in a living language. The only unfailing guide should have been the Arabic, although many of the sounds were utterly unfamiliar to one whose native tongue was not Arabic. But even this could be confusing, when the Arabic was written only after a transliteration that could only approximate the original sound. And this, too, was dependent on perfect knowledge of the names of places on the part of the guide, perfect hearing of his intonation by Niebuhr, consistent rendering of those sounds in the system of transcription he had chosen, and perfect repetition of those sounds to the later copyist who would be assumed to know the names of the villages or places in order to spell them correctly. And this further assumed that there *was* an accepted Arabic spelling. The rest was a matter of convention, of systems of orthography in the several foreign languages in which the map would ultimately be produced. There were many systems: the French spelling of Arabic words differed from the English, which differed from the German. Even today, there is not a single, universally accepted system of transliteration from Arabic to English.

A second problem was the determination of direction and distance. To pursue the nautical example, this would be dead reckoning: the plotting of a track on a chart based on a compass heading and average speed over time. The problem of direction might seem the simplest: the magnetic compass had been use for centuries. Niebuhr carried a small pocket compass that he used on the road, and even in the streets when making his maps of cities. But even here there were problems. In the first place, the magnetic compass needed correction to true north, although this was not a great problem, given the approximations with which Niebuhr was dealing. In fact, he implicitly recognizes that giving compass bearings in degrees would give a spurious accuracy to his map. Instead, he gives headings in points of the compass.

By the sixteenth century northern Europeans had settled on a thirty-two point compass as the standard, each point of 11.25 degrees. The mean bearing of north-northeast, for example, was 22.500 degrees, but it represented everything from 16.875 to 28.125 degrees. Even here, it appears that Niebuhr's compass was not standard, being graduated in units of sixty degrees marked

---

20. Also *kafr*, as in *Kafr al-Sheikh*, or "village of the Sheikh," and *azba*, a small rural cluster. In Upper Egypt villages are generally called *naga*.

in Roman numerals, with intermediate graduations of five units.[21] This meant that the headings Niebuhr read were each of a mean value of twelve degrees. This differed slightly from the standard compass rose, although his headings are given in the standard thirty-two points. Niebuhr occasionally added greater refinement by headings such as "S.S.W. by a little west," which was something greater than S.S.W. but less than S.W. by south; or "E by S 1/2 S," which presumably meant something greater than 101.25 degrees but less than 112.5 degrees. These headings represented general headings of tracks or paths, which probably had minor deviations along the way. Some of the deviations, depending on the scale, probably represented less than the width of a pencil line on a map. But Niebuhr had to be very careful about taking compass headings. In many cities it was downright dangerous, nowhere more so than Cairo. Even today, the use of a compass in the streets in Cairo excites interest, if not suspicion, and Niebuhr remarks that his direction finding would not have been permitted in many parts of Europe.

The next variable was that of distance. Here, there were two problems. The first was measuring the distance, in itself difficult enough, and the second expressing that measure in some commonly accepted standard. At sea, the log and log glass would be used to determine the speed made good over the water. But on land, the pace of a man or an animal was often the only measure. This could be static, as with Niebuhr's map of Cairo. The street from the Bab Zuwayla to the Bab al-Nasr, Shari Muez li Din Allah in the old Fatimid city, for example, is shown as approximately 5,000 Danish feet, or 1,100 double paces, in Niebuhr's map of the city. This was surely measured by stepping or pacing it off and counting the number of steps or paces. This assumed a standard double pace of just under five feet[22] depending on traffic in the street and the ability to maintain a consistent stride. The problem would be different with, for example, the length of the walls of St. Catherine's Monastery in the Sinai, where crowds were not a factor.

However, this static measure did not depend on time: 1,100 geometric paces were 1,100 geometric paces, regardless of the amount of time it took to pace them off. But just as Niebuhr could not travel with his compass constantly in hand, he could not be expected to count every pace he or an animal took. This required that he develop a standard measure of distance as a function of time. A rapid pace of a man walking was four English miles per hour, but this could not be maintained over a period of many hours, and would have to be adjusted for a long journey. More useful was the average hourly pace of an

21. A photograph of the actual compass carried by Niebuhr is shown in Stig Rasmussen's *Carsten Niebuhr und die Arabische Reise 1761–1767*.

22. It appears that the number of actual strides would have been divided by two to allow for variation in stride length.

The Delta

animal, a camel or a donkey, on a level surface. The problem would later be more difficult on tracks in the mountains of the Yemen, some of which were very steep. There, the pace was much slower and distance was not "overland" but represented a combination of horizontal and vertical vectors. But this was the only reasonable means of determining distances during a long journey. Once any of these average paces was known, time could be measured and the distance estimated. Niebuhr carried a pocket-watch by means of which he measured the elapsed time. The problem was different on the flat-bottomed boat on the river, when Niebuhr made his map of the two major branches of the Nile. However, the problem with the Nile was made easier by the fact that he could establish the location of various cities and villages by taking astronomical sightings. The trend of the Nile was generally north and that meant that differences in latitude were close approximations of the distances between places. Those distances could then be calculated.

Once a method of estimating distances was established, there was the problem of the standard in which they would be expressed. Niebuhr's map of Cairo is calibrated in "Geometric paces" and "Danish feet." His map of the Delta is measured in "German miles" and "French leagues." The scale in his map of the Yemen is expressed in "German miles" and two different "French leagues." This illustrates the widely recognized problem that, before the French Revolution and introduction of the metric system, there simply was no uniform standard in Europe. Each European country—and often different regions of the same country—had its own set of measures and this led to differences that were not unlike the problems we have seen with spelling. In a small book published in 1787, John Whitehurst stated the problem as he perceived it in the late eighteenth century:

> It is therefore obvious that weights and measures being thus divided and established, can have no more relation or agreement with each other, than the languages of the several nations by whom they were adopted; and their being divided into a diversity of smaller parts renders them still more incoherent and embarrassing to those nations who have any intercourse in matters relative to science or commerce ... For although their relative proportions may be truly known, whereby the quality of space, magnitude, or weight ascertained by the measures of one nation, may be found correspondent to a certain number of measures of another, yet it must be owned that such operations not only demand an experience of time, but are also liable to error. It is therefore evident, that the diversity of weights and

measures established by different nations, is no small impediment to the progress of science.[23]

The fact that several scales were often laid out side by side in Niebuhr's maps makes it relatively easy to determine, at least, their relationship to one another. For example, from the scales in the map of Cairo, we can establish with some degree of accuracy that 1,000 Danish feet is the equivalent of 210 geometric paces, giving 4.76 Danish feet to the geometric pace. That leaves the problem as to the value of a Danish foot, as expressed in terms of some more familiar measure. As shown below, it was slightly shorter than the English foot. The detailed calculations would probably be of interest only to a very few, although one is shown below.[24] For the purposes of understanding the distances used by Niebuhr, then, the following conventions have been adopted. The calculations can be verified using methodology similar to the example above drawing from, among other sources, the Whitehurst book.

Measure	English Equivalent	Metric Equivalent
Danish Foot	11.7 inches	29.46 cm
German Mile	4.61 miles	7.42 km
German League	3.25 miles	5.23 km
French League	3.46 miles	5.57 km
Geometric or Double Pace	4.59 feet	1.40 m

The league is particularly interesting measure of linear distance. Dating from Roman times, its length varied from 1,500 Roman paces, or nearly one and a-half modern English miles, to a figure that was closer to three English miles at the time of the Norman conquest of England. The above figures show the inflation to have continued, and the 3.25 to 3.46 miles represent the distance a man would cover at a moderately rapid pace in an hour. Indeed, the German word used by Niebuhr, or *Stunde*, is also the word for hour.

23. Whitehurst, *An Attempt towards Obtaining Invariable Measures of Length, Capacity & Weight*.

24. An example of the calculation from the *Travels*, vol. III, Plate X would be as follows: ten German miles are the equivalent of forty minutes of longitude at 31° of latitude. Using the Smithsonian Geographical Tables, one degree of longitude is equal to 69.167 English miles at the equator. Correcting for the latitude at 31° (40/60 = .67) yields a figure of 46.12 English miles to the 10 German miles, or 4.612 English miles to the German mile. As a check, the *British Weights and Measures, a History from Antiquity to the Seventeenth Century*, lists the English equivalents of miles in Bavaria, Brunswick, Hamburg, and Prussia of respectively 4.61, 4.60, 4.68 and 4.68 miles. Richard Lepsius in his *Egypt, Ethiopia, and the Peninsula of Sinai* in 1853 (516) lists 776 English miles as 161 German miles, or 4.82 English miles to the German mile.

The Delta

The care we have seen in Niebuhr's maps of Cairo and the Delta was repeated in other cities, from Constantinople through Rashid, Damietta, Suez, and Jidda; Loheia, Beit el Fakih, Taizz, San'a, and Mocha in the Yemen; then Bombay, Surat, Muskat, Bushire, Shiraz, Meshed Ali, Baghdad, Diarbakr, Aleppo, Antioch, Jerusalem, Adana, Konya, and Brusa. It is clear that Niebuhr himself felt his greatest responsibility on the expedition would be his map of the Yemen. Included as Plate XXV of his *Description of Arabia*, it is the largest and most detailed of the maps he produced. So slight was the interest in this remote corner of the Arabian Peninsula, however, that his map remained for a century the only detailed map of the country in existence. This was not the case of Egypt, and European travelers had gone there for centuries. There already were maps showing the general outlines of the country. None, however, was as detailed as Niebuhr's and it remained the standard for nearly half a century. As mentioned above, the map was reissued virtually unchanged in 1796, curiously entitled "Carte Physique and Politique de l'Egypte," as part of Herbin's *Conquêtes des Francais en Egypte*. There, an inset "Carte Particulaire et Detaillée du Delta" incorporates detail so close to Plate X of the *Travels* that it might have been traced from the Niebuhr original. Indeed, it appears that it was.

11

Manners and Customs

> It is of no great importance to know how Orientals spend their leisure hours. But the little games that are popular among the people are for the most part of great antiquity; and since they might perhaps serve to illustrate today the expressions of the ancient authors, I will report here what I saw of the exercises and diversions of the Orientals. However, I must confess that have not delved deeply into the subject. (*Travels,* vol. I, 168)

SEVENTY-FIVE YEARS AFTER THE Danish expedition left Cairo, Edward William Lane[1] would publish a seminal sociological study of the manners and customs of the Egyptian people of his day. We have seen that Lane's task of careful observation was made immeasurably easier by the relatively enlightened despotism of Mohammed Ali, where Europeans were accepted as useful agents in the modernization of Egypt. Europeans in Cairo in 1761–62, on the other hand, were very circumspect observers, if they were observers at all. They were tolerated by the native Egyptians within very narrow limits, beyond which they dared not trespass. In the Lane opus all but some fifty of the 571 dense pages were devoted to subjects dealing with human social behavior: their character, infancy and early education, religion, domestic life, superstitions, games, music, public recitations, death, and funeral rites.

Carsten Niebuhr was a mathematician and cartographer, subjects that might seem dull, not to say sterile, to the general public. His abridgers appeared to share this opinion, and have spared generations of readers most of the latitudes and occasional longitudes, compass headings, and place names that constitute so much of the *Travels.* But we should have learned already to be wary of Niebuhr's penchant for understatement, of his apology for entering into subjects in which he had little formal education or training. Would that those professing greater expertise in these matters had been half

---

1. Lane, *Manners and Customs of the Modern Egyptians.*

Manners and Customs

as observant. And the quixotic assignment of the Royal Danish expedition was nothing less than to learn as much as possible about the manners and customs of the ancient Hebrews, so as to assist in the explication of the Hebrew Bible. In fulfilling that responsibility, there was hardly a subject in the land of the "Orient" that could not be examined with profit by inquiring European minds. Niebuhr's was such a mind, and not one to be troubled by the fact that many of the subjects he dealt with in the year in Cairo were the responsibility of other members of the expedition. He leaves us a record of everything he thought was of interest and we would agree that if Niebuhr thought it was interesting, it probably was.

Oriental Dress

Twelve pages of the first volume of the *Travels* are devoted to the "Dress of the Orientals." It included comments on dress in general, the product of his nearly seven-year odyssey through the East. But much of the coverage was devoted specifically to Egypt where, as elsewhere in the Orient, dress made the man or the woman. Clothing was often a badge of ethnicity or religion, the covering of the head, in particular, the key to an understanding of a man's status and his station in life. In general, the Eastern preference for loose dress contrasted sharply with the European upper-class fashions of the middle of the eighteenth century, where men wore elegant chemises, waistcoats, and tight-fitting breeches, and the women featured elaborate corsets and great round skirts. The latter wore their hair in pompadours and the men in powdered periwigs. The dress of Europeans of both sexes—not to mention their relatively easy intercourse with one another—was indecent in the eyes of the Orientals, the low-cut bodices of the women's dresses revealing more than a hint of breasts and the men's breeches outlining parts that were better left concealed.[2] Monkeys in Cairo were often dressed, mockingly, in European clothes, with the tail of the animal protruding behind the waistcoat and lending the appearance of the sword. But Europeans were *not* expected to go native, and they were not respected when they did. The members of the Danish expedition, however, appear to have done so, at least in the matter of dress. In the first place, the loose Oriental clothes were more comfortable in hot climates, and more conducive to social exchanges in a society where divans and cushions, not chairs, were the rule.[3] In Cairo a man removed his slippers

---

2. Burton would later remark that, however indecent, the tight-fitting trousers of the Englishman in India were the mark of authority and, so, inspired respect among the natives.

3. Lane also wore Oriental dress in order, we suspect, to enter as completely into

and reclined on the floor, whether chatting, smoking a water pipe, drinking a tumbler of coffee or tea, or taking a meal. European clothes and shoes were both uncomfortable and unsuitable under these circumstances.

In the second place, in a country where Muslim fanaticism and dislike of Christians lay very near the surface, and it was safer not to call attention to oneself, and most Europeans adopted local dress when they could. Niebuhr gives us a description of his own "Turkish" attire. First came great wide linen breeches over which fell a shirt like the shifts worn by women in Europe. His feet were covered with linen stockings, then very fine leather slippers called *terlek*s followed by *mest*s or heavier outer slippers. Wide red pantaloons called *shaukshir* completed the underclothing. The layers of outer clothing then followed. A doubled linen *entari* fell about two hands below the knee. Over the *entari* was worn the *caftan* which reached the feet. A wide belt or sash allowed the caftan to be tucked up, so as to display the *entari* and *shaukshir*. A dagger, or *khansjar*, was frequently worn in the belt. Over the caftan was worn a short-sleeved *juppe*, fur-lined in the winter. And finally, completing the ensemble, a *benisch*, or coat that fell to the ground covered the *juppe*. This ensemble competed with upper class European dress in complexity, if not in display. But Niebuhr deconstructs it for us. Among the lower classes, only the pantaloons and the shirt were worn.

In Cairo even Christians and Jews could wear whatever color they wanted, with the exception of green, which was reserved for Muslims. However, their slippers had to be of red, black, or blue leather. The madmen of Egypt dressed according to their own fancy and some wore nothing at all, their madness being considered a mark of divine favor and their nudity consequently tolerated. Niebuhr's chaste mention of naked madmen in Cairo would contrast with Flaubert's later, and lurid, descriptions of women masturbating mad *marabouts* in the street, the symptoms of unnaturally-acquired syphilis in patients at Qasr al-Aini hospital and of a man copulating with a monkey in Shubra. Ninety years was the difference in time between Niebuhr and Flaubert, but the difference in temperament could be measured in eons. Flaubert slept his way to Luxor and back, and never became used to the Egyptian woman's habit of depilation. But his was a different time, when the biblical associations of the Orient had long since given way to a Europe that reveled in its color and diversity, and when young Europeans came to Egypt to indulge their fantasies.[4]

---

and move as seamlessly as he could through the society he was describing. Even David Roberts in 1838 adopted Oriental dress, and the portrait that hangs in the Scottish National Portrait Gallery shows him "masquerading" in his Turkish clothes.

4. See *Flaubert in Egypt*, 65.

Manners and Customs

The covering of the head was especially important in the Muslim world, where a glance at the style served to identify a man by religion, ethnicity, and occupation. Where Europeans wore wigs, Muslims generally shaved the head, with the exception of a tuft or topknot.[5] Niebuhr says that the Orientals took a kind of "sensual pleasure" in being in the hands of a barber and this, along with their indulgence of the public bath, probably contributed to the general European impression of Oriental idleness, sensuality, and luxury. An entire cottage industry would grow up in the next century, where European artists would devote themselves to the depiction of "Oriental" scenes, with the bath, the slave market, and female nudity prominently featured. Niebuhr does not mention visiting a bathhouse, we suspect because he was not circumcised. On the shaven head was worn a skullcap or *fez*, after the city in Morocco where they were famously made. The hat, or cap, which surmounted this inner covering was a mark of religion, scholarship, or rank. Over the fez, European merchants wore the *kalpak*, a cap trimmed in fur, which was the distinguishing mark of the European dragoman in Constantinople. This was a recent requirement of the authorities, and previously the merchants had worn turbans, which allowed them to blend more easily into the crowds in the street. Niebuhr wore a brown turban, shown as figure two in Plate XIX.

It differed from the Muslim turban only in color, the latter generally being white. Volume I of the *Travels* includes five plates showing head coverings, probably sketched by Baurenfeind. They consisted of forty-eight examples of hats of three major kinds: the Turkish *kaouk*, the Arabic turban, and the *kalpak* worn by Oriental Christians, European merchants, and Tatars:

> In short, if one has the time, the desire, and the opportunity to see all the different kinds of caps worn by Orientals, enough material could be found to fill an entire volume. Since this article of dress is what distinguishes not only different nations, but also different stations in lands under the Turks; and since it seems to be most subject to change, I have therefore chosen to show the principal kinds of headdresses. But my collection is far from complete ...[6]

Although the Jesuits and the members of the Society for the Propagation of the Faith wore the local garb, the Capuchins and Franciscans wore their customary dress. As we have seen, to the simple habits of the latter, particularly the dirtiness of their sandal-shod feet, the Muslims felt an extreme aversion. The ritual cleanliness of the Muslims, where the extremities

---

5. Burton says this was so that the severed head could be picked up by an unbeliever without defiling it by inserting his fingers into the mouth.

6. *Travels*, vol. I, 159.

Tab. XIX.

Manners and Customs

were washed five times a day, contrasted sharply with the habits of the Europeans who had still not learned to bathe on a daily basis. Facial hair was another statement of a man's station and religion. In the Orient, a beard was a man's glory and among the first acts of members of the Danish expedition in the Yemen was to permit their beards to grow. The Greek ecclesiastics also let their facial hair grow, and the Jews were known by the hair "which they leave on the upper part of the beard, as did Abraham."[7] It is said that the Sunni practice of clipping the mustaches—removing the hair on this "upper part of the beard"—was a result of an order by the Prophet for the Muslims to distinguish themselves from the Jews of Medina, who allowed their mustaches to grow long.

The common women's dress in Cairo consisted of wide breeches, a loose shift, and the jewelry that constituted her dowry. It also included, most notably, the veil. Niebuhr comments on what, to a European, was the strange behavior of an adolescent girl. She was naked except for the facial veil, which she believed covered her "shame." A European girl would have covered her nether parts. It was impolite to stare fixedly at a woman, even if she were veiled. For the dress of the higher classes, Niebuhr refers us to the "admirable letters of Lady Montague."[8] Her husband, Edward Wartley Montagu, had been appointed Ambassador to the Porte in 1716 and she accompanied him to post. Her "Letters" in part cover two years in Constantinople and Adrianople where she learned a little Turkish and interested herself in the customs of the Turks. Niebuhr remarks that this clever lady was more interested in describing the things that were *not* reprehensible about the Turks, while everyone else seemed to concentrate on the things that were. European attention to the faults of the "unspeakable Turk" was building, to reach a climax in the early nineteenth century with the lionization of that other ancient people who, with the Israelites, had formed the twin bases of European civilization, the Greeks.

Upper class women of Cairo, like those of Constantinople, walked very awkwardly, being used to sitting or riding with their legs drawn up beneath them. A women of quality never walked if she could be carried. The same was true of prominent Christian and Jewish women, who did not suffer from the same discrimination as their husbands and were not required to dismount in presence of a Turk. Relations between the sexes were carefully monitored, at least among the upper classes, and we have seen that the members of the expedition had been warned to avoid all contact that might excite the "jealousy and cruel vengeance" of the Orientals. Aside from the

7. Niebuhr, *Description of Arabia*, 58–59.
8. *Travels*, vol. I, 165.

dalliance with the slave girls on the voyage from Rhodes to Alexandria, they had put the caution into practice, although they would be present at several performances by dancing girls in Cairo. In the Sinai, as we will see, Niebuhr would talk relatively openly and freely with the wife of a sheikh. In spite of a year's stay in the first city of the Muslim world, it would be the first respectable Muslim woman with whom he had a conversation.

Diversions

Twenty-one pages of vol. I of the *Travels* are devoted to the "Exercises and Diversions of the Orientals in the Hours of Leisure." Here, Niebuhr continues his excursion into what would become Lane's territory, and away from strictly scientific and linguistic reportage. But there was more than mere sociology involved: it was felt by Michaelis that an understanding of the everyday concerns and habits of the Orientals would shed light on those same concerns and habits among the ancient Hebrews. And many of the games Niebuhr describes had a mathematical basis that exercised some of the best minds in Europe.

The "Osmanlis," Niebuhr reported, were fond of riding and their mounted games—*jerid* and *bardak*—which, if not a preparation for war, were at least designed to hone the skills that would be used in battle. The same was true of practice with the bow and matchlock, the arquebus not being suited to firing at the gallop. Later in the century at the Battle of the Pyramids these same men, still disdainful of firearms, would break themselves on the French squares. The audacious, splendidly accoutered Mamluks would come face to face with European science—although Napoleon's army was composed of the sweepings of France—where they would effectively pass out of history as a fighting force.

The leading men of Cairo gathered twice a week in the *Mastabe*, open ground to the west of the Birket al-Fil and between the *khalig* and the Nile.[9] There, followed by an army of slaves and servants on horseback, they tilted vigorously, with some suffering broken bones in the tests. The common people wrestled and fenced with staves, and later observers would comment on the sheer physical strength of the *fellah* and his suitability, when well led, for the military profession. When not tilting and when the Nile was high, the beys would relax by boating on the Nile or the large ponds in the city—al-Azbakiyya, al-Fil and ar-Roteli—often attended by fireworks. With these jaunts, at least, the Europeans would have been familiar, George Frederick

---

9. From Niebuhr's map, it is at the approximate location of the present-day Midan al-Tahrir.

Handel (that son of Halle, born some eighty miles from Michaelis's university at Göttingen) having written his three "Water Musics" for excursions by the king on the Thames between 1715 and 1736, and his "Music for the Royal Fireworks" in 1749. His oratorio "Israel in Egypt" appeared in 1738. Beginning with "The ways of Zion do mourn," followed by story of the release of the Israelites from bondage, and finally the song of celebration and thanksgiving of "Moses song," the oratorio was an epitome of the odd connection that existed, at least in some minds, between the Europeans of the Enlightenment and the little-understood Hebrews of the Orient.

In the evenings the beys were occupied with the *harim*, the very word conjuring visions of lasciviousness in European minds. Niebuhr remarks that European travelers "cannot see how they amuse themselves there,"[10] the uncertainty only feeding notions of luxury and licentiousness. The common people were extremely fond of public celebrations and amusements on saints' days, the opening of the *khalig*, and the departure of the pilgrim caravan, and on those days "machines D, E, and F" in Plate XXV were set up for children. They are primitive swings, and carousels, not too different from those seen in the poorer neighborhoods of Cairo to this day. As for children's games, they seemed to be the same the world over, from Egypt to Lower Saxony. With regard to the games of adults, there were chess—at which some men passed the entire day—*tawila*, or backgammon, and *dama*, or the game of draughts, or checkers. The game of *dris et talata* consisted of a board with twelve holes, six on a side for each player. Into the six holes the players played six small stones, with a player taking all the stones in a particular hole, depending on the number played.

It was the mathematical probabilities in the game that particularly interested Thomas Hyde in his *Syntagma Dissertationum et Opsuscula*. The object was to capture as many of the opponents pieces as possible, and it was the same game as *trip, trap, and trul* in Lower Saxony. With apologies for having tried the reader's patience, Niebuhr concludes his description of Oriental board games with descriptions of the games *dris et tissa, lab el kab*, and *tab wa duk*. They were played with a variety of pieces, generally uneven numbers such as seventeen, nineteen, and twenty-one, depending on the locale, and these also were of interest to Hyde. European cards were not in evidence, except among the Greeks of Cairo, although the Arabs had a kind of thick and clumsy set of cards with which they played a game called *lab wa qamar*. Muslims were forbidden to gamble for money but human nature being no less irrepressible among the Muslims than the Christians, gambling was widely practiced out of the view of the puritanical.

10. *Travels*, vol. I., 170.

Niebuhr in Egypt

Coffee had been introduced into Egypt in the sixteenth century by Yemeni Sufis of the Shadhiliyya order, and by 1762 the coffee shop had became a favorite haunt of Cairenes. There, a man could smoke a pipe, imbibe a beverage that was not forbidden, and drink in the tales of the storytellers. There, we imagine, was played out the timeless appeal of the *Alf Leila wa Leila*, or the *Thousand Nights and a Night*, even in the "debased brogue of Egypt or rather of Cairo."[11] For all of the coarseness of the *Nights*, it was a coarseness of language rather than of thought, and its moral lessons were patent: "Poetical justice is administered by the literary Kazi with exemplary impartiality and severity; 'denouncing evil doers and eulogising deeds admirably achieved.'"[12] The Egyptian pipe was long-stemmed and wrapped in silk or linen in the better establishments. In the hot weather water was poured over the stem to "cool" the smoke. Men spent entire evenings in these establishments, in quiet enjoyment of their *kaif*, or leisure.

Music, on the other hand, was the province of the poor throughout the Orient, the upper classes believing that "they dishonor themselves in learning music and dance."[13] Niebuhr found the "natural music" of the chanting of the Koran very pleasant, but thought the average Oriental musician unskillful. No one appeared to understand musical notation and it was perhaps here, as much as anywhere, that the difference between East and West manifested itself. The lusty bellowing of the Egyptian sailors, accompanied by the striking of tambourines and the concussive Egyptian manner of clapping the hands, held little attraction for this particular German amateur violinist. On the other hand, European music was not pleasing to Arabs or Turks:

> In Cairo we gave a concert in which several merchants, monks, Mr. Baurenfeind, and I played. As we returned contented to our lodgings, believing that we had played very well, we came upon an Egyptian in the gloom of the street who sang a song, accompanied by a flute. This so moved one of our servants from Sennar that he cried: 'By God, that is beautiful, God bless you!' We were greatly surprised at this, and we asked him how he liked our concert. 'Your music is a harsh and disagreeable noise,' he said, 'in which no serious man can take pleasure.'[14]

---

11. Burton, "Translator's Preface" to *The Book of a Thousand Nights and a Night*, xxi. For all of his inveighing against the Egyptian language, Burton loved the *Nights* and the translation, particularly the poetry, footnotes, and terminal essay were his real *magnum opus*.

12. Ibid., xvii.

13. *Travels*, vol. I, 176.

14. *Travels*, vol. I, 176.

Manners and Customs

The European system of major and minor scales with twelve steps to the octave, harmonies, and elaborate counterpoint could be particularly heavy to some ears in the middle of the eighteenth century. It is no wonder that the Orientals, used to a more supple seventeen-step octave, did not find it to their taste. While Niebuhr did not delve deeply into Oriental music, he did provide a plate showing instruments used throughout the East.[15]

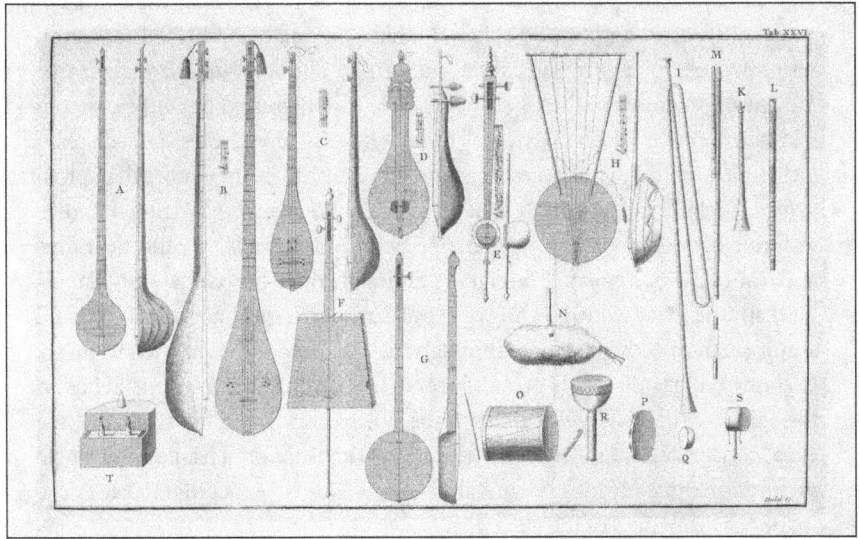

They consisted primarily of strings, drums, winds, and reeds, and include the Egyptian single-stringed instrument called the "marabba" and the two-stringed viol called the *semenje*. These were undoubtedly the more familiar *rabab* and the *kamanja*. Martial music consisted largely of that produced by drums, oboes, and a kind of trumpet, all of which "make a disagreeable noise to the ear of a European."[16] The Egyptians, says Niebuhr, favored loud music but the *Berberis*[17] from Dongola preferred a kind of harp, which, whether plucked or struck, produced a softer sound. And did not the last, he thought, resemble the harp of David?

> And it came to pass when the evil spirit from God was upon
> Saul, that David took an harp, and played with his hand: so Saul

---

15. M. Villoteau, who accompanied the French expedition, would publish his *Musique des Egyptien et des Orienteaux* and include descriptions of instruments and "signes ou notes en letters arabes." He also includes remarks on the "chants religieux des Juifs d'Egypte" and some interesting commentary on the differences between the Rabbinites and Karaites and the areas inhabited by each.

16. *Travels*, vol. I, 179.

17. The term still used in Cairo to denote black Africans.

was refreshed, and was well, and the evil spirit departed from him. (1 Samuel 16:23)

As we have seen, dancing was considered to be in poor taste for a respectable man, but not so for his wife. Oriental women danced not only to please their husbands, but also for their own pleasure. It was not uncommon at gatherings of wealthy ladies with finery and their most beautiful servants and slaves, for the participants to change clothes ten times in the course of an evening, each outfit more elaborate than the one before. Needless to say, Niebuhr was not a witness to these gatherings, although he did suggest that Europeans, concerned that their wives were too interested in clothes, should take comfort from the fact that they did not live in Cairo. Professional dancing girls—*ghasiya* in Egypt—were considered by some to be little more than prostitutes. Niebuhr remarks only that they "make no pretense at being the most virtuous of Mohammedan women."[18] They lived apart from other Muslims and, on occasion, married foreigners although their husbands were usually blacksmiths. They were like the gypsies of Europe. Respectable Muslims could admit them into their homes, although the men had to be married to do so. Niebuhr reports that their greatest profit came from the Europeans living on the banks of the *khalig*, where they were able to perform in the street that the canal constituted for most of the year. In the time-honored tradition of single men in foreign countries the members of the Danish expedition took solace in the performances of the *ghasiya*:

> To dispel our fears of the upcoming journey, we tried to distract ourselves by as many means as possible, among them by having a troupe of these dancing girls sing, dance, and play in the dry canal, which then served as a street. And although, at first we were not very amused by this kind of spectacle, given that the music, both instrumental and vocal, was very bad, and the women made every kind of indecent gesture . . . in the end we listened and watched them with as much pleasure as if we were being entertained by the best singers and dancers of Europe.[19]

He does not mention the common practice of men dancing in the garb of women. But he had already seen the phenomenon in Istanbul, catered by Greeks and appealing to "debauched Turks." His stern Saxon soul was moved to the strongest expression of disapprobation:

> Such Christians, who supply the Mohammedans with entertainment of this nature, by prostituting the innocence of their

---

18. *Travels*, vol. I, 183.
19. *Travels*, vol. I, 183–84.

Manners and Customs

coreligionists and providing the opportunity for the vice to which it leads, surely deserve to be classed with these Gypsies.[20]

And it was not only in the matter of dance. The freedom enjoyed by European women was a kind of standing indictment of Europeans in general. In Turkey both sexes behaved as if they were in Europe, with elaborate balls and masquerades in Pera at carnival time. They were despised by the Turks for carrying on in a manner that characterized the rabble in the East.

Baurenfeind was not idle In Cairo and he left many sketches that were later engraved for inclusion in vol. I of the *Travels*. They included a sketch of the performance of a troupe of dancing girls,

included by Niebuhr as Plate XXVII. He also sketched a marriage procession, with the elaborate dress, display of jewelry and furniture that constituted the dowry, the whole accompanied by the din of the music, guns being fired, and the ululations of the women in the company. Funerals were also elaborate affairs, with hired female mourners to assist, given that unrelieved lamentation was sheer, hard work. The veil conspired in the charade, since one couldn't tell whether or not a woman was really weeping. Muslims as well as Christians employed hired mourners to lend numbers, if not solemnity, to the occasion. There was the other side of the coin as well,

20. Ibid., 184.

Niebuhr in Egypt

Cairenes being as notorious for their sense of humor as for their hostility to outsiders. Troupes of comedians—made up of Muslims, Christians, and Jews—performed for a pittance, and the members of the Danish expedition were invited to a performance in the home of an Italian merchant. Niebuhr, for one, was not edified. Puppets were a great street favorite in Cairo, and one of the favorite topics of the shows was the ridicule of the manners and dress of Europeans. We have already seen that monkeys were often dressed in tight European clothes. Tricks of sleight-of-hand, accompanied by much buffoonery, were another popular favorite, although the charlatans earned very little from their performances.

Finally, there were the beggars, including men who had supposedly suffered outrages at the hands of the Europeans and now appealed to the charity of good Muslims. For every Joseph Pitts of Exeter—who chronicled his capture by Barbary pirates, conversion to Islam, and pilgrimage, before escaping from the barbarity of the religion and its prophet, a "bloody impostor"—there was a Mustafa with an equally sad tale of mistreatment at the hands of the Christians:

> I can only say here, in passing, that in Cairo many times I saw a man sitting in the street in enormous chains that had been put on in Malta, recounting in a pitiful voice to his countrymen the evils he had suffered during his captivity, i.e., being required to keep the pigs during the day and to sleep with them at night in their sty, etc. Sensible Mohammedans regard this kind of mendacity with indignation. But the beggar was ordinarily surrounded by a crowd of people who were moved by his sad tale, not only to provide copious charity to him, but especially to pour out imprecations against the alleged barbarians of Europe.[21]

In some respects, not a great deal has changed in the 250-plus years since the Danish expedition was in Cairo.

With the tale of the beggar Niebuhr brings to a close his sociological review of the inhabitants of Cairo. These were interesting matters. However, we suspect that they were a distraction from more serious work. With its wealth of Islamic and Pharaonic monuments, Cairo probably provided more raw material for historical and linguistic research than any other city in the Orient. Niebuhr was not immune to its attractions during the year of his stay. We will now turn to his careful treatment of the antiquities of Egypt.

---

21. *Travels*, vol. I, 190.

12

The Antiquities of Egypt

> Among the many scholars of Europe surely there are some with the patience and skill to study the ... ancient Egyptian inscriptions. So if travelers provide them with a sufficient number, I am certain that they will be able to clarify many matters, especially if ... they have a good understanding of the Coptic language spoken before the arrival of the Greeks; for this seems essential to an understanding of the hieroglyphs. (*Travels*, vol. I, 201)

THE YEAR THAT THE members of the Danish expedition spent in Egypt was in some respects a regrettable diversion from their ultimate goal, Happy Arabia. However, for eighteenth-century scholars, the Orient was a mine to be worked and there were many other things to be seen and described along the way. As we have already seen, Niebuhr would contribute to an understanding of the political and economic situation of the country in the middle of the eighteenth century, not to mention its geography, with his detailed maps of the city of Cairo and the Delta. He was the cartographer on the expedition, responsible primarily for maps of the areas they visited. But he was also an interested observer of everything he saw in the Orient, and the time in Egypt was very profitably spent. The "antiquities of Egypt," a subject which occupies thrity-one pages in Niebuhr's account, was theoretically beyond the purview of the expedition, although the explication of the Hebrew Bible was so all-encompassing a task that there was hardly a subject that could not throw light into its hoary recesses.

After all, according to the Bible, the Israelites had sojourned 430 years in Egypt. The story of Joseph's sale into slavery to a band of Ishmaelites, his rise to a position of prominence in Egypt, his brethren joining him, the suffering of the children of Israel under the yoke of Pharaoh, and their ultimate escape occupied many chapters in Genesis and Exodus. By the time of the Exodus, according to the Bible, the original seventy souls that issued from the loins of Jacob and accompanied him "down" to Egypt had grown into more than 600,000,[1] not counting women and children. After a long contest with Pharaoh, whose heart was hardened against them—although from the account it seems that it was his head, not his heart, that was hardened—the children of Israel had been permitted to go. Pharaoh's second thoughts had led to the swallowing up of his host in the waters of the Red Sea. It was one of the most stirring and memorable accounts in the Hebrew Scriptures, acceptance of whose broad outlines was shared by the three "people of the Book," all of whom were represented in Egypt in 1761–62.

Surely, there were aspects of the biblical account that could be examined by specialists in Egypt. It would be unusual if an expedition, whose brief was expressly biblical, would spend a year in the country and not speculate on the purported sojourn of the children of Israel as bondmen to the Pharaoh. Given what we have seen of the *Fragen*, we can imagine the questions with which Michaelis might have equipped the members of the expedition if he had known that they would spend a year in Egypt.

What was the frequency and average length of famine in the Nile Valley? Was a famine of seven years unusual, and had such a famine occurred within the memory of the inhabitants? Where were the corn storehouses of Joseph located? Where was the Land of Goshen? Where were the sacred cities of Pi'-thon and Ra-ma-ses? What was the Egyptian method of making bricks? Were swarms of locusts generally borne into the country on an easterly wind?

Fortunately, with the exception of speculation about the path the children of Israel took on the Exodus (see Chapter 13 below), the members of the expedition appear to have resisted the urge to examine Egypt's ancient history within a specifically biblical framework. This was fortunate because it was the account in the Bible itself, among other influences, which stood in the way of serious scholarship about ancient Egypt.

Egyptology

The study of the antiquities of Egypt, or "Egyptology" as we know it today, was still in its infancy in 1761. In the absence of an understanding of the hieroglyphs, the written language of the ancient Egyptians and the key to an understanding of the civilization, scholarly knowledge was still dependent on the sources that had defined western knowledge of ancient Egypt for over a millennium: the classical authors and the Bible.[2] The authors—Strabo, Ptolemy, Arrian, Pliny, Diodorus Siculus, Herodotus, and Flavius Josephus—were

---

1. Exodus 12:37.
2. See Wortham, *British Egyptology 1549–1906*.

the most reliable sources, although all were either flawed or gave partial or tendentious accounts of the history of the country. We have already seen the contributions of some of them above. But the greatest contributions available to scholars were those of Herodotus and Manetho. A brief review will remind us of the state of knowledge about Ancient Egypt in 1761.

Herodotus (484–425 BC) was the earliest of the classical sources but it is useful to remember that even his account, dating from the mid-fifth century BC during the Persian period in Egypt, was written more than two thousand years after the building of the pyramids at Giza. His account of Egypt, largely comprising Book Two in *The Histories*, covers plant and animal life, the rise of the Nile, religion, the hieroglyphs, the major monuments, and quasi-historical anecdotes. Many details in his account of Egypt we now know to be accurate: his careful description of the sacred ibis, the widespread use of barley beer, the eleven fathoms of muddy bottom a day's sail from the mouth of the Nile, the "black and friable" earth of the alluvium, the "sacred and common writing," the lower half of the pyramid of Mycerinus being cased with red granite, among many others. His history—presumably learned from the priests of Hephaestus at Memphis—is confused, although he has the broad outline and many of the details right: his identification of the first king as Min, is commonly accepted today, in its variations, as the name of the first Pharaoh of the 1st Dynasty. But his chronology is convoluted, covering "three hundred and thirty monarchs in the same number of generations," before arriving at "Sesostris, who succeeded them."[3] He correctly identifies the 4th dynasty Pharaohs Cheops, Khephren, and Mycerinus as the builders of the three major pyramids at Giza. And his list of kings—the "three hundred and thirty monarchs"—when corrected and reconciled with other lists, would be the key to the history of the country.

The list, or variations thereof, was apparently kept by Egyptian priests, and in far greater accuracy than that transmitted by Herodotus. It was also the source of Manetho's *Aegyptiaca*, written a hundred years later. As we have seen, Manetho was an Egyptian priest of the fourth century BC, a native of Sebennytus in the Delta, who wrote in Greek. His history was an expressly Egyptian work in an age that was otherwise Greek. His principal aim, among others, was to correct the misinformation of Herodotus, among which the latter's suggestion that the pyramids at Giza were built by slave labor and that prostitution was widespread in ancient Egypt seemed particularly egregious errors. Manetho's sources were probably papyri in temple archives, hieroglyphic tablets and inscriptions, including information that appeared in the royal lists of Abydos, Karnak, and Saqqara, as well as what would later be identified as the Palermo stone and the Turin papyrus. The last contained the

---

3. See Herodotus, *The Histories*.

names of over 300 kings in order, with the lengths of their reigns recorded in years, months, and days. These combined sources listed kings from Menes, or the Min of Herodotus, through Rameses II of the 19th Dynasty.

Unfortunately, as we have seen, the authentic Manetho had been lost and the *Aegyptiaca* has come down to us only in fragmentary and distorted form through Jewish and Christian apologists, each of whom had similar, although slightly different, polemical interests. The Jews, of whom Josephus was the foremost representative, were interested in the history because of their own tradition of the bondage and Exodus from Egypt. The period played a major role in Jewish history, the bondage in Egypt being no less than "the cradle of the Jewish nation."[4] In Manetho they were able to perhaps identify their ancestors with the Hyksos,[5] and the Exodus with the expulsion of those invaders, but the story told by Manetho was very different from that told in the Hebrew scriptures: it turns the captivity story in the Bible from one in which an oppressed, servile people escaped through the intervention of God into the story of the expulsion of a hated, alien race after a century of dominance. As we have seen, the "efforts of Jewish apologists account for much re-handling, enlargement and corruption of Manetho's text, and the result may be seen in the treatise of Josephus . . ."[6] Later Christian chronographers, primarily Africanus and Eusebius, were equally interested in correlating Egyptian history with the Bible, and they ingeniously related the lunar years of the Egyptians to biblical incidents such as the Flood, demonstrating that Egyptian writings at odds with "our divinely inspired Scriptures are really in agreement with them."[7] They were not the first to do so, and the Egyptian monks Pandorus and Annanais wrote in the late fourth century, attempting to harmonize Egyptian chronology with that of the Hebrews. Bending facts to the dictates of biblical history does not appear to be a practice confined to Europeans of the eighteenth century.

But even with access to the admittedly bowdlerized Manetho, there was little that scholars of the eighteenth century could have made of him:

> most of the royal names, especially Greacized, have been so mutilated by non-Egyptian scribes, who did not understand their form, as often to be unrecognizable, and the regnal years given by him have been so corrupted as to be of little value unless confirmed by the Turin Papyrus or the monuments.[8]

4. Fargeon, *Les Juifs en Egypt*, 25.
5. See Redford, *Egypt, Canaan, and Israel in Ancient Times*.
6. *Manetho*, xvi.
7. Ibid., 13.
8. Hall, "Egyptian Chronology," 260.

The Antiquities of Egypt

This confirmation would require an understanding of the hieroglyphs, or the sacred language of the Egyptians. For Europeans of the middle of the eighteenth century, that breakthrough was still nearly a century in the future, if by breakthrough a working knowledge of the script is meant. Serious investigation of the history of Egypt would not take place until the language of the hieroglyphs was deciphered and Europeans could examine the some of the same sources as Manetho himself used in his *Aegyptiaca*.

It should be noted that even these sources—Herodotus and Manetho—were relatively recent if measured by what we now know of the dates that encompass the civilization of ancient Egypt. If there was some knowledge, through Manetho, of what later came to be known as the dynasties, there was no comprehension of the time periods involved.[9] Even in the early twentieth century, there was still dispute as to the dates. Flinders Petrie wrote in 1906 that "no account of the present knowledge of Egyptian chronology is generally available," so he provides one. His use of astronomical data is ingenious, if incorrect, and he places the beginning of the Ist dynasty at 5510 BC, rather than the commonly accepted figure of 3050 BC today. It would be his most serious error in print. But in earlier periods, Europeans were not reluctant to fill in the gap in western knowledge of this and other ancient civilizations, again using the Bible as one of the authorities. In 1658, James Ussher, the Archbishop of Armagh and Primate of Ireland, had published his *The Annals of the World, Deduced from the Origin of Time*, in which he had determined precise dates in the history of the world. His sources, sacred and profane, included the Scriptures as well as many of the classical authors listed above. In the "Epistle" to the reader that precedes the tables, he states:

> I incline to the opinion that from the evening ushering in the first day of the World, to that midnight which began the first day of the Christian aera, there was 4003 years, seventy days, and six temporarie hours ...

About other key dates he was equally certain. In the "Year of the World" 1655:

> Now in the second month of the year, upon the 10 day thereof (answering to the 30 of our November, being Sunday), God commanded Noah ... to enter into the Arke.

Later, in 1491 BC, or the "The year of the World" 2513 according to Ussher, the Exodus took place.

---

9. See Petrie, *Researches in Sinai*.

The Bible as History

Mention of the *The Annals* is not out of place. The biblical account of history was still immensely important as Europeans approached the new discipline that would become Egyptology. But, whatever truth it may have contained, the Bible simply was not a repository of historical fact and the persistent attempt to treat it as such led to grotesque errors. The Bible was not the only source of error in European knowledge of ancient Egypt, but even a century later Egyptology would still suffer from its baleful influence. We have already seen how a respected scientist, the Scottish national astronomer Charles Piazzi Smyth, would suggest in 1867 in the *Life and Work at the Great Pyramid*, that an Old Testament figure must have been the architect of the "Great Pyramid," that it was a perfect structure and its dimensions contained the sacred cubit of the Israelites, that it held the key to an understanding of the history of man, etc. The nonsense overshadowed much useful work by Smyth, including the most accurate measurements of the pyramids to date and photographs of the interior. His book was what first prompted Flinders Petrie to visit Egypt. If he had done nothing else, this would have cemented Smyth's contribution to the discipline of Egyptology.

Much of the furor provoked by Smyth had to do with the anti-evolutionist movement in England and the belief in the literal truth of the Bible. The search for evidences of the cities of Rameses and Pithom of the bondage, of the corn storehouses of Joseph, or of the route of the Israelites as they crossed the Red Sea into Sinai, occupied travelers and scholars, from Rabbi Benjamin in the eleventh century, to Flinders Petrie in the twentieth. The errors lay not so much in the derivation of incorrect information, as in the squandering of scholarly time and energy in correlating what was known with the sketchy account in the Bible, thereby "proving" its accuracy:

> Scholars expended substantial effort on questions that they had failed to prove were valid questions at all. Under what dynasty did Joseph come to power? Who was the Pharaoh of the Oppression? Of the Exodus? Can we identify the princess who drew Moses out of the water? Where did the Israelites make their exit from Egypt: via the Wadi Tumilat or by a more northerly point? One can appreciate the pointlessness of these questions if one poses similar questions of the Arthurian stories, without first submitting the text to a critical evaluation. Who were the consuls of Rome when Arthur drew the sword from the stone? Where was Merlin born?[10]

---

10. Redford, *Egypt, Canaan and Israel in Ancient Times*, 260.

In fact, for all of the length of the account and the literary charm of the story of the bondage and escape from Egypt, there was virtually nothing in the historical records of the country itself that made mention of the events recorded in the Bible. The only reference to the Israelites in Egyptian records appears to be a stele of Merenptah (1224–1214 BC) that refers to Israel in Palestine, not in Egypt. It was discovered by Flinders Petrie in December 1895, on the temple ground to the west of Thebes, and the story of its discovery is telling:

> The great discovery was the large triumphal inscription of Merenptah naming the Israelites . . . I had the ground cut away below, blocking up the stones, so that one could crawl in and lie on one's back, reading a few inches from one's nose. For inscriptions, Spiegelberg was at hand, looking over all new material. He lay there copying for an afternoon, and came out saying, 'there are names of various Syrian towns, and one which I do not know, Israr.' 'Why, this is Israel,' said I. 'So it is, *and won't the reverends be pleased*,' was his reply. To the astonishment of the rest of the party I said at dinner that night, 'This stele will be better known in the world than anything else I have found, and so it has proved.'[11]

What were the biblical accounts of Egypt that largely constituted the baggage with which Europeans entered the country in the eighteenth century? In fact, they were remarkably slender. The country is first mentioned in Genesis 12:10 when it is said that Abram went "down into Egypt," although there is no information given about the land before he went "up out of Egypt" in chapter 13. In Genesis 37:25–27, Joseph is sold into slavery to the Ishmaelites going down into Egypt "bearing spicery and balm and myrrh," and thus begins the tale of the bondage. In the remaining 13 chapters of Genesis and the first 15 chapters of Exodus the familiar story is played out, but it contains almost no information about the country. We learn in Genesis 41:29 that the country was subject to famine; in 45:21 that the Egyptians used wagons; in chapter 46 that shepherds were an abomination to the Egyptians and that the kinsmen of Joseph were settled in the land of Goshen; in Genesis 47:26 that Pharaoh kept the fifth part of the produce of the land, except for that of the priests; and in chapter 50 that physicians embalmed Jacob. The story continues in Exodus with the tale of the bondage, of Pharaoh's anxiety about the growing numbers and influence of these outsiders, of his forbidding that straw be given to the Israelites with which to make the bricks that were their bond, and with the familiar tale of the plagues visited upon Pharaoh—the turning of the waters to blood, the frogs, the lice, flies, the "grievous murrain," boils and blains, hail and fire, and, finally, the slaughter of the firstborn—before Pharaoh relented.

---

11. Petrie, *Seventy Years in Archaeology*, 160. My italics.

Stirring stuff, and a tale to inspire generations with its saga of suffering, faith, and deliverance. But there is hardly a fact—historical, demographic, anthropological, sociological, or religious—that would be of much use to a scholar of ancient Egypt. By comparison with Herodotus, the Bible is virtually silent on the subject of Egypt.[12]

This paucity has not kept believers from attempting to correlate Egyptian history with the biblical account, many of the attempts ingenious efforts by otherwise sensible people. It was Rabbi Benjamin who first suggested that the pyramids at Giza were the storehouses of Joseph, and the notion persisted through Sir John Mandeville in the fourteenth century at least until John Greaves suggested in his *Pyramidographica* in 1646 that a pyramid makes a poor granary. As for evidence of the presence of the Israelites in the cities of Rameses and Pithom,[13] it seemed to depend on the religious belief of the investigator. And acquaintance with Manetho did not eliminate error, even in the middle of the nineteenth century. Another early giant of Egyptology, Richard Lepsius, would put the new found knowledge of the king-lists to work in an effort to achieve agreement between the Egyptian records and those in the Old Testament.[14] This eminent man of science would put his considerable forensic skills to use in a review of the chronologies—those in Manetho and the Old Testament (purged of obvious repetitions in Genesis, 1 Chronicles and Numbers)—side-by-side, in the process arriving at, among other things, the date (1314 BC) and the pharaoh ("Menephthes, the son of the great Ramses, in the 19th Dynasty") of the Exodus. His starting and ending point was a belief in the fundamental historical truth of the Bible and he used it to lend authority to Manetho's account. However, he suffered from a number of handicaps: he got his Manetho through Josephus and although he was rightly skeptical about the latter's editorializing, he still was forced to accept much that was dubious. And without the advantage of later lists with which Manetho could be compared, there were still too many

---

12. Later references to Egypt in the Bible are not without historical interest, their accuracy, it seems, a function of the contemporaneity of the accounts. In 2 Chronicles we read of "Shishak king of Egypt," surely one of the five Shoshenks of the 22nd Dynasty who ruled between 945 and 735 BC. And in 2 Kings 23:29, 2 Chronicles 35:20 and Jeremiah 46:2 mention is made of the invasion of Syria by the Pharaoh Necho, we now know to be the 26th Dynasty Necho II who invaded in 610–605 BC. But these references were many hundred years after the purported events mentioned in Genesis and Exodus. Knowledge of the dynasties was, of course, unknown to eighteenth-century biblical scholars and would remain so until the deciphering of the hieroglyphs.

13. Lepsius believed that Tell al-Maskhutah in the eastern Delta was the site of Rameses, one of the cities that the Israelites were supposed to have built for Ramses II. However, no archeological evidence has emerged to support the contention.

14. See his *Letters from Egypt, Ethiopia, and the Peninsula of Sinai*.

gaps in Manetho's account. His own "Tables of Egyptian Dynasties" was an impressive early bit of scholarship, although his beginning date was off by nearly a millennium (3893 BC for the beginning of the reign of Menes).[15]

The evidence of the Israelites in Egypt during the purported period of the captivity, we now know to be virtually nonexistent in Egyptian sources. Taking the numbers allegedly involved in the Exodus—some 600,000 men, not to mention women, children and flocks—an event of such magnitude in an Egypt with a population of a few millions, would have been cataclysmic, especially if the Israelites were disproportionately a servile population. Yet, Egyptian sources are silent on the matter.

Ideally, historical truth would be arrived at through the comparison of information from various sources and its eventual reconciliation in a commonly accepted version. This has been done, for example, with the various Egyptian king lists that have come down to us, whether in the lists from Manetho, Abydos, Karnak, Saqqara, or in the Turin Papyrus or the Palermo Stone. The version that we accept today is a product of this process of comparison, purged of obvious errors, lacunae, and absurdities. With the Bible, on the other hand, the preferred methodology seemed to be acceptance of the text as the "truth," or as the "word of God," followed by the search in historical or archeological records for plausible evidence that the events recorded there may, in fact, have taken place. Since the Old Testament was not created out of whole cloth, but preserved, however inaccurately or incompletely, information available to its authors or compilers, we would expect to find evidence of accuracy in some of the accounts.

For example, Flinders Petrie reasons[16] that since it says in Exodus (3:18; 8:27) that the Israelites repeatedly asked to be allowed to go "three days into

---

15. We will later see Lepsius in the Sinai, when he weighs in on the location of the mount where the Law was given: it was surely in the vicinity of the Wadi Faran, he believed, not fifty miles east at the present location of the Greek monastery as many others believed.

16. Petrie, *Researches in Sinai*, 203. Petrie grew up in a deeply religious, if slightly nonconformist household. His original interest in Egypt came from reading Piazzi Smyth's *Our Inheritance in the Great Pyramid*, which contained much nonsense about the sacred cubit etc. The families—the Petries and Smyths—had known each other in South Africa and they shared many scientific interests, although Petrie did not share all of Smyth's enthusiasms and was not one of the British "Israelites" who believed that the British were one of the ten lost tribes of Israel. But his mother was an amateur writer in her own right, and in the 1840s had published several anonymous essays, poems, and short stories including the treatise *The Connection between Revelation and Mythology illustrated and vindicated* "in which she attempted to show that the mythologies of ancient nations—Egyptian, Persian, Indian, Phoenician, Chinese, Scandinavian, and classical Greek—all reflected a single body of tradition derived from the Old Testament, which she held to contain the primeval truth, 'the earliest facts of Sacred History.'" See Drower, *Flinders Petrie*, 8.

the wilderness to sacrifice to their God," and that one familiar with the waterless three-day journey from Suez to Wadi Gharandel would know that this meant Sinai, the biblical account was "proven." But this reasoning proves nothing, except that the author or authors of *Exodus* may have had some understanding of Egyptian geography. Accepting Petrie's testimony makes the account perhaps more plausible, but it says nothing about whether the Jews ever went there at all. There is simply no other record with which the account can be compared. Petrie's comment on the controversy is instructive:

> The whole question of the direction of the journey and the position of Sinai has been much disputed of late years. The first step is to see what the direct narrative shows, *then to examine it if any other indications are discordant with that.*[17]

But the only evidence given, other than geographical details of the eastern Delta and Egyptian names that may or may not be identified with names in the biblical account, is the Bible itself. That is, depending on the assumptions, the account in the Bible is not implausible and the Bible is not manifestly inconsistent with itself. Petrie is no ordinary researcher, and his account includes evidence that purports to show that no dramatic change in the climate of Egypt took place over the preceding five millennia, making the sources of water little different at the time of the Exodus from those of his own time. In addition, since the peninsula would probably been capable of supporting the same number of human beings then as in 1906, he speculates that the numbers of the children of Israel were actually 600 *families* rather than 600 *thousand*, not including women, children and flocks of Numbers. The error lay in a misreading of the Hebrew word *alaf* for "thousand" rather than "family." Petrie's reasoning was as follows:

> As bearing on this, observe the size of the region from which they came. The land of Goshen was at the mouth of the Wady Tumilat, a district of about 60 or 80 square miles, as it did not include the great city of Bubastis. This is about a hundredth of the whole Delta; and this, on the basis of the population before the present European organization, would hold about 20,000 people. This estimate is reckoned on an agricultural basis, whereas the Israelites were a pastoral people ... If the numbers of the population stated in Exodus and Numbers were correct, the 600,000 men would imply at least 3,000,000 people, which would equal the whole population of the Delta on an agricultural basis; and there is no trace of a depopulation of the Delta at this period.[18]

17. Petrie, *Researches in Sinai*, 203, my emphasis.
18. Ibid, 208.

Petrie's creativity was not confined to proving the plausibility of the biblical account from the historical record. He also contributed to the discussion a distinction[19] between *non-natural* and *co-natural* events. A non-natural event was an event that flew in the face of nature, in effect a "miracle." But a co-natural event was a natural event—the rising of a strong east wind, for example—that would explain what otherwise appeared to be a non-natural event. Michaelis's suggested explanation of the crossing of the Red Sea by the Israelites was, to use Petrie's term, co-natural: that is, a strong wind created an ebb upon ebb tide, which exposed dry land and enabled the Israelites to cross an area that had previously been covered by water. As we will see, Niebuhr, after much thought and examination of the tides and the bottom at the northern end of the Red Sea, concluded otherwise. The whole tale seemed so non-natural, again to use Petrie's term, that divine intervention—a miracle—was the only plausible explanation. It would be surprising if a mid-eighteenth century expedition with an explicitly biblical purpose could have come to any other conclusion.

The Hieroglyphs

In short, eighteenth-century European attitudes about the history of ancient Egypt were a combination of credulity, religious provincialism, and ignorance of such written sources as were at hand. But they were not the only reasons for the lack of understanding. The ancient Egyptian writing, or hieroglyphs, were still thought to be an esoteric system of writing that made use of symbols to impart cosmic philosophic truth, known only to initiates, and the obvious difficulty of decoding or deciphering such a system stood in the way of any serious attempt. The notion that hieroglyphs did not represent a system of writing in the accepted sense went back at least as far as Diodorus Siculus in the first century BC. Instead, it was believed that the script contained "figurative meaning" an understanding of which depended on the memory of the initiate. A genre of "understanding" the hieroglyphs grew up, which bore little relation to the study of language or script in the commonly accepted sense. The *Hieroglphika* of Horapollo, compiled in the fourth or fifth century AD, was an epitome of this thinking and it had an unfortunate influence over later opinion and research on the hieroglyphs. The notion that the Egyptian and Chinese systems of writing were essentially the same, and that China was a colony of Egypt, had an early vogue. It led to some grotesque speculation: if the Psalms of David were translated into

---

19. Ibid., 203.

Niebuhr in Egypt

Chinese and written in the ancient characters of that language, Egyptian hieroglyphs would be reproduced![20]

Copies of the hieroglyphs had, of course, been available to Europe for centuries, if only on the obelisks that had been carried off during Roman times. Along with the history of other ancient peoples, matters Egyptian had received renewed attention with the Renaissance. In the sixteenth century, after a millennium of neglect, many of Rome's obelisks were unearthed and erected in the prominent piazzas of the city by a series of Popes. By 1761, the city of Rome with its thirteen standing obelisks contained more of these monuments than the rest of the world combined, Egypt not excepted. But even with these examples close at hand, the reproductions of hieroglyphs by Europeans were crude and show evidence of the belief in their esoteric nature. They have none of the grace and simplicity of what is perhaps the most beautiful system of writing ever devised. But such was the nonsense that surrounded the hieroglyphic script in particular, and ancient Egypt in general, that serious study of either was difficult if not impossible. In fact, ancient Egypt may have been the prototype of the "Orient"—mysterious, arcane, and inscrutable—that had simultaneously intrigued and repelled Europe for centuries.

The German Anastasius Kirchner (1602–80), otherwise a linguist of great ability and the author of the first Coptic grammar and vocabulary, was probably the greatest representative of the breed of symbolic interpreters. His translations of hieroglyphs, "based entirely on notions as to their symbolic functioning, are wholly wide of the mark, to the point of absurdity."[21] This inability to understand the structure of the language apparently influenced the ability to write the hieroglyphic script: European copyists, even with the originals in front of them, were unable to reproduce the symbols with much accuracy. And it is here that Niebuhr made his modest contribution to the study of Egyptology. The mere act of copying the hieroglyphs over a long period of time led him to several key insights into the script, including the difference between the script properly so-called and other symbols, and the direction in which animate objects faced as a key to the direction in which the text should be read. Surely, his suggestion that the need for greater familiarity with the script could be met by copyists such

---

20. M. le Comte de Palin, *Essai sur le moyen de parvenir a la lecture et a l'intelligence des Hieroglyophes Egyptiens* in *Memoires de l'Academie*, vol. 29, 1764. Quoted in Budge, *An Egyptian Hieroglyphic Dictionary*. Ernest Alfred Thompson Wallis Budge was curator of Egyptian and Assyrian Antiquities at the British Museum from 1894 to 1924, and an indefatigable worker in the field of Coptic, Arabic, Syriac, Ethiopian, and hieroglyphic texts, as well as a field worker in Egypt, among other places. His best-known work was probably the translation of *The Egyptian Book of the Dead*.

21. Ibid., 48.

The Antiquities of Egypt

as himself, was as fundamentally sound as it seems obvious today. And his statement that "We would be much better acquainted with the history of this interesting country if we knew how to read the inscriptions of the ancient inhabitants..."[22] was surely as unexceptional as it was true.

But a key development in the mid-eighteenth century had pointed the way to the future. In 1761, the year of the arrival of the Danish expedition in Egypt, the Abbé J. J. Barthélemy had demonstrated satisfactorily that the "ovals in Egyptian inscriptions which we call 'cartouches' contained royal names."[23] Another laborer in the field, Georg Zoega (1756–1809), accepted this view and pointed to what he believed to be the alphabetic nature of the letters. Unfortunately, so arcane seemed the subject that this promising lead was not pursued, and the existence of a hieroglyphic alphabet would await the work on the Rosetta stone many decades later. Then, in the early nineteenth century, Akerblad and Silvestre de Sacy would successfully produce demotic[24] vocabularies and relate them to Coptic letters. They were even able to read the demotic equivalents of Greek royal names, but they failed to apply the method to the hieroglyphic inscriptions.

Demotic was the vernacular of Late Egyptian, the everyday language of the New Kingdom and an intermediate stage in the development of the language, from Early Egyptian, Middle Egyptian, and Late Egyptian to Coptic, its final stage. The demotic script, from the Greek *demotika* or "popular," was a secular, cursive script barely recognizable in relation to its hieratic intermediate phase, let alone its original hieroglyphic form. However, the Coptic script was an almost total break with the past with the adoption of twenty-four letters of the Greek alphabet, supplemented by six demotic letters to render sounds not known in the Greek. This made the relation between the Coptic and demotic—and eventually the hieroglyphic—scripts, at first, very difficult to recognize.

It was only after the work of Young and Champollion that the fundamentally alphabetic nature of the hieroglyphs was finally recognized. But even then, as we have seen, the persistence of the notion that the hieroglyphic characters embodied arcane meaning was strong and enduring. The serious study of the hieroglyphic script had begun in scholarly circles in Europe in the middle of the eighteenth century. The key to opening its secrets had been unearthed, if it long remained unrecognized. Men like Akerblad, Young, and Champollion would later lay the foundation of the science of Egyptology in the first quarter of the nineteenth century. Dedicated men

---

22. *Travels*, vol. I, 201–2.
23. Budge, *An Egyptian Hieroglyphic Dictionary*, v.
24. See Davies, *Egyptian Hieroglyphs*.

followed and elaborated what they had begun, tireless laborers in the field like Birch, Lepsius, Brugsch, Chabas, Goodwin, de Rouge, and Wallis Budge. However, there were predecessors as well, and if Barthélémy, Zoega—and Carsten Niebuhr—were not giants, at least they were exceptional men, on whose shoulders those who followed them stood.

Niebuhr and the Hieroglyphs

Niebuhr's appreciation of the key to an understanding of the hieroglyphs was as sound as Wilkinson's, and then Petrie's 120 years later:

> For a complete enough collection of hieroglyph texts to be obtained, so that an explanation by scholars is possible, a traveler would have to spend a long time in Upper Egypt, and to copy not just fragments but entire inscriptions, of which I believe the ancient temples are full . . . Where is the European who, until now, has taken the trouble to gain the friendship of the common Arabs, and thus be able examine everything with them at a leisurely pace?[25]

Between 1821 and 1856, John Gardner Wilkinson would devote himself to just that task, making careful copies of inscriptions, reliefs and paintings that would form the basis of his monumental three-volume *Manners and Customs of the Ancient Egyptians*. Petrie's later success was in no small measure due to his willingness to work with the natives. "I am *ibn el beled*, a son of the country," Petrie explained, "all the Arabs are my friends and I know them all."[26]

Given the uncertain duration of their stay in Egypt and the firm instruction that they should not divert themselves from the ultimate goal of Happy Arabia, Niebuhr did not visit the monuments of Upper Egypt. But he spent the better part of the year in Cairo copying the inscriptions he found. The philologist on the expedition, von Haven, was not idle during the year in Egypt, and we will see that he made copies of the Arabic inscriptions on several bridges over the branch of the Nile leading to the pyramids at Giza. But there is little to suggest that he took much interest in the hieroglyphs. Niebuhr, on the other hand, took an early and keen interest. In his own words:

> I was not, strictly speaking, charged with research into antiquities. But when I was making the map of Cairo and I saw hieroglyphic inscriptions on many occasions, I decided to copy them to satisfy my own curiosity. The first inscription took me a great

25. *Travels*, vol. I, p. 201.
26. Quoted in Drower, *Flinders Petrie*, 50.

The Antiquities of Egypt

> deal of time, since the signs were unfamiliar to me. With the second I had less trouble, and in the end the hieroglyphic signs had become so familiar to me that I could copy them as easily as Greek or Cufic characters.[27]

His suggestion that the Coptic language was another key to an understanding of the hieroglyphs was also correct. As with the cuneiform inscriptions at Persepolis (see Chapter 14 below), however, his contribution would not be in his ability to decipher the language, but rather in providing a portion of the raw materials for scholars who would pursue the study.[28] His ambition was explicitly no more than that. For some explorers, to be first is everything and to be second nothing. Niebuhr, on the other hand, was quite willing to be a part of a larger enterprise whose cumulative effort would ultimately result in the deciphering of the script.

In vol. I of the *Travels* Niebuhr provides sixteen copper plates (Plates IV, XI, XXX–LXI, LXV, LXVI) with examples of hieroglyphs. The first, on Plate IV, were on the obelisk in the Hippodrome in Constantinople and we suspect that the copies were not made during his first stay in the city in August of 1761. But by the time he returned in February of 1767 he was well versed in the practice of copying inscriptions in the presence of a querulous and often hostile public. As we have seen, it was the red granite obelisk of Tutmoses III (1479–1425 BC), celebrating his crossing of the Euphrates in northern Syria.[29] Plate XI contains hieroglyphs on small scarabs and Plates XXXVII–XL are of inscriptions on mummy cases and small alabaster urns, or painted on wood or stone. But in between there are more substantial objects. Plate XXX shows "Hieroglyphen *auf einem Kasten von schwartzen Granit bey* Kalla el Kabsch in Kahira," or hieroglyphs on a black granite sarcophagus near the same area in Cairo. The sarcophagus had been shown by Pococke in his *Description of the East*[30] and Maillet had called it *la fontaine des amoureaux*.[31] It lay under the steps of the *madrasa* of Qaytbey near the ruined fort above the mosque of Ibn Tulun. Only one side was visible, and it was filled with water, being used as a watering trough for animals.

---

27. *Travels*, vol. I, 202.

28. See Budge, xxxii where, discussing the state of Egyptology in 1838, he states that "What was most wanted was good copies of texts on which scholars in every country could work..."

29. See chapter "To the Orient."

30. Pococke, *Description of the East*.

31. Maillet, *Description de l'Egypte*.

Niebuhr in Egypt

Niebuhr spent a great deal of time, and went to a great deal of trouble, to make the copy that appears in the plate. This was not only because of the difficulty of reproducing the signs, but because of the hostility of the authorities in the area. He took a *mulla* with him on the first attempt, but even this was not enough to save him from abuse by the *sarraj* in the quarter, and he was forced to beat a hasty retreat. It was only after repeated visits and small gifts that he was able, in piecemeal fashion, to complete the drawing. His conclusion from this experience stood him in good stead throughout the remaining five years of his travels: when approaching a task, he should avoid officialdom whenever possible and rely on his own courage, discretion, and self-control. The authorities were generally venal and while the lower orders such as this *sarraj* were capricious, at least they could be managed with humor and a judiciously-applied gift.

What appeared to be this same sarcophagus was mentioned by Lane, several years after its removal. Writing in 1835 on "The Periodical Public Festivals etc.," the following passage appears:

> Some of the people of Cairo say that a party of genii, in the form and garbs of ordinary mortals, used to hold a midnight "sook" (or market), during the first ten days of Moharram, in a street called Es-Saleebeh, in the southern part of the metropolis, before an ancient sarcophagus which was called 'el-Hod el-Marsood' (or the Enchanted Trough). This sarcophagus was in a recess under a flight of steps leading up to the door of a mosque adjacent to the old palace called Kal'at el-Kebsh; it was removed by the French during their occupation and is now in the British Museum.[32]

And so it is. The black granite sarcophagus of Hapmen, Twenty Sixth Dynasty (about 600 BC), sits in Room 25 of the British Museum, just as Niebuhr drew it. The following label is attached:

> Exterior reliefs and texts include figures of Isis and Hephthys at foot and head, Anubis (shown twice) and the four sons of Horus, *udjat* eyes and part of Chapter 72 of the Book of the Dead. The floor has the body of the goddess Nut, on the inner walls is a frieze of major Egyptian deities. Gift of His Majesty King George III, 1802.

Plates XXXI–XXXV show various views of another black granite sarcophagus[33] which, according to Niebuhr, had been brought from Upper

---

32. Lane, *Manners and Customs of the Modern Egyptians*, 422.

33. It also appears to be of the Late period and was probably brought from Saqqara. A figure similar to the large figure in Plate XXXIII appeared as Plate 33 of Perry's

Egypt by Osman Kikhya about twenty years before the Danish expedition arrived in Cairo. It was also to be used as a watering trough but, as it was being landed in Bulaq, it had shattered and the pieces placed around a tree near the port. Niebuhr noticed that there were differences in the attitudes of the figures in the hieroglyphs: some faced to the left and some to the right, which led him to speculate on the direction in which the text should be read. We now know this to be a key insight. He also had trouble making this drawing, including a first attempt, which was seized by the *sarraj* and not returned. But with the present of a small coin he was allowed to return and copy as long as he wanted. Plate XXXVI contains two broken, black schist obelisks shown as "A" and "B." Niebuhr says that "A" was in the threshold of a mosque in the Citadel of Cairo. It is of Nectanebo II[34] (360–343 BC), and is now in the Egyptian Museum. The figure "B" in the same Plate was of granite and appears to be of the 26th dynasty, perhaps of Psammetikos II or Apries. Niebuhr says the fragment he sketched was used as a step in front of a house near the "Kantaret Jedid." But it is now in London, also a gift of His Majesty King George III. The British Museum displays several other similar gifts of the same provenance, donated by George III. They were seized by the British after the defeat of the French in 1801 and include the Rosetta Stone.

Niebuhr, however, was not content to merely copy the hieroglyphs. As his familiarity with the signs and proficiency at reproducing them grew, he was led to speculate on the organization of the writing system.
Plate XLI contains his attempt to reduce the hieroglyphs to some formal order or system:

> I wish that scholars would pay particular attention to the characters, which form the hieroglyphic writing. Those who would apply themselves to this work should begin by collecting all the different characters and figures of which the script is composed. Since the inscriptions I copied were not equally well preserved, I have gathered together in Plate XLI all the letters or characters that appear clearly in my sketches. It would be easy to add to the list other well-drawn inscriptions; and if travelers in Upper Egypt were diligent enough to copy these ancient inscriptions, we would soon know all the different characters which

---

View of the Levant, although Niebuhr says that the accompanying text was different. The Egyptian Museum today has several fragments similar to these, although an exact match is not possible.

34. The last native ruler of Egypt, Nectanebo dedicated the obelisks to "his father, Thoth," the titular deity of Ashmunein, the capital of the fifteenth nome of Upper Egypt near present-day Mallawi. The fragments in the British Museum would seem to fit with a fragment in the Egyptian Museum in Cairo. See Habachi, *The Obelisks of Egypt*.

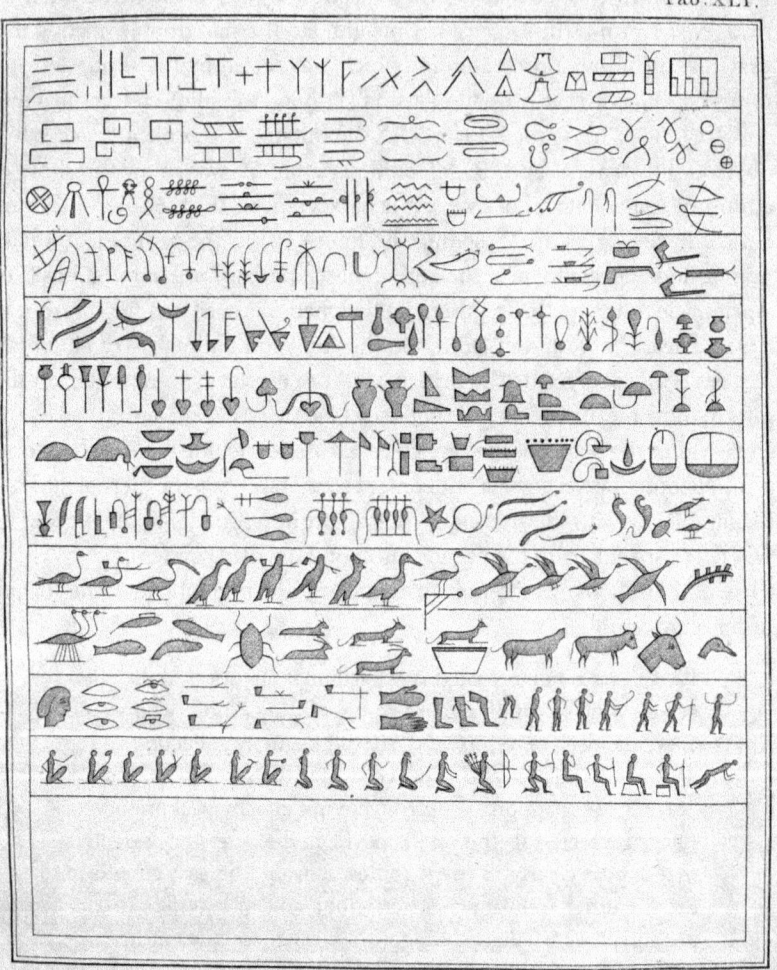

Verschiedene Zeichen und Figuren der hieroglyphischen Schrift.

constitute the hieroglyphic script . . . All this must guide the scholar who has the goal of deciphering the hieroglyphs.[35]

Even though he was only an amateur "Egyptologist," the mere effort to copy the hieroglyphs had already yielded important fruit, leading to his speculation on the rules of the language. As we saw above, while copying the sarcophagus at the *madrasa* of Qaytbey, he noticed that the same text was often written in a mirror image and that the direction in which, for example, a bird was facing held the key to the direction in which the script was to be read. He also noted the difference between the large figures, three of which appear in the sarcophagus near the *madrasa* of Qaytbey, and suggested that they were different from the smaller figures that constitute the script itself. This is, of course, correct as the large figures are representations of the gods Isis, Hephythis, and Anubis and the smaller characters are texts from the *Book of the Dead*.

Plate XLI is Niebuhr's presentation of the characters he copied, and reduced to a system. He was not a trained artist and the copies have little of the grace of the originals. But, from bottom to the top of the Plate, Niebuhr has done what Birch,[36] Champollion and others would do seventy-five years later with the first dictionaries of hieroglyphs. Champollion died before the dictionary could be published, but his editor followed his reasoning in adopting

> a methodical, or so to say, natural classification,' that is, he grouped into sections the figures of men, human members, animals, birds, fish, reptiles, plants, etc. This method was a modification of the system of arrangement of the words in their Vocabularies by the Copts, for Champollion argued that if the Copts, who are racially the descendants of the ancient Egyptians, and whose language is substantially the same as that of the ancient Egyptians, arranged their Vocabularies in this way, they must be reproducing a system that had been in use among their remote ancestors thousands of years earlier.[37]

Budge would later choose an alphabetical arrangement for his own dictionary, since he did not believe that Champolion's arrangement was suitable for a working dictionary, where the reader with a limited knowledge of hieroglyphs would find it difficult to get much help with even an ordinary historical inscription. An alphabetic arrangement, while it may not have been as "rational" or "natural" as Champollion's, nonetheless was seen as more useful to the student. However, Budge added just such a "List of Hieroglyphic Characters"

35. *Travels*, vol. I, 208.
36. See Budge, xxv.
37. Ibid., xxvi-xxvii.

as collected by Champollion—and Niebuhr before him—in the introduction to his dictionary. They include twenty-five classes of characters, from "I. MEN (Standing, Sitting, Kneeling, Bowing, Lying Down," through "IV. MEMBERS OF THE BODY," to "XXV. STROKES AND DOUBTFUL OBJECTS." Included in the twenty-five classes are 1,445 individual signs.

Niebuhr's Plate XLI is, on a smaller scale, and with admittedly less erudition, an attempt at the same system of organization. Of the twelve horizontal groupings or classes in the plate, the lowest, or first, includes men, sitting, kneeling, and lying down. It included nineteen signs, with significations we now know to be as determinatives or ideographs such as "invoke, address," "offering," "captive" and "soldier." The second grouping includes parts of the body such as eyes, arms, legs, hands and standing men. The third grouping is devoted to animals, although it includes a bird, fishes, and a scarab. The fourth is devoted to birds, although the last of the fifteen signs may or may not be "backbone," a part of the body. In the fifth, Niebuhr's classification system begins to break down as it includes signs gathered by Budge under the heading of "sacred vessels," "trees," "furniture," "ships," "birds," "buildings," and "weapons."

In Niebuhr's sixth grouping or class the signs have become inanimate and abstract, although some are still recognizable objects such as vessels, weapons, and parts of buildings. The seventh and eighth are much the same, with milk pots, jars, and crowns. The ninth through twelfth are increasingly linear and abstract, with a range of significations, when recognizable, that are too wide to permit generalization. What Niebuhr has done, however, is to collect together over 300 individual signs, arranged in a "natural" order in twelve classes. Most of the 300 can be collated to the 1,445 listed by Budge, although some are duplicates and others are doubtful or unrecognizable.[38] However, it was a liberal reading of his responsibilities that this traveling cartographer would make the attempt to contribute to the collection of signs necessary for the scholars of Europe to decipher the hieroglyphic language. When Budge states that "The first printed list of Hieroglyphs was pubished by Champollion in his Grammaire Egyptienne, Paris, 1836, and contains 260 hieroglyphs,"[39] he probably meant lists with corresponding meanings. But if Niebuhr's effort did not lead directly to decipherment, his recognition that Coptic was the key to the language and his arrangement of the signs in categories that intuitively mirrored the Coptic lexicons surely pointed the way to the future. We will see below that his contribution to an understanding of the cuneiform script was not so modest.

38. Of the 319, I was able to identify 270 among Budge's categories.
39. Budge, lxx.

The Pyramids of Giza

Niebuhr's interest in the antiquities of Egypt was not confined to the hieroglyphs. He also visited the Giza plateau with the intention of applying his surveyor's skills to a measurement of the two largest pyramids. On his first attempt, he and Forsskal went as far as the little village of Giza with a group of Europeans who had rented a country house in the area. We should not be surprised to learn that only Forsskal among the members of the expedition would accompany him to the site of these "prodigious masses." Taking two Bedouins they had found on one of the bridges[40] as guides, they set off along what is today the Pyramids Road. It is shown in the bottom right corner of his map leading to the west and off the page. The Bedouins were mounted on horses, Niebuhr and Forsskal on asses. Arriving at the foot of the plateau on which the pyramids stood, Niebuhr set up his astrolabe and, laid off a base of 203 feet and, by means of triangulation, calculated that the base of the first pyramid was approximately 200 feet above the surface of the river.

This first visit was not without drama. Their guides were not from the area and a horseman, "a young Sheikh," soon arrived and demanded that he be their official guide. After rancorous discussion with the sheikh, during which they flatly refused to pay what he demanded, Niebuhr and Forsskal decided that retreat was the better part of valor. However, not wanting to leave empty-handed, the Bedouin first took Forsskal's turban and then attempted to seize Niebuhr's astrolabe. If anything could rouse the ire of this most rational of men, it was a threat to his precious instruments. Niebuhr describes the sequel and the dangerous pass to which it brought him:

> Not quite so coolly as Forsskal, I seized him by the large, loose cloth he wore around his body, and as he kicked his horse without holding onto the reins, the animal started and the Arab fell to the ground. This would have placed me in great danger; for the young man believed himself to be so affronted at being seen by peasants to be thrown from his horse by a Christian, that he took one of his

---

40. Von Haven later copied the Arabic inscriptions on several of the bridges over the "branch" of the Nile flowing between Giza and the pyramids and they were later translated by Reiske. It appears that there were four bridges over the river in the vicinity of Giza: two large ten-arched bridges of sixty and fifty double paces (or 291 and 243 English feet) in length, and connected by a masonry dam 1,500 double paces in length. Niebuhr saw two other smaller bridges in the vicinity of Giza, one of five and the other of three arches. From the inscriptions on one of the smaller two, it had been built in 1087 AH (or 1676 AD) by a certain Hussein Pasha. Von Haven also copied the inscriptions on the larger, and they indicate that both were dedicated in the reign of Qaytbey, one in 880 AH (or 1475 AD) and 884 AH (or 1479 AD). The Arabic versions of the three inscriptions are provided by Niebuhr as footnotes.

pistols and placed it against my chest. I could do nothing, and at first thought that I was near my end. But the pistol was probably not loaded. The other Arabs sought to appease the Sheikh, and at last I managed to do so by giving him half a crown.[41]

They later returned in the company of a French merchant resident in Cairo and a large retinue of Europeans. Niebuhr was able to make a cursory survey of the "first" and the "second" pyramids,[42] although without the precision he would have wanted. He established that the four sides of the first were oriented to the four cardinal points of the compass. But with little time and surrounded by onlookers and curious Arabs, he could measure the sides only by pacing them off. After establishing a survey chord, he determined that the distance between the nearest corners of the two pyramids was 565 feet, Then by triangulation he established that the base of the two pyramids differed by thirty-four feet in height. After a series of calculations he came to the conclusion that height of Cheops was 440 feet and that Khephren is 443 feet. The accepted figures today are 481 feet for Cheops and 471 feet for Khephren. Niebuhr's overall error was probably due to the hurried nature of his survey. However the fact that he shows Khephren to be higher than Cheops may be due to the fact that the accumulated debris near Cheops—according to Niebuhr "a great pile of refuse and sand"—was greater than near Khephren, and the side he measured correspondingly shorter. That would lead to a correspondingly shorter calculation of the height. The relationship between the two is shown graphically in the figure "E" of Plate V. The more mathematically sound way would have been to measure the angle at one of the corners where the debris did not cover the limestone plateau that constituted the base. However, this would have required an "extended" corner so that he could have seen the both the top and the limestone base at the same time. This, again, was not possible due to time constraints.

Niebuhr's measurement of a side of the second pyramid, "141 double paces or 705 feet" is in agreement with the 215 meters accepted today. However, the first pyramid's "142 double paces or 710 feet" is shorter than the accepted 755 feet. This would lend credence to the probability that his measurement was taken from a place above the limestone base of the structure.[43] So, the figures did not have his customary precision, and he was tempted not to publish them. He would have preferred to begin from the limestone

---

41. *Travels*, vol. I, 192.

42. Although Herodotus correctly identified the builders as Cheops and Khephron, his information was not confirmed until Egyptologists were able to read the names in texts.

43. For the dimensions, see Lehner, *The Complete Pyramids*.

base at one of the corners, rather than a hastily stepped-off estimate of the length of the sides. However, in the interest of providing yet another opinion, he provided the calculations in his account, with an apology for their inherent limitations.[44]

Unlike most visitors, who confined their interest to the first pyramid, Niebuhr climbed nearly to the top of the pyramid of Khephren to satisfy himself that the material of the casing was the same limestone of the rest of the structure. Several accounts had reported that the pyramids were originally encased in marble, and at least one that they were covered with a coat of plaster. Niebuhr brought back with him a small piece of the limestone casing that still exists near the top of the pyramid. Then, like most visitors, he made the ritual climb of Cheops and afterward entered the King's Chamber, with an apology for not having found the chamber that Davison later discovered.[45] Near the smaller, or third pyramid, of the Giza complex, later confirmed as that of Mycerinos, he noticed the large number of granite blocks but did not see evidence, such as the oblique dressing, which suggested that they once served as the outer casing.[46] As we have seen, Herodotus had already called attention to the fact that the lower portion of this pyramid was encased in red granite.

Niebuhr also noticed a phenomenon that had an indirect bearing on his biblical research: small snail-shaped "petrifactions" in the limestone blocks that made up the pyramids. They were similar to those he had seen in the Muqattam range to the east of Cairo. Earlier travelers had also noticed the petrifactions, including Strabo[47] who thought they were the remains of the lentils fed to the workers on the pyramids. But Niebuhr was led to questions that were not those of the average traveler:

> This is cause for reflection on the antiquity of Egypt. For, how many years must have passed before a sufficient number of little

---

44. Niebuhr, incidentally, shared his training and instincts as a surveyor with none other than Flinders Petrie, who would contribute so much to the study of ancient Egypt in the next century. It was Petrie's early excursions to the monuments of England, including Stonehenge, that prepared him for work in Egypt and gave him "the ability to measure quickly and accurately and to gauge distances by eye, and a visual memory that were all developed in him to an unusual degree." Drower, *Flinders Petrie*, 24.

45. Niebuhr cannot be serious in this apology, but seems only to be saying that he was careful and cannot be accused of careless observation. He was nothing if not careful, but serious pyramid exploration was not among his priorities. Davison discovered the lowest of the five stress-relieving chambers in the Great Pyramid, that of Cheops in July 1765, before publication of the *Travels*. See Lehner, *The Complete Pyramids*, 45.

46. Sixteen courses of undressed red granite lay at the bottom of the pyramid. The upper part was cased in limestone. See Lehner, 134.

47. Strabo, *Geography*.

snails were born and died, for these mountains to have reached their present height? How many years must have passed before Egypt became dry, especially if in ancient times the water receded as slowly from the coast as in the last thousand years? How many years must have passed before Egypt was populous enough for one to think of building the first pyramid? How many years must have passed before the number of great pyramids that we still see in Egypt were built?" And yet we still don't know with certainty in which century, and by whom, the last was built.[48]

These were important geological and historical questions and it is clear that Niebuhr possessed the scientific temper and the scholarly instincts that animated the best of the Enlightenment seekers. Without an understanding of the principles of petrology, he still wondered as to the length of time involved in the formation of the sedimentary rock with organic constituents, such as that from which the blocks of the pyramids were quarried. The question, seemingly naive in its formulation, was surely intuitively perceptive in that most of the limestone occurring throughout the world is the product of decomposed shell or coral imbedded in inorganic sedimentary deposits. And even without a knowledge of modern geology, of stratification or upthrusting by tectonic forces, he would here—and later in the vicinity of the Red Sea—speculate on the presence of marine fossils in areas that were far above the current level of the sea.

It was a problem whose solution seemed to some in the eighteenth century to be satisfied by the "diminution of waters" theory: that is, if the biblical flood had once covered the face of the earth, the waters had since receded and this would explain the evidences of marine life in areas that were now dry. The seventh chapter of Genesis had been clear:

> 12 And the rain was upon the earth forty days
> and forty nights . . .
> 19 And the waters prevailed exceedingly upon
> the earth; and all the high hills, that were
> under the whole heaven were covered.
> 20 Fifteen cubits upward did the waters prevail; and the mountains were covered.

So, there appeared to be nothing inherently contradictory between these fossil remains and the biblical account. But the times involved in the process were unknown although, as we have seen above, the Bishop of Ussher had determined that the cataclysm took place in late November 2349 BC. It would be another sixty-nine years before Lyell published the first of the three volumes

48. *Travels*, vol. I, 200.

The Antiquities of Egypt

of his *Elements of Geology* in which he suggested that the processes involved were slow and gradual, rather than cataclysmic, and that the time spans involved should be measured in millions, not thousands, of years. But with this speculation Niebuhr is already questioning the accepted wisdom. We will see the same questions later in his investigation of the tidal movements of the Red Sea and the story of the passage of the Israelites out of Egypt.

Niebuhr's musings about the development of Egyptian civilization and the antiquity of the pyramids are at the other end of the spectrum of geological time and dealt more with the pre-history of the Nile valley. The pre-dynastic period in Upper Egypt is dated by Hoffman[49] at approximately 5500 BC, with the proto-dynastic period beginning around 3100 BC. Today, we accept the period of the building of the Pyramid complex at Giza as that of the 4th Dynasty, from 2575–2465 BC. But Niebuhr deals also with more recent phenomena, such as the location of certain ancient cities of the country. As indicated above, an ongoing debate[50] raged in England about the location of the ancient city of Memphis. To Niebuhr, based on his reading of Pliny, Abul Feda, Sherif ed-Dris, and the more recent testimony of Maillet, there was really no question that Memphis was located on the west bank of the Nile, a short day's journey to the south of *Masr*, or Cairo. He and they were right.

Michaelis and the Language of Ancient Egypt

A puzzling, although perhaps understandable, lapse in the instructions to members of the Danish expedition was the absence of questions about the ancient language of Egypt. Michaelis was immensely erudite in Oriental languages—Hebrew, Chaldean, Aramaic, Syriac, and Arabic—but seemed to have little curiosity about other languages that must surely have had some influence on the Hebrew that appeared in the Bible. He can be forgiven for being unfamiliar with the Old, Middle, and New Egyptian and perhaps demotic stages of the language, but, as we have seen, a Coptic grammar and vocabulary had been available in Germany for nearly a century. The avowedly biblical nature of the expedition may explain this curious myopia: there was a sacred language—Hebrew—and its cognates, and then there were the profane languages that merited, at best, cursory attention. But to a serious scholar of the Hebrew scriptures, surely the purported sojourn of the children of Israel in Egypt would have led to borrowings that might have appeared later in the Old Testament. Michaelis, however, confines his attention to the "Arabic, Ethiopian, Syriac, Chaldean, Hebrew, and Samaritan"

49. See Hoffman, *Egypt before the Pharaohs*.
50. See Chapter 4.

that were necessary, he thought, to an understanding of the totality of Oriental languages. We would probably say that this was enough, and more than enough, for a group of travelers most of whom were amateurs in the study of languages. The hieroglyphs were another problem altogether, and it is tempting to suggest that the reason for Michaelis's lack of interest is that they were profane, and therefore not a part of God's plan for mankind. Not all Europe shared this religious provincialism and we have already seen that in 1761, a contemporary of Michaelis, the Abbé J. J. Barthélémy, was demonstrating satisfactorily that the cartouches contained royal names. However, it would be another seventy years before this key, along with a knowledge of demotic, was used to unlock the secrets of the language.

However, this is probably a case of being wise after the fact. Not only was there little understanding of the stages through which the language of the ancient Egyptians had passed, but the script in which the language was written was undeciphered and unknown. Still, one would have expected a man as schooled in language as Michaelis, with his knowledge of the relationships between some of the languages of the ancient Near East, to show some curiosity about a language that must have played a part in the development of the languages of the neighboring Semites. Here again, the limitation that belief imposed in the thinking of the Christian scholars of Oriental languages may have been at fault. It was not many years before, as we have seen, that Hebrew was considered to be the language of God, the first language of mankind, the most perfect of all languages, and not susceptible to the same analysis as other languages. Michaelis had moved beyond these notions, but his belief in the sacred nature of the Bible appears to have been a tether to those earlier ideas that he was unable to break.

Even in the early part of the twentieth century the relationship between the language of the Egyptians and the Semitic languages, though of great philological importance, had not been adequately examined. This was because the scholars in the separate disciplines lacked the knowledge of the others to make the exchange fruitful. For a period of time toward the end of the nineteenth century, the belief that Egyptian was a Semitic language acquired currency among scholars. In 1892 Erman presented a paper in which he made a systematic analysis of the two languages and a provided a list of words common to both. Brockelmann and Brugsch accepted this hypothesis, while concluding that Egyptian had gone its own, separate way thousands of years before. A troubling factor was the apparent lack of the triliterality, so familiar to the student of Arabic. Budge's own conclusion, based on a lifetime of work in the Egyptian hieroglyphs, was that, while there were undoubtedly many Semitic words in the language, Egyptian was fundamentally not a Semitic language. Instead, it was a Hamitic language developed by the African people who inhabited the Valley of the Nile. The words used to express the fundamental beliefs and relationships, feelings and beliefs were African and foreign to those of the Semites, in much the way that Farsi remains structurally an Old Persian and Indo-European language, with heavy borrowings, along with a script, from Arabic, a Semitic language.

The terms used in the discussion—Semitic and Hamitic—may be passing out of fashion, and it is tempting to think that it is because of their biblical origin. They were artificial constructs that may once have had utility. But, based as they were, on the biblical tale of the dispersion of the sons of Noah after the flood, especially with the politically explosive curse on Ham (Genesis 9:24–25), they are now subsumed in other, more benign constructs and labels. Both Egyptian and the Semitic languages are now considered to be Afroasiatic languages. The new labels may reflect a better understanding of the structural differences as well as the relationships between the languages of the ancient Near East, including that of Egypt. A problem with the new grouping may be its scope: it includes languages spoken from Lake Chad in West Africa, to the Berber regions of North Africa, to Egypt, then east and south to Ethiopia and Somalia in the Horn of Africa down to the foot of Mt. Kilimanjaro, across the Red Sea to the Arab peninsula and then, finally north into Western Asia including Palestine, Syria, and Iraq. New languages are continually being discovered and the classification scheme is continually being updated in response to "an increasingly amorphous heap of unassimilated information."[51] But even with the new scheme, some old labels remain: "Semitic," after the oldest son of Noah, "Cushitic," after a son of Ham, "Canaanitic," after another son of Ham, and even the term "Hamitic" itself still occasionally appears.

It was men like Niebuhr, laborers in the field, who showed the way to the future. It was not the first time that Niebuhr's curiosity carried him beyond the narrow frame of reference of the Danish expedition, nor would it be the last. His conclusions were often diametrically opposed to the preconceived ideas of Michaelis. This cannot have been welcome to the older man. So it is probably no surprise that, in the end, Michaelis took little interest in the labors of the investigator he had been so instrumental in sponsoring.

---

51. See the website on The Afroasiatic Index Project.

13

To Suez and Sinai

> The Arabs in the vicinity of Tor . . . permitted no caravan from Suez to Cairo to pass peacefully . . . Such was the feebleness of the Government of Egypt, and of their mighty sovereign, the Sultan . . . that one small Arab tribe dared interdict the traffic between the cities in the area. This kept us in Egypt for almost an entire year. (*Travels*, vol. I, 210).

ON THE 27TH OF August 1762 a cannon shot rang out from the Citadel in Cairo announcing that the returning pilgrim caravan had reached the Birket al-Hajj. It was a sign, eagerly awaited by the little band of northern Europeans, that the feud with the Arabs had been reconciled and that it was, therefore, probably safe to travel. We have already seen the dangerous state of relations with the Bedouins of the Hejaz. But there was another complication as well. The year before, the Arabs in the vicinity of Tor, who normally participated in the carrying trade between Suez and that city, had been unable to resist the prospect of plunder when a ship carrying corn to the Holy Cities had put into the port to take on water. They had seized the boats of the fishermen in the area, captured the ship and carried off the cargo. The act caused uproar among the merchants of Cairo. They demanded the return of the cargo, which was of course refused. And as long as the dispute continued, no caravan was permitted to pass undisturbed overland, and no ship was safe in the Gulf of Suez, since Tor was a regular stop for water. But a solution was clearly in the interests of all concerned: of the government, anxious about supplies to the Holy Cities; of the merchants, whose trading activities had been curtailed; and, not least, of the Arabs themselves whose actions had put an end to their own carrying trade. Besides, the booty had nearly all been consumed. "In a word, everyone wanted peace."[1] The Amir of the returning Hajj caravan apparently brought the parties together and brokered the peace. The merchants

---

1. *Travels*, vol. I, 210.

To Suez and Sinai

of the city were back in business and they advised the Europeans that they could now make preparations for departure.

Whatever the reason for the extended stay in Egypt—whether because they were awaiting word on their own internal problems from Copenhagen as Hansen suggests, or because of a serious problem with the Bedouins of the Hejaz as Jabarti reports, or because of this affray with the Arabs of Tor as Niebuhr states (and it may have been a combination of all three)—the expedition had been in Cairo for nearly a year and some members were anxious to be on their way. They moved off in some haste on the 27th to the gathering place of the caravan to the northeast of the Citadel, although it was not until the afternoon of the following day that movement of other groups indicated that departure was immanent. In spite of the prolonged stay in Egypt, they soon discovered that their preparations were inadequate to travel in the desert. They had provided themselves with the usual paraphernalia recommended by other travelers: ample food, a tent, beds, tinned copper cooking utensils, candles, condiments, leather bags for drinking water, and, a commodity that did not form part of the kit of most travelers, great glass flasks carrying their wine. They were to learn that the flasks, each containing the contents of about twenty bottles, were too easily broken in the jostling of the animals as they were loaded and unloaded. Waxed skins served much better, all the while not imparting the least unpleasant taste to the contents. But, more importantly, they did not take a native servant with them and had only Berggren, their Swedish servant, a Greek cook, a Greek interpreter who was also a renegade, and a young Jew from Sana who had taken the opportunity of traveling with Europeans to accompany them to that city. Neither Berggren nor either of the Greeks had the slightest knowledge of the desert and the renegade, in spite of having embraced Islam, suffered from the suspicion that always attaches to those who have abandoned the faith of their fathers. The Arabs supremely despised the Jew.

Their route would be direct to Ajerud, just short of Suez, over the well-beaten caravan path. Due to the hurried departure from Cairo, the caravan was relatively small, consisting of only 400 camels carrying mainly corn and material for the shipyards of Suez. Niebuhr and his companions tried always to stay in the middle of the caravan, as stragglers ran the risk of being robbed by Bedouins along the way. The Arabs in the company were poorly armed with a motley collection of matchlocks without wicks and broken or rusty sabers, and so afforded little protection. Forsskal, von Haven, Kramer, and Baurenfeind, now outside of Cairo, were mounted on horses. Niebuhr alone chose a dromedary and, laying his mattress over the saddle, rode in much greater comfort than the others.

They traveled until late that night before pitching camp at a little place called "El firn bebad." They were up again before dawn the next morning, marching for over five hours and stopping only long enough to eat. They were back on the way after the short break and marched, Niebuhr informs us, a total of seven and one quarter leagues before pitching camp just before sunset. Assuming three and one quarter English miles per league, the seven and one quarter leagues in approximately nine hours would be the equivalent of about two and one half miles per hour. This would be the pace of a loaded camel. Those who were riding horses would have to accommodate themselves to this pace and, in places where it was considered safe, could ride ahead and wait for the remainder of the caravan to catch up with them. This first experience of travel in the desert would be important as Niebuhr would later measure distance by the average pace of the slowest animal in the company. He provides little detail other than that the first stop was near a mountain called "Wehbe" and the second near a mountain called "Taja." In fact, except to connoisseurs of dry watercourses and low-lying ranges, the track between Cairo and Suez is relatively featureless. To the south, there was only an extension of the Muqattam culminating in the looming presence of Jebel Ataqa within a few miles of Suez, and to the north only an occasional elevated plateau. As his skills in desert travel developed, Niebuhr would gradually acquire the ability to recognize geological and geographical features at a glance, and the painstaking habit of mind that this required.

On the 30th of August they were up just after midnight, and by 5:30 had reached Ajerud, a ruined Turkish fort where, nonetheless, potable water could still be found. It was the last place the pilgrims camped in Egypt on their way into Sinai, and the first place they stopped on their return.[2] At this natural boundary, a kind of checkpoint separating Egypt from the wilderness, Niebuhr's mind was fixed elsewhere than on the progress of this little caravan in 1762. Given the purposes of the expedition, it would be unusual if he were not aware of the fact that they might be retracing the steps of the children of Israel in their flight from Egypt, one of the most memorable stories in the Old Testament. He devotes a chapter in the *Description of Arabia* to "The Desert of Mount Sinai" where he collects his thoughts on the event as recorded in the Bible:

---

2. Pococke says that there was a Turkish garrison there in 1739. In 1816, Burckhardt spent three days inside the walls to avoid being plundered by Bedouins in the area. In 1839 Robinson and his party only had a glimpse of *Ajrud*, "a square fortress with a well of bitter water two hundred and fifty feet deep, built for accommodation and protection of pilgrims on their way to and from Mecca." It has recently been excavated by the Egyptian Ministry of Antiquities and the circular pit of the well and the rough outline of the walls of the *qalaa* have been exposed.

> There can be no doubt the children of Israel actually crossed the Red Sea; but as several thousand years have elapsed since the event, and before careful research on the place where they passed was even considered, it will be difficult to determine the place with precision.[3]

As we might expect, his speculations were those of a practical man: where the Israelites gathered—probably not far from the place where caravans assembled in 1762; how they must have been led by Moses, their *caravan-bashi*—the company must have included men who knew the routes and would not allow themselves to be led "willy-nilly" to a crossing point that made no sense; where they stopped—Ajerud is a candidate; and where they crossed—to the north of Suez, not south and near the Wadi Bedea as some European scholars, Michaelis among them, maintained. To Niebuhr, there was no question as to the truth of the biblical story and, although the topographical detail as it is told in Exodus and Numbers was scanty, he nevertheless speculated on the relevance of the Scriptures to this journey in 1762. At Ajerud, this border post at the beginning or end of the wilderness depending on the direction of the caravan, he thought of Numbers 33:6:[4]

> And they departed from Suc'coth, and pitched in E'-tham, which *is* in the edge of the wilderness.

and Exodus 13:20:

> And they took their journey from Suc'coth, and encamped in E'-tham, in the edge of the wilderness.

They watered the animals and, unlike the pilgrim caravans that turned north from Ajerud to pass around the upper reaches of the Gulf before striking straight across the Sinai to Aqaba, the little caravan turned southeast toward the brackish wells of Suez. Like the children of Israel, they turned towards the water (Exodus 14:2):

> Speak unto the Children of Israel that they turn and encamp before Pi–ha–hi'–roth, between Migdol and the sea, over against Ba'–al–ze'–phon: before it shall ye encamp by the sea.

They covered the remaining distance to the city in an hour. By Niebuhr's calculation, the caravan had covered thirty-two German leagues

---

3. *Description of Arabia*, 348.

4. The citations here are in the English of the King James Bible, although Niebuhr would probably have thought in terms of Luther's German Bible. Niebuhr was not a student of Hebrew and this Bible "of the people" would probably have been more familiar to him than the original Hebrew of the scholars.

and forty minutes, or twenty-three German miles from Cairo. The distance was equal to today's 106 miles, or just under 171 kilometers. He found the latitude of Suez to be 27° 57', or 8' 37" south of that of Cairo.

Suez

Located on the western shore of the gulf, Suez in 1762 was a squalid little town, completely open to the east or gulf side and protected to the north and west only by poorly maintained walls. The primary industry was shipbuilding although the materials had to be brought overland from Cairo, greatly increasing the cost. The vessels built in Suez, called "Cairo ships," trafficked only as far as Jidda and some twenty vessels each year carried corn and passengers to that city, often with a stop in Yanbu, the port of Medina. With the exception of a few public *khans*, Suez was meanly built and the small population included a few Greeks and Copts who subsisted primarily on the fish they caught in the waters of the gulf. Everything else came from Cairo, Gaza, or Mount Sinai. The lack of good water left the inhabitants—not to mention the bey and his ragtag collection of household troops—at the mercy of the Arabs in the surrounding areas. The Arabs could bring the city to its knees simply by refusing to provide the water that was their main transport commodity, or by preventing the residents of the city from approaching the outlying wells. The surrounding area was sterile, utterly devoid of ". . . plants, much less trees, gardens, meadows or fields."[5]

It had not always been so humble. The ruins of Kulzum[6] ("the sluice" in Greek) since it was at the head of the canal which led to the Nile and the *khalig* (see chapter on Cairo) lay just to the north of the city. As we have seen, it was dug by the Pharaoh Necho in the sixth century BC, and after the Arab conquest it had been repaired by Amr ibn al-Aas to carry corn from Fustat to the Red Sea and the Holy Cities. But there was later concern about its possible use as an invasion route into Egypt, and the canal had gradually been allowed to fall into disrepair. By 1762, according to Niebuhr, all traces of it were gone, unless a long, narrow wadi between the city and the Bir Suez might have once marked the bed. It filled with water after the infrequent rains in the area, and the inhabitants of the city were able to briefly fill their water skins. Afterward, there was a brief riot of desert vegetation before the wadi, again, reverted to its dry state.

5. *Travels*, vol. I, 222.
6. There is still a Shari tel Kolsum in Suez today but it is all that remains of the name. The *tel*, or mound, marking the ruins of the old city, was leveled years ago, along with the Jewish *madafin*, or tombs, near the town.

Suez would be their base of operations for the next month and the expedition made arrangements for an extended stay. From Suez, there was no question of their proceeding overland—as Christians they could not travel through the Hejaz—and they would have to go by sea to Jidda before proceeding onward to the Yemen. It was now the end of August and the last ship for the south would leave in early October to catch the tail end of the prevailing northerly winds. But during this short interval of just over a month there were two tasks that would occupy them, both related to the biblical research that was the primary responsibility of the expedition. The first was a careful examination of the northern reaches of the Red Sea, where Scripture stated that Moses crossed with the children of Israel and where the waters swallowed up the host of Pharaoh.

The question as to where they had crossed the Red Sea had exercised scholars for centuries, and Michaelis hoped that careful research conducted by the expedition might shed some light on the details of the actual crossing. As we have seen, Michaelis himself was certain that Wadi Bedea, south of Suez, was where the crossing had taken place. Question II in the *Fragen* had framed the questions that Michaelis sought to answer: "On the ebb-tide at the northern extremity of the Red Sea. On the timing and measure of this ebb tide. On the depth & sea floor at the place where the Israelites passed. On coral." As we might expect, he sought natural explanations for the extraordinary events reported in Exodus 14:21, 22 and 14:27:

> 21. And Moses stretched out his hand over the sea; and the LORD caused the sea to go *back* by a strong east wind all that night, and made the sea dry *land*, and the waters were divided.
> 22. And the Children of Israel went into the midst of the sea upon the dry *ground*, and the waters were a wall unto them on their right hand and on their left . . .
> 27. And Moses stretched forth his hand over the sea, and the sea returned to his strength when the morning appeared; and the Egyptians fled against it; and the Lord overthrew the Egyptians in the midst of the sea.

Niebuhr was to observe and record the timing of the high and low tides, inquire of the inhabitants or sailors about the possible occurrence of a double ebb tide, look into the existence of an underwater isthmus in the vicinity of Suez, and look for coral and its possible effect on the feet of the sandal-shod or barefoot Israelites. Affirmative answers to these questions—there *was* the possibility that the parting of the waters took place naturally in the shallow, northern reaches of the Red Sea through a combination of wind and tides; an underwater isthmus or sand bar *might* have eased the passage of the

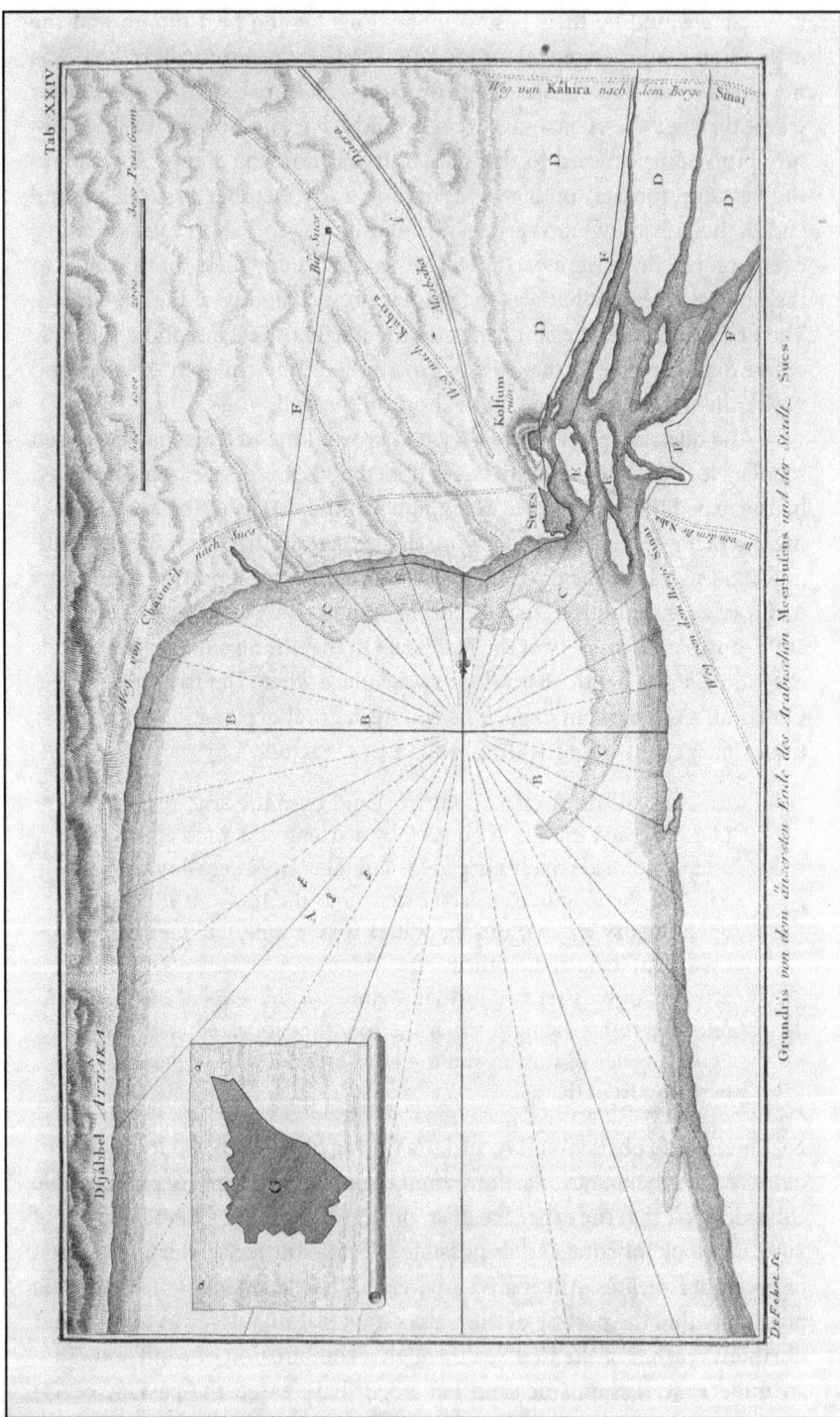

Israelites; and the absence of coral *would* have made the passage less arduous—would lend scientific plausibility to an event which Exodus suggests was brought about only by divine intervention. The congruence of science and religion would, thereby, be demonstrated.

The second task had to do with reports of unknown inscriptions in Sinai that had circulated in scholarly circles in Europe for several decades. They had caused an immediate stir, as it was felt the inscriptions might be the rude beginnings of the square Hebrew script, scratched by the Israelites during their wanderings in the wilderness. The existence of inscriptions had been reported in 1743 by Richard Pococke in his *Description of the East*[7] and, most recently, by Robert Clayton, Lord Bishop of Clogher in Northern Ireland. Clayton had not seen the inscriptions himself but had seen the journal of the Prefect of the Franciscans in Cairo who had seen them in 1722. Clayton had written to the Society of Antiquaries in London, enclosing his translation of the journal and expressing the hope that the Society might look into the matter:

> Being possessed of the original journal from Grand Cairo to Mount Sinai mentioned by my worthy friend Dr. Pococke in his Travels Through the East, which was written by the Prefetto of the Franciscans in Egypt, who set out from the Convent de Propagande fide at Grand Cairo A.D. 1722, I think proper to communicate to you a translation of it, in hopes of exciting you who are now erected into a Society of Antiquaries, to make some enquiry into these ancient characters which, as we learn from it, are discovered in great numbers in the wilderness of Sinai, at a place well known by the name of Gebel el Mokatab, or the Written Mountains, which are so particularly described in this Journal, that it is impossible for an inquisitive traveller to be at a loss in his searches after them. By carefully copying a good quantity of these letters, I should apprehend, that the ancient Hebrew characters, which are now lost, may be recovered.[8]

Such a possibility was clearly was of interest to Michaelis and Article three of the "Instructions" in the *Fragen* directed that the expedition should proceed "across Egypt to Mount Sinai, to the country in the vicinity of the Gebel el Mocatab . . ." Once there, they should make detailed copies of the

---

7. Pococke, *Description of the East*.

8. Clayton, *A Journal from Grand Cairo to Mount Sinai and Back Again*. Clearly, the report, if not the translation, was available to scholars prior to the 1810 date of publication. The translation was published as part of a little volume also containing a *Journey from Aleppo to Jerusalem at Easter AD. 1697* by Henry Maundrell and *A Faithful Account of the Religion and Manners of the Mahometans* by Joseph Pitts. Pitts's tale of capture and eventual escape from the Muslims had captivated Europe.

inscriptions and, if circumstances permitted, plaster casts should be prepared by Professor von Haven, with the aid of Baurenfeind the artist (Article Twelve). If the importance Michaelis attached to this task was not already clear enough, Article Forty-three directed Baurenfeind to prepare copies of the inscriptions "under the inspection of Professor von Haven, with the greatest fidelity and exactitude." The results, if satisfactory, would be of interest not just to Michaelis but to all of Europe.

There was also great interest in the sojourn of the Israelites in the wilderness of Sinai. Michaelis had included several specific questions about Sinai in the formal list prepared for the travelers. They included queries about the springs of Sinai and the purported stream in the valley of Rephidim (Question XX), the various sorts of "Arab manna" (Question XXVI), the manner of preparing manna (Question LIII), and the height of various mountains in Sinai (Question LXV). But even without specific queries, it was historic ground over which the expedition would pass, and the impressions of seasoned travelers—if nothing else, reports on the terrain, the availability of water, the way the Bedouins lived—could not help but contribute to an understanding of Sacred Scripture. Finally, when they reached the southern part of the peninsula, there was the reported collection of Oriental manuscripts in the Monastery of St. Catherine. Then, as today, the library in the monastery had a collection of medieval manuscripts that was second only to that in the Vatican Library. The *Fragen* called particular attention to the importance of written documents:

> the writings which we principally have in view are those dealing with Natural History, Geography, and History; old manuscripts of the Greek and Hebrew Bibles, or in every case old Arabic versions of the holy book.[9]

One of the responsibilities of von Haven was to purchase manuscripts and, as we have seen, he had already bought a large number of Arabic documents in Istanbul and Cairo. There could be no question of purchases at the monastery, but he was also to look for versions of the Bible with which the translations into European languages could be compared. Von Haven had been instructed to note variants in the texts (Article Forty-one of the "Instructions") and, if there was any question as to their meaning—or any doubt as to the importance of his research—he should have his notes countersigned by another member of the expedition. The results would have obvious importance in understanding the document as it was originally written.

However, time was short. The members of the expedition would have to find the "Gebel el Mocatab" and make copies of the inscriptions, travel to

---

9. *Fragen*, XXVI.

the monastery, allow enough time to examine the manuscripts in the library, and somehow look into these other matters along the way. All this had to be done in a month, before the departure of the ships in early October. It was a tall order for a group of travelers who were just cutting their teeth in the desert. The most immediate need seemed to be arrangements for the journey to the southern part of the peninsula. Since Niebuhr would accompany the travelers and prepare a map of the route, it meant that he could devote little time to the study of the tidal movements at this most northerly part of the Red Sea. A table provided in the *Description of Arabia*[10] shows that on the 4th of September, a few days after their arrival in Suez, Niebuhr made a measurement the tides, correlated to the day of the month, the passage of the moon through the meridian, the time of high and low water, and the difference in height between the two. There is then a gap of twenty-six days until their return from the journey to Mount Sinai, when the same information is provided for the 30th of September and the first four days of October. He would save his conclusions until he had time to analyze this admittedly scanty information.

Meanwhile, there was the journey to the south. Shortly after their arrival they made contact with the Greeks of Suez, whose travels to the monasteries at Tor, Wadi Feran, and Mt. Sinai made them natural sources of information about the features, topographical and otherwise, of the peninsula. But none of them had ever heard of the "Gebel el Mocatab" reported by the prefect of the Franciscans. The Greeks took them to the sheikh of the Sa'id, one of the three major tribes in the peninsula, but he also failed to recognize the name. However the next day they brought the sheikh of a second tribe, the Leghat, and he said he knew of rocks that bore inscriptions. The word was now out that the foreigners were looking for a "Gebel el Mocatab," and if that was what they wanted, that is what he would give them. Niebuhr placed little faith in his assurances that such a place existed or, if it did, that it was known by that name. His doubts were later confirmed: he found many evidences of crude scratchings on the soft rocks along the way, but no "Gebel el Mocatab" as such, although he was led to a discovery that was almost as interesting, as we will see below. But now there was a further complication: since the journey to St. Catherine would take them through the territory of all three tribes, they needed a *ghafir*, or guide, from each of the three: the Said, the Leghat, and the Sualha. Negotiations ensued and a contract with the three was drawn up in the presence of the *qadi* and the *sanjak*, the governor of Suez. The latter had been exiled from Cairo and asked Niebuhr to consult the inscriptions to learn when his exile would end. As we saw above (see chapter on government), he was probably Ahmed Bey as-Sukari, the son of a Cairo sugar merchant and born a Muslim.

10. "Observations on the sea in the Arab Gulf made in 1762 and 1763," 369.

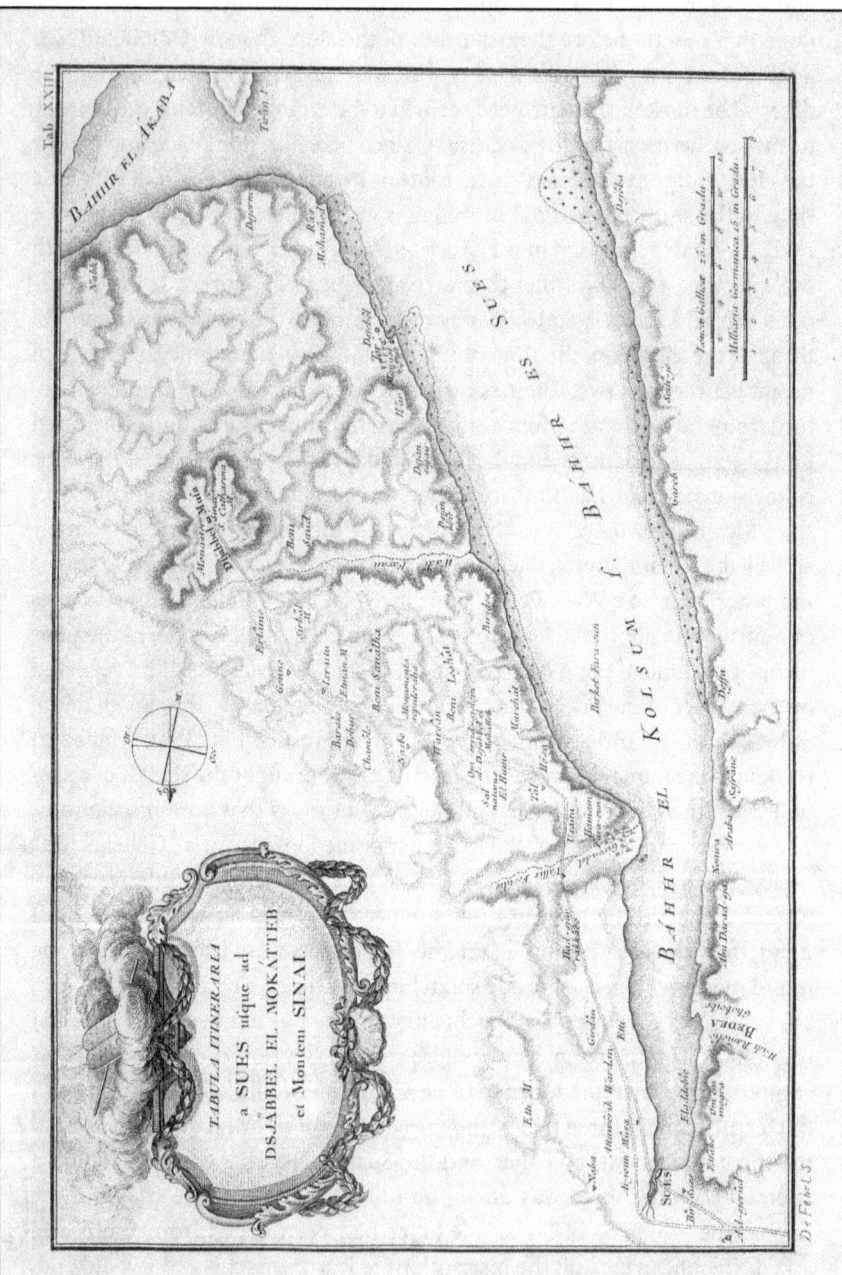

To Mount Sinai

They would be a much-reduced party that set off into the wilderness. The artist Baurenfeind had suffered greatly on the journey to Suez and had been felled by a "malign fever" shortly after their arrival. The presence of large bodies of standing water in Cairo after the summer rise of the Nile in 1762 could have created conditions for the transmission of malaria, and Baurenfeind may have already been infected with one of the less virulent strains of the disease. As we have seen, his work would be critical in recording the inscriptions they found. But there was no question of his fitness to travel, and the others had even begun to despair of his recovery. Kramer, the doctor, was needed to tend to the stricken man. Forsskal, whose assistant Baurenfeind nominally was, and who depended on the artist for the illustrations of flora and fauna he collected along the way, also refused to leave his side. This left only Niebuhr and the temperamental philologist, von Haven, to pursue the researches in Sinai. But if anyone other than Niebuhr should have set off into the wilderness, it was certainly von Haven. In addition to the inscriptions, he would be able to examine the manuscripts at the monastery of St. Catherine. So, for this part of the journey, we have two detailed records, that of Niebuhr whose general outlines we will, as usual, follow here, and the diary of von Haven, in handwritten form and still kept in the Royal Library in Copenhagen.[11]

Hansen suggests that the expedition to Sinai was von Haven's particular responsibility, as important in its potential results as the work in the Yemen itself. He suggests that the Sinai, the storied ground over which the Exodus took place, held even greater promise than the exploration of Arabia itself.[12] We know with hindsight that the expedition would find little of a biblical nature in the Yemen, but this was not known to Michaelis at the time and, in terms of the expectations that were entertained in Europe, Hansen probably overstates the case. As we have seen, Happy Arabia was the ultimate goal of the expedition and the fact that they reached the Yemen through Egypt and the Sinai rather than Tranquebar was accidental. On the other hand, given that they would now be able to trace the path of the Israelites on their way out of Egypt, there were targets of opportunity that presented themselves, such as the inscriptions in the desert and the library of the monastery. Hansen's further statement the monastery was renowned for its collection of

---

11. The original of the diary includes fifty-one loose-leaf, handwritten pages entitled "*Relation of a journey from Suez to Jebel el-Mokatteb and Jebel Musa from the 6th to the 26th of September, 1762.*"The complete diary was published in 2006 as *Frederik Christian von Havens Reisejournal fra Den Arabiske Rejse 1760–1763*, edited by Stig T. Rasmussen and Ann Haslund Hansen).

12. Hansen, *Arabia Felix*, 166.

"ancient Hebrew manuscripts" is probably also misleading. The collection was, and remains, primarily of medieval manuscripts, most of them Greek and Arabic, but including Aramaic, Persian, Amharic, and Syriac texts. Perhaps the oldest, the Codex Siniaticus, was a fourth-century Septuagint that became the basis for the revision of much of the Old Testament. But it was written in Greek.[13] While the expectation of what they would find in the Sinai may be debatable, the fact that Michaelis held high hopes for the visit to the monastery of St. Catherine is not. Von Haven would now be asked to produce a return on the time and expense that had been lavished on him in preparation for the expedition.

The two men crossed the head of the Gulf in a skiff on the evening of the 6th of September and set off for the supposed Gebel el Mocatab on the morning of the 7th. They were accompanied not only by the *ghafirs* but also assorted hangers-on who expected to travel at the expense of the foreigners. The little vexations with their traveling companions seemed to occupy von Haven to an extraordinary degree, and his diary is full of complaints about the behavior of the Arabs as well as the primitive nature of the traveling arrangements. We would expect this from a man who had spent the previous year in the most comfortable circumstances that Cairo could afford, and one for whom we suspect that this journey into the wilderness was an unpleasant and unwelcome duty. But the cost of von Haven's shortcomings to the Danish expedition is arguable in this case. As difficult as von Haven apparently was, Hansen still makes him out to be a thoroughgoing villain, not only of the trip into the Sinai but of the entire expedition, whose refusal here to cooperate with the Arabs was responsible for the failure to find the true Gebel el Mocatab.

In fact, as we will see, there *was* no mountain of that name, but only a series of names and petroglyphs scratched in the soft rock along the several routes to the places of pilgrimage in the south of Sinai. Von Haven was probably no more than what he appears to Niebuhr—an unpleasant traveling companion whose behavior did nothing to ingratiate himself with the Arabs. Niebuhr, on the other hand, found that little gestures of the kind von Haven was constitutionally incapable of making, paid large dividends in terms of information. The Bedouins had their own priorities, which were primarily to reach their tents in the south in as little time and at as little cost to themselves as possible. These priorities did *not* include catering to the wishes of a pair of Europeans, regardless of how pleasant or unpleasant

---

13. It was "borrowed" from the monastery by the German theologian Constantin von Tischendorf in 1859 and never returned. Presented as a gift by Tischendorf to Tsar Alexander II, it was later purchased by the British and is now in the British Museum. See Hobbs, *Mount Sinai*.

they may have been. Niebuhr's copies of what inscriptions they found were probably the best the expedition could have hoped for, given the time and circumstances involved in what was, after all, a hurried and poorly-planned detour on the way to Jidda.

The little band traveled southeast then south-southeast before reaching the intriguingly named Aiyun Musa, or Wells of Moses, a few miles from Suez. Only one of the wells, of which Niebuhr counted five, produced drinkable water. The area was particularly important as it was "... one of the most noteworthy in the Orient, since Moses has solemnized the journey of the Children of Israel in this desert. So I was careful to take measurements as exactly as I could along the way..."[14] After the short trip from Cairo to Suez, it was Niebuhr's first experience of systematic map-making in the desert and he soon developed the habits that would serve him over the next five years. He noted the bearings along the way with a pocket compass, which, surprisingly, excited no suspicion among the unlettered Arabs. Distance was a function of time. After they had settled into a rhythm, depending on the kind of animal they rode and the average distance each made over time, he calculated the distances by measuring the elapsed time. As usual, the most difficult problem was that of names. Only their guides knew the names of the prominent features along the way, and they were notoriously uncommunicative and unreliable. But after Niebuhr befriended one of the Arabs of the company and allowed him to sit on the crupper of his saddle, he seemed to get straight answers. Von Haven would not lower himself to this kind of familiarity and, as a result, received answers "that were sometimes erroneous, sometimes offensive."[15] The map of the journey from Suez to St. Catherine and the return is included as Plate XXIII of the *Description of Arabia*.

They followed the coast for two days, before turning inland on the 9th of September, south of Wadi Girandel and the Hamam Faraoun, or Pharaoh's baths. Niebuhr speculated that the wadi, with the presence of water a few feet below the surface, might be the El'-im of the Scriptures (Exod, 15:27):

> And they came to El'-im, where there were twelve wells of water, and threescore and ten palm trees: and they encamped there by the waters.

Even though there were passes through the mountains on both the eastern and western side of the gulf, it seemed too wide and deep at this point for the passage of the children of Israel as some scholars had suggested. Niebuhr set

---

14. *Travels*, vol. I, 226.
15. Ibid., 226.

up his astrolabe, established a base and hurriedly calculated that the gulf was approximately five German miles (or just over twenty-three English miles) wide. As they moved inland, the coastal plains turned into rugged mountains, and the farther east they went the harder and blacker the rocks became. They made little progress on the 10th because they were near the tents of their Leghat *ghafir*, although they pitched camp in the evening near the supposed Gebel el Mocatab. Niebuhr was, naturally, anxious to see the inscriptions:

> I had hoped on that day to finally have the good fortune to see the famous inscriptions, which were the object of our journey into the desert, especially since I assumed they were near our route, and appeared in such abundance that the mountains were covered with them. But the words of our Arabs now made me doubt whether our journey had been entirely profitable. He told us that he knew of no other inscriptions than those which were at the top of a very high and rugged mountain, to which he could not or would not lead us until the next day.[16]

Serabit al-Khadem

So, on the morning of the 11th, Niebuhr, von Haven and several Arabs set off and made their way to the foot of a promontory at the top of which the inscriptions supposedly lay. After a strenuous climb and then a long trek along a plateau that lasted an hour and a half, they found what their Leghat *ghafir* had promised. There Niebuhr was " . . . not a little surprised to find, here in the wilderness and at the top of such a steep mountain, a superb Egyptian cemetery."[17] He knew he had found something of interest, even though it bore little resemblance to the mountain described by the Franciscans in Cairo. He speculated that this might, instead, be the sepulchers of the covetous, mentioned in Numbers 11: 34:

> And he called the name of that place Kib-roth-hat-ta'-vah: because there they buried the people that lusted.

Or, perhaps it was Mount Hor (Numbers 33:38):

> And Aaron and the priest went up into mount Hor at the commandment of the LORD, and died there, in the fortieth year after the children of Israel were come out of the land of Egypt, in the first *day* of the fifth month.

16. Ibid., 231.
17. Ibid., 235.

Tab. XLIV.

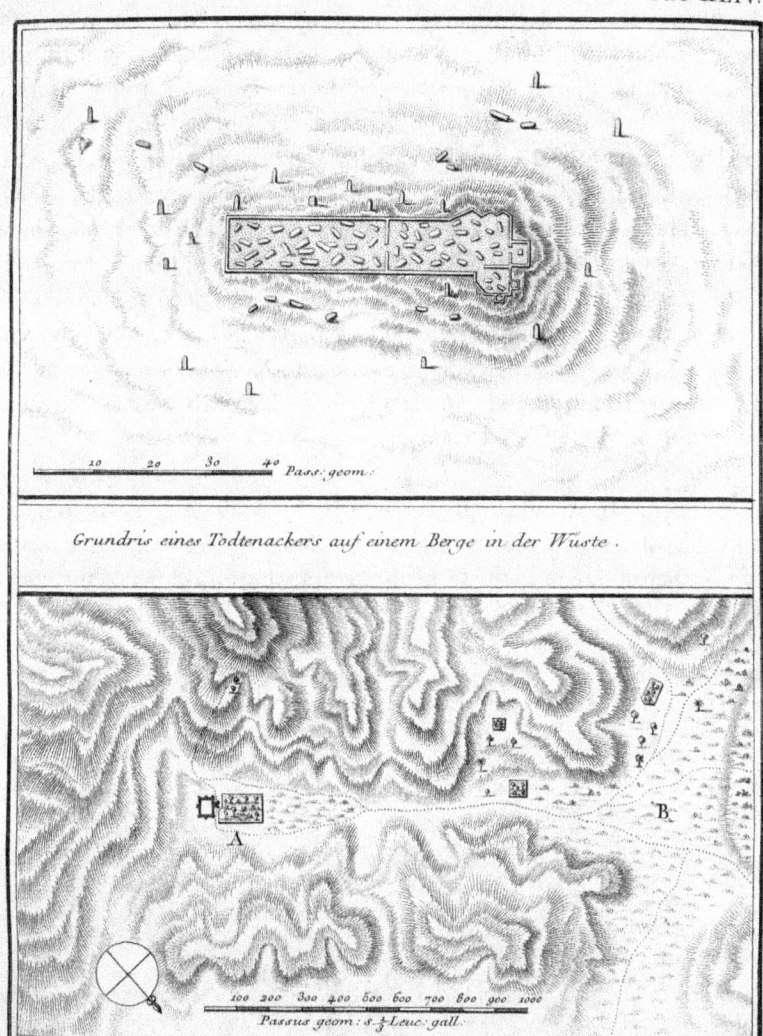

Grundris eines Todtenackers auf einem Berge in der Wüste.

Lage des Klosters am Berge Sinai.

Niebuhr in Egypt

If it was ancient inscriptions that were the purported precursors of Hebrew he was looking for, his reservations were well founded. His familiarity with the hieroglyphs, formed during months of painstaking copies of monuments in Cairo, suggested to him that this was Egyptian site. What they had, in fact, stumbled on, the first Europeans to do so, was an Egyptian temple now known as Serabit el-Khadem[18], although it was later shown to have strong Semitic characteristics. Niebuhr was wrong in believing it to be a cemetery, although even today a view of the upright steles from a distance still gives it the look of a country churchyard. Niebuhr and von Haven wandered over the site, inspecting the standing and broken steles, entering what was later identified as the sacred cave, and inspecting the hieroglyphs that covered the walls and pillars. They were carved in what Niebuhr called hard soapstone, unlike the granite monuments he had seen in Egypt, and contained many figures of goats, rather than the larger horned animals on the hieroglyphs he had seen in Cairo. But the *ghafirs* were absolutely insistent that he could not make copies of any of the inscriptions, the fear among the Arabs being that Europeans, like Maghribis, could thereby discover treasure and spirit it away to their own country.

Niebuhr suspected the real motive was money, but even that did not move the sheikh. So he secretly agreed with an Arab of the party to make another visit on their return from Mount Sinai. It was then, later in the month, that he made the plan of the temple and drawings of the steles that appear as plates XLIV, XLV, and XLVI in the *Travels*. Later scholars

---

18. The name *Serabit el Khadem* does not appear to be entirely Arabic in origin, and perhaps is not Arabic at all. Sometime between its discovery by Niebuhr in 1762 and the arrival of Burckhardt (*Travels in Syria and the Holy Land*) the name *Serabit* was applied to the temple, although Burckhardt called the site *Sarbout el Jemel*, as if *Sarbout* were the singular of *Serabit*, on the face of it reasonable enough. Not a great deal had been learned by the Spring of 1816 and Burckhardt's passage:

> The burying ground seen by Niebuhr near Naszeb, which as I have already mentioned, I passed without seeing, and missed on my way back, by taking a more southern road, appears to have been an ancient cemetery of the same kind, formed at a time when hieroglyphical characters were in use among all the nations under Egyptian influence. (482)

Richard Burton, in *The Land of Midian [Revisited]* refers to "a *Sarbut*, or rock said to be written over" (vol. I, 81) and "... a Sarbut, or 'upright stone'" (ibid., 304) without his usual etymology. This would suggest a Bedouin origin for the word, although the plural form of *Serabit* would suggest a more learned derivation. Raymond Weill also called it *Sarbout*, which he rendered as "mine." But I can find no justification for Weill's complete version— "the mines of the fortress." No Arabic, Persian, Turkish, or hieroglyphic dictionary I have seen contains the words *Sarbut, Sarbout* or *Serabit*. *El-Khadem* does appear to be Arabic, meaning "the servant." But even *khadem* is suspect, the possible substitution of a known word for an unknown but similar one. In spite of several interesting speculations, the meaning of the name has not been established with any certainty.

would decipher their meaning. Niebuhr's Plate XLV shows the stele of Hor-ur-ra of the 13th dynasty. It is still standing, the upper hieroglyphs still sharp after some 3,500 years, although in the lower parts the sandstone has "rotted," the result, Flinders Petrie speculated, of water damage. Raymond Weill collated copies of all the inscriptions in Sinai in his *Recueil des Inscriptions Egyptiennes du Sinai* and identified the "Hieroglyphen *auf zwey* Leichen Steinen *in der Wuste* on Niebuhr's Plate XLVI as a paean to Pharaohs in whose reigns expeditions took place. They were both of the 12th dynasty. The stele on the left of the plate is still standing and sharp, although a visitor has deeply cut the year "1825" in the face at the top. The stele on the right is of one of the Amenemhats.

Unable to make further progress that day, they descended from the plateau and the party resumed its way to the south. Niebuhr hoped that the discovery would be of interest to travelers and that the site might eventually be excavated. His hopes were rewarded in both cases. Publication of the *Travels* led to a regular stream of visitors of the hardier sort in the next century, including the Scottish artist David Roberts and the American divine Edward Robinson within a month of one another in 1839. Roberts saw the site on February 16th of that year although his lithograph "Temple on Gebel Garabe" displays so little of his customary architectural exactitude that one wonders if was really Serabit that he saw. The large mesa he shows in the left-center of the print is certainly nonexistent. Edward Robinson visited the site a month later in March of 1839 and published his impressions in his *Biblical Researches in Palestine, Mount Sinai and Arabia Petrea*. He gives a brief description of the site in general agreement with Niebuhr's, before giving a short history of the visitors since Niebuhr:

> This spot was first discovered by Niebuhr in 1761 (sic), who inquiring for the inscriptions in the Wadi-el-Mukatteb, was brought by his guides to this place as one of still greater interest and wonder; or rather, as it would seem, from ignorance on their part of the real object of his inquiries. The next Frank visitor seems to have been the French traveler Boutin in 1811, who was afterwards murdered in Syria; he was followed Ruppell in 1817. Many other travelers have since been here on their way to Sinai. So Lord Prudhoe and Major Felix; and after them Laborde and Linant, who have given drawings and views of the place and several of the monuments."[19]

19. Robinson, *Biblical Researches in Palestine*, vol. I, 78–79. Robinson was a sober and informed biblical scholar and later professor at the Union Theological Seminary in New York. He was the translator from the German of Gesenius's *A Hebrew and English Lexicon of the Old Testament*, and seems to be one of the few writers in English who was familiar with Niebuhr's work in the original German.

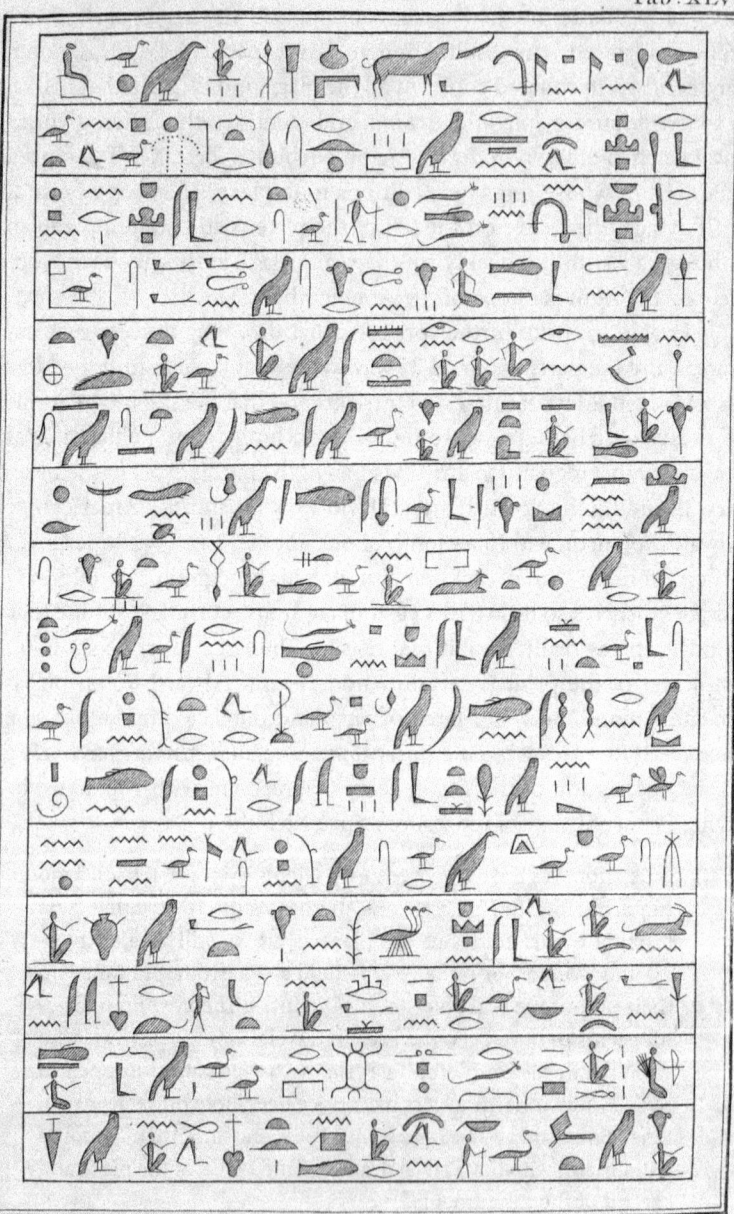

Hieroglyphen *auf einem Leichen Stein in der Wüste.*

To Suez and Sinai

Robinson goes on to doubt that the site was a cemetery, a doubt confirmed by Flinders Petrie. The definitive archaeological work on the site would come with the arrival of Petrie in 1904.[20]

Petrie identified the temple, Niebuhr's "cemetery," to be a place of worship for Egyptians working the nearby turquoise mines. He found that the earliest part of the complex, the sacred cave, dated from the Old Kingdom and the reign of Seneferu (2575–2551 BC), although the working of the mines took place mainly in the Middle and New Kingdoms (2040--1100 BC). The long axis shown in Niebuhr's Plate represented the successive additions to the temple over the 1,500 years that it was in use. The greatest builder had been Hatshepsut (1473–1458 BC). But there was an interesting Semitic aspect to the temple: it also appeared to be a place of worship for the deities of the area, some of whom were not Egyptian. This gave the complex its Semitic aspect. But perhaps Petrie's most important discovery was

> that of a new and unknown script; alongside the inscriptions in hieroglyphics left by Egyptian officials who had visited the area with the mining expeditions, others—he supposed them to be natives of Sinai—had left their own records in small, roughly-shaped stone figures or in the shafts of the turquoise mines . . . The script became known as proto-Sinaitic . . . early attempts to write a proto-Canaanite" (i.e. Semitic) language. Petrie's dating of them to the early eighteenth dynasty . . . is nowadays generally accepted.[21]

The fragmentary nature of these inscriptions made it difficult to draw immediate conclusions. But Petrie offered his usual perceptive speculation, including the thought that the script was most likely alphabetic and Semitic. His dates seemed to be off by several hundred years, the scripts more probably dating to the late Middle Kingdom or Second Intermediate Period.[22] But in late 1999 it was announced that Egyptologists had discovered evidence of an alphabet by "workaday people that simplified and democratized writing" in an ancient track to the west of the Nile. The script, identified as Semitic with Egyptian influences, was dated tentatively to between 1900 and 1800 BC and, so, several hundred years before what had been the previously earliest-known alphabetic writings. Petrie's dating of the inscriptions at Serabit was early 18th dynasty, or the sixteenth century BC. The discovery seemed to confirm speculation that the earliest alphabetic writing had been developed by Semites in an Egyptian context—something that could

20. See Petrie, *Researches in Sinai*.
21. Drower, *Flinders Petrie*, 292.
22. See Davies, *Egyptian Hieroglyphs*.

certainly be said of Serabit—but suggested that the date should be pushed back by some 300 years.

Later, the archaeologist and biblical scholar William Foxwell Albright would work closely with the texts and add his speculation as to the meaning of the discovery:

> there was presumably little change in the nature of Semitic life in Egypt during the two centuries between our inscriptions and the Exodus . . . there was normal overland intercourse between the Hebrews and Canaanites in Egypt and their relatives in Palestine . . . These documents throw much light on the continuity of language, practices, and ideas. Many Hebrew words, idioms and personal names appear already at Serabit . . . Our Serabit vocabulary throws light on the background of the patriarchal religious tradition . . .[23]

Although not all of his conclusions are accepted today and his translations suffered from a lack of understanding of the clues that signaled word breaks, Albright's work was an important step forward. Subsequent work by Itzhaq Beit-Arieh in 1982, Benjamin Sass in 1988, Yeshua'yahu Lender in 1990, and James Harris and Dan Hone between 1994 and 1997, on a greatly expanded body of signs, resulted in the suggestion that they had characteristics that made them more likely precursors to Hebrew rather than to Arabic scripts.

So maybe Michaelis was not so far wrong in looking for contemporary evidence of a written Semitic language at the time of the events of the Exodus. The eighteenth dynasty would place these scratchings just before the beginning of the reign of Ramesses II. On the other hand, as the putative precursor of the square Hebrew script, this surely would have been because Hebrew was itself only one of several northern Semitic languages, although there were important linguistic differences between the proto-Sinaitic and biblical Hebrew scripts. It was not because it was *the* language, the oldest and most perfect language, taught by God to Adam and Eve in the Garden of Eden, as some eighteenth-century sacred philologists were wont to believe.

If the Gebel el Mocatab reported by the Franciscans of Cairo was not Serabit al-Khadem, what then was? Some indication had already been given to Niebuhr as they made their way into the mountainous area. Shortly after departing on the morning of the tenth, they had halted under a small rock outcropping "on which many Greeks, who make the pilgrimage to Mount Sinai, have carved their names."[24] For centuries, pilgrims to the holy places of southern Sinai—Christian and pre-Christian alike—had traveled these

23. Albright, *The Proto-Sinaitic Inscriptions and Their Decipherment*, 13.
24. *Travels*, vol. I, 231.

same routes and had carved their names in the soft stone that lined the wadis. There would be many of these carvings that Niebuhr would see, both around the monastery and on the return journey when he would make the copies that appeared as Plates in vol. I of the *Travels*. In all probability, it was these inscriptions, many in the Nabataean script, which had been reported as the "Written Mountain."

Wadi Feran

The need to reach the monastery and return before the end of the month was still their overriding concern, and on the twelfth they moved on to the southeast and towards Mount Sinai. They moved first to pick up the sandy scrub and the "smooth valley called *Chamele*." But instead of turning to the southwest, through Wadi Seih to its confluence with Wadi Sidr and then south to its junction with Wadi Feran, they continued on to the east by south and then southeast. They covered five and a-quarter German miles, or just under twenty-four English miles that day before camping in "the valley of *Genne*." The next morning, they continued in the same general direction towards the Wadi Feran where their *ghafirs* would visit their families. They appear to have made good time: the twenty-four miles meant steady marching for eight to twelve hours, at between two and three English miles per hour. It is clear, then, that the party did not pass through what would later be called the Wadi Mokatteb on their way to the monastery. Just a few kilometers north of the track through Wadi Feran, but well to the west of the oasis, Wadi Mokatteb is probably the most concentrated area of inscriptions in the peninsula, with names and petroglyphs lining the rocks above the sandy floor, off and on, for several hundred yards. Since they simply retraced their steps on the way back to Suez, they did not do so on the return journey either. The map Niebuhr provides as Plate XXIII of the *Description of Arabia* (see above), confirms this: their track was generally east by south and then southeast from the coast as they approached the monastery, quite different from that of the metalled road of today.

On the 13th of September they exited from the interior fastness and entered the Wadi Feran.[25] Its name had not changed since the time of Moses, and Niebuhr's description suggests that its appearance had not changed much since his time either:

---

25. Lepsius would find Rephidim here in Wadi Feran, and not at the more celebrated "Mount Sinai" at the foot of which the Greek monastery sits. There were ample supplies of water in Feran and it was unlikely, he believed, that the Israelites would have decamped from there to occupy a place where there was almost none.

> The mountains that enclose the valley on both sides are very steep, of sandstone, mixed here and there with coarse granite, speckled with red and black. The valley was then dry; but after a heavy rain, so much water from the surrounding heights pours into the valley, that the Arabs in the area must seek refuge with their tents on higher ground . . . According to the Arabs, near the camp of our *ghafirs* there were enough date palms to feed a thousand souls.[26]

Niebuhr heard of the existence of ruins in the area, probably that of the Greek Orthodox monastery that had once served as headquarters of the Church in the peninsula. But the *ghafirs* were now near their tents and interested only in seeing their families, and no one would take him there. The enforced idleness gave Niebuhr time to reflect on the domestic arrangements of these Arabs and he was able to speak, for the first time, with a respectable Muslim woman. Although they had been in Egypt for a year, their only previous female contact had been with the dancing girls in Cairo. Here, she was the first wife of the sheikh and the first dame of the camp. But the sheikh had taken a second, younger wife and she was banished from his bed, if not from his affections. Not surprisingly, she found favor with the European practice of permitting only a single wife at a time. Her husband spent most of his time on the road to Suez and Cairo, which she referred to only as the *ryf*, applying the Arabic term for "countryside" or "cultivated land" to the entire country. While von Haven talked with the dame, Niebuhr—now pressed into duty as an artist—sketched her clothing, particularly the veil and earrings that would appear in Plate XXIII of the *Travels*.

On the 14th they were underway again to the southeast, camping in the evening near mountain springs whose water gave "a pleasure that the best wine could not afford in Europe."[27] During the day Niebuhr saw more inscriptions in a wholly unknown character and resolved to make copies on the return journey. They appear in Plates XLIX and L of vol. I of the *Travels*. On the 15th of September, they arrived before the monastery of St. Catherine after, by Niebuhr's calculation, twenty-eight and three-quarters German miles (or just over 132 English miles) from Suez.

---

26. *Travels*, vol. I, 240.
27. Ibid., 243.

Mount Sinai

It was the realization of one of Michaelis's fondest hopes. Given the time available, however, it is difficult to see how the two men could have accomplished much in the monastery's library. It had taken the party nine days to reach the monastery and they could look forward to at least the same amount of time on the return journey. If they were to be back in Suez by the end of the month, this left less than a week in the library. It would probably take that long simply to catalogue the most interesting manuscripts. Von Haven had already done something of the kind in the library of the mosque of al-Azhar in Cairo. However, careful textual examination of the kind that could yield important information, not to mention any copying of the manuscripts, would seem to have been out of the question. Unfortunately, not even an opportunity for a cursory examination was given to him. And it was here that the drama that occupies so prominent a place in the Hansen book—the responsibility of von Haven for the failure of much of the original intent of the Danish expedition—was played out again.

In this case it appears to be justified. Entry to the monastery required a letter from the bishop of Mount Sinai in Cairo.[28] Von Haven had contacted him while they were in the city but, unfortunately, that prelate had left unexpectedly for Istanbul and had not provided them with the necessary letter. They did have a letter from a deposed patriarch residing in Istanbul,

---

28. As we have seen, his residence and church were in the Fatimid city, just inside the Bab al-Nasr.

one Christopher of Macedonia, says von Haven, which had been secured through the good offices of the English ambassador in the city. It was this letter that they passed to the monks through a small aperture in the wall. But it was not enough. The monks were cordial, seeing that the visitors were Europeans, and even sent several bunches of fresh grapes to slake their thirst. But there could be no question of entering the monastery without a letter from their own bishop. About this they were adamant. The letter from the patriarch was passed back, unopened and unread, through the same aperture by which it had entered.

The rebuff had to be a bitter disappointment to von Haven and Niebuhr, although Niebuhr's account suggests that he swallowed his regrets and resolved to make the most of the situation. Hansen, again, faults von Haven for his passiveness, for his failure to "insist on entry" or to "demand any bible manuscripts."[29] There was, of course, no hope that either would have had the slightest effect on the monks ensconced, as they were, behind nine-foot thick walls. The fault was surely von Haven's, but it lay not in his lack of aggressiveness at the monastery, but rather in having failed to secure the necessary letter on one of his several visits to the bishop in Cairo. Even more fundamentally, there was simply not enough time in the Sinai to have accomplished much of scholarly value. But, as we will see below, if von Haven was disappointed at the rebuff, his disappointment was as nothing when compared with that of the authorities in Copenhagen when they learned of the failure to gain entry to the monastery.

Niebuhr decided to see as much as he could of the surrounding area. This was, after all, sacred ground. Von Haven was indisposed, and seemed unwilling or unable to exert himself. He and the rest of the party moved a quarter-of-an-hour down the Wadi ed-Deyr towards the plain of Raha so as not to be a source of molestation to the monks. The Europeans had, of course, been seen when they arrived and the Bedouins flocked to the camp in an attempt to extort bread from the monks or money from the visitors. Niebuhr paced off the exterior dimensions of the monastery and sketched it from two perspectives, one from a point that appears to be halfway up Jebel ed-Deyr to the northeast, and the second from the head of the valley looking southeast (Plate XLVIII). Like many pilgrims before and after him, he climbed the mountain, up the steps to the Basin of Elijah where he saw the two tall cypresses that still mark the site today. The Bedouins who accompanied him devoutly kissed the images of Christ and the Virgin Mary in a small chapel—probably that of Our Lady of the Storehouse—as they had undoubtedly seen Greek pilgrims do many times before. But they were

---

29. Hansen, *Arabia Felix*, 180.

dealing here with a northern European and a Lutheran, and Niebuhr could not help but remark that they could have saved themselves the trouble if they thought they would please him by "feigning adoration of these images."[30]

At the Basin of Elijah, the Arabs said that they were at the summit of the mountain. Niebuhr knew from Pococke that there were an additional thousand steps to the top and he could see some of them before they disappeared over the rise. But being unable to persuade any of the Arabs to accompany him further, he retraced his steps to the bottom and rejoined the party at the head of the valley. He speculated on the meanings of the names of the mountains as used by the Arabs. Jebel Musa seemed to be the entire range of mountains that had accompanied them through Wadi Feran, and Tur Sina the mountain at whose base the monastery stood. It is odd, however, that while he speculated that the area in which Jebel Musa stood was capable of accommodating the host of the children of Israel, he saw nothing in the vicinity of Tur Sina that could have done so:

> So, even if near Mount Sinai narrowly so-called there would not have been enough room for the entire Israelite encampment, still, there are perhaps broader plains on the other side, or else they may have camped around Jebel Musa, and even in part in the valley of Faran.[31]

He did not see the square mile of the plain of Raha as the likely assembly point. Von Haven was in agreement:

> I had already come to the conclusion that it was impossible that we were at Mt. Sinai. The monastery lay in a narrow valley, the valley whose dimensions I had already noted. It was not large enough for a small army to camp within, much less the 600,000 men Moses had with him and who, with the wives and children, must have amounted to more than 3,000,000 people... The entire range of Jebel Musa extends over many cross branches, and this branch ends in just the small and narrow valley... through which we came. It is not the Arabs who call this mountain Mt. Sinai. It is only the monks and other travelers who cling to the belief that Mt. Sinai must lie here, the monks, perhaps in order to give the place and the pilgrims that come here greater prestige...[32]

The comment is borne out in the twenty-first century. A regular stream of Christian pilgrims visits the site, Africans, Americans, Europeans, and

---

30. *Travels*, vol. I, 247.
31. Ibid., 248.
32. From *Frederik Christian von Havens Reisejournal*, 357.

Asiatics. The nearby "people of the book"—Muslims and Jews—seem, in general, indifferent to its attractions.

It is interesting to compare these observations with the very different impressions of another biblical scholar, Edward Robinson, sixty-five years later:

> We measured across the plain . . . we may fairly estimate the whole plain at two geographical miles long, and ranging in breadth from one third to two thirds of a mile; or as equivalent to a surface of at least one square mile. This space is nearly doubled by the recess . . . The examination of this afternoon convinced us, that here was space enough to satisfy all the requisitions of the Scriptural narrative, so far as it relates to the assembling of the congregation to receive the law.[33]

Return to Suez

The return to Suez was over the same ground they had covered in their outward journey. The party left the monastery on the 17th but were delayed again at the camp of their *ghafirs* in the Wadi Feran. While there, they were taunted by a young Arab who appeared to be intoxicated:

> When he heard that we were Europeans and Christians, he sorely tested our patience by mocking us, rather the way an insolent and drunken young European might behave towards a Jew. It appears that the Bedouins of this area make wine. I thank the heavens that Mohammed forbade the people of the Koran the use of strong drink.[34]

It was not until the 19th that they were on their way again. At Serabit on the 21st of September, Niebuhr climbed the mountain and was able to make the copies he had been denied on the outward journey. This was followed on the 22nd by copies of the "unknown inscriptions" of what he now called "Jabbel el Mocatteb." What Niebuhr saw and recorded that day was in an area they had passed through at night on the outward journey. It was another cliff face probably in the vicinity of Um ar-Riglain, the most prominent geographical feature to the west of Serabit, on which travelers had scrawled their names from time immemorial. Since Serabit—Niebuhr's cemetery—itself was not the Gebel el Mocatab described by the Prefect of the Franciscans, he must have reasoned, the crude inscriptions on this cliff face must be.

---

33. Robinson, *Biblical Researches*, vol. I, 95–96.
34. *Travels*, vol. I, 249.

> On the 22nd of September, we passed during the day through an area where, on the outward journey, the Arabs had made us travel by night. In one place where the path was hemmed in on both sides by steep cliffs, in addition to the rough incisions in the stone, I saw unknown inscriptions of the same sort as I had already copied on Jebel Musa. I quickly dismounted from my dromedary to examine the inscriptions more closely and to copy them. Our Arabs thought this was an unnecessary loss of time; however, Mr. von Haven, by a combination of cajolery and scolding, caused us a short delay; and in this short time I quickly copied inscriptions E, F, G, H, I, K, L, M, and N of Plates 49 and 50.[35]

They were crudely executed, only scratched in the soft stone, and Niebuhr speculated that they were only the names of travelers. He did not think they merited the attention of European scholars. He was right about the fact that they were names, but certainly wrong in that they were not of interest. Like the inscriptions in the Wadi Mokatteb to the south—and, for that matter, on rock faces near trade and pilgrimage routes all over the peninsula—they *were* the names of travelers. They were probably from the early Christian era and many were in the Nabataean script, which occupied an intermediate place between the Aramaic and the eventual Arabic script. The earliest identified Nabataean inscriptions date only from 250 AD, a century after the destruction of the kingdom of the same name by the Romans. The script outlived the political structure, but remained relatively undeveloped until the sixth century when distinct and recognizable Arabic forms began to appear. With the rise of Islam, the need for a written form of the language received a strong impetus. By the end of the eighth century, in a spectacular artistic flourishing, the classic monumental and cursive scripts had all appeared. The unknown inscriptions in the Sinai, then, represented early examples in the development of the Arabic, not the Hebrew, script.

---

35. Ibid., 249–50.

Inschriften am Wege von Sués nach dem Berge Sinai.

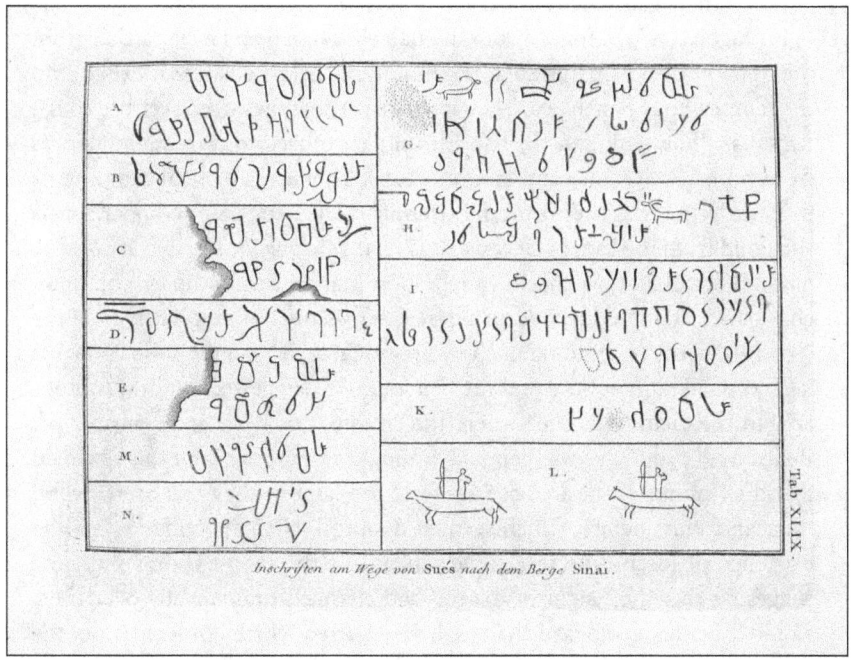

Inschriften am Wege von Sués nach dem Berge Sinai.

On the 25th they reached the vicinity of Suez. This time the tide was out and they crossed at once, not waiting for boats, the Arabs on foot and Niebuhr and von Haven on dromedaries. Baurenfeind's health was much improved and they could now think about the voyage to Jidda. But it was not until the 5th of October that they would board the larger of the two little ships recommended by the merchants in Cairo.

Preliminary Conclusions

And what, then, of the other task they had been assigned before they left for the south? As we have seen, Niebuhr spent the better part of the month on the journey to St. Catherine's and there was little time to devote to research into tidal movements near Suez. But after a gap of nearly a month, he resumed the observations begun on their arrival in Suez. Between the 30th of September and the 4th of October, and he found the mean difference in tides to be three feet six inches, very modest by comparison with tidal movements in other parts of the world. The high tide occurred between 11:15 a.m. and 12:56 p.m., as the moon passed through the meridian. The low tides were approximately six hours later. The data was hastily gathered and hardly exhaustive. But, even so, nothing that Niebuhr saw suggested to him that there was much similarity between the situations of Suez, 300 German leagues (or 975 miles) from the Bab el Mendab and even farther from the open sea, and Cuxhaven at the mouth of the Elbe where it exited into the tempestuous North Sea. The North Sea, a shallow cover over the continental shelf on which the British Isles sit, is exposed to extreme influences from both the Atlantic and Arctic oceans. This leads to tidal movements that are dramatic.[36] The northern extremity of the Red Sea, by comparison, is sheltered from the Indian Ocean and is relatively placid. The two areas were quite different, and Michaelis's suggestion that an extraordinary ebb upon ebb—such as sometimes occurred at Cuxhaven—was responsible for the "parting" of the waters did not appear reasonable to Niebuhr. Based on what he had seen, both in Suez and further south, he suggested that it was probably in the vicinity of Suez where the crossing took place, not near Wadi Bedea to the south, as was generally thought in Europe. Even this assumed that the features of the area of Suez were the same as they had been several thousand years before, which seemed doubtful. Niebuhr remarks that the existence of fossil crustaceans well inland of the then-existing shoreline suggested that the height of the sea had changed dramatically over time. Could it be that geological changes had occurred which rendered moot the question as to the place where the Israelites had crossed? The lapse of time between the purported events of the Exodus and 1762 AD was, geologically speaking, the blink of an eye. But if Niebuhr was thinking in terms of thousands not millions of years, he surely was on the right track. Changes in the area had clearly taken place, some of them relatively recent—the disappearance of the canal, for example—and the attempt to relate the movement of the tides in 1762 to those at the time of the purported exodus of the Israelites seemed unprofitable at best.

Niebuhr did not come to this conclusion lightly, however. It is clear he wrestled with Michaelis's issues at some length and he devotes the better part of the chapter on "The Wilderness of Mount Sinai" in *The Description of Arabia* to the most probable circumstances of an event that he believed surely took place, the passage of the children of Israel into the Sinai and the swallowing up of Pharaoh's host in the waters of the Red Sea. He compared the accounts in Exodus and Numbers with the places on his itinerary, and suggested that the head of the Gulf was the most likely place for the crossing. As for the possible existence of an isthmus or the nature of a smooth sea floor, both of which might have eased the passage, these seemed fruitless speculations. His interpretation differed from that of Michaelis. But Niebuhr was not

---

36. The mean difference in the tides at Cuxhaven is over three meters, or nearly ten feet, with extremes that are much greater.

one to let orthodoxy stand in his way, and he publicly differed with his patron on this point. Very sensibly he concluded that there was little science could do to unravel the mystery. If the passage of the Israelites over the Red Sea was a miraculous event, the search for a natural cause seemed to be pointless:

> I say that all these circumstances appear to me as miracles. If all this happened naturally, I no longer know what scholars mean by the word *miracle,* and I am willing to accede to the opinion of Mr. Michaelis.[37]

The conclusion cannot have been welcome in Göttingen.

Niebuhr did speculate on several other biblical references, although he neglected to ask about manna in the vicinity of Mt. Sinai. But he later saw what he described as manna near Merdin in southeastern Turkey and found that it consisted of "small, round yellow grains and so the same as the Israelites described it in Exodus, XVI, 14."[38] A kind of gluey exudation like sap, it attached itself to certain trees as "a kind of meal or flower." But, again, the feeding of the host of Israel year round with manna must have been miraculous since the substance appeared only at a certain time of the year. And might not the city of Tor, one reason for their long delay in Egypt, be the *Sur* mentioned by *Moses* in Genesis, 25:18, and the coast from Tor to Hammam Faraoun be the desert of *Sur* mentioned in Exodus, 15:22[39] And there did appear to be much that the life and practices of the Bedouins could tell latter-day scholars:

> If some intelligent and judicious theologian lived for a period of time with the wandering Arabs, he would learn perhaps many things about the daily life reported in difficult passages of Holy Scripture.[40]

But Niebuhr was not a theologian and his time in Sinai had been limited.

With regard to copies of the unknown inscriptions and the inspection of the manuscripts in the library of St. Catherine's monastery, the results were hardly more satisfactory. Instead of plaster casts of the former made by the expedition's artist, they had only Niebuhr's hastily sketched copies, made while von Haven distracted their guides. The "Gebel el Mocatab" of the Prefect of the Franciscans appeared to be nothing more than rude scratchings on a cliff face. They *had* discovered something of interest, but

---

37. *Description of Arabia*, 356. Michaelis would later say it was all a mistake and that he believed in the supernatural explanation of the event. See chapter on "Results" below.
38. Ibid., 125.
39. Ibid., note, 341.
40. Ibid., 149.

another Egyptian temple was not what would interest the foremost sacred philologist in Europe. As for access to the monastery, it had ended in utter and abject failure. When Bernstorff read von Haven's diary[41] dispatched from Suez, he was furious at the lack of preparation that seemed to have made the entire journey into the wilderness a waste of time. But he was angrier still with the tone of the account, the petty caviling at the behavior of the Arabs, the complaints about the food and water, the problem with von Haven's foot and his fever.

This was not what the Danish authorities expected and the fiasco at the monastery, when revealed to the eagerly awaiting scholars of Europe, would expose them to ridicule. Not only von Haven, but all of them, were at fault for the failure not only of this particular undertaking, but for the slender results of the expedition to date. Bernstorff reprimanded each of the members of the expedition in writing. But he saved his wrath especially for von Haven, the most learned of the group, the philologist and most prominent Dane in the expedition, whose failure in Sinai this most particularly was. He would translate von Haven's diary into German and send it to the eagerly waiting Michaelis, who could only be as disappointed as he was with its contents. However, he was certain that this was only a provisional account, of which a fuller and more substantive version would be forthcoming when von Haven had the time to prepare it. And the letter contained a veiled threat that, if this were *not* the case, von Haven would have forfeited an opportunity to render a great service to science and to win the good will of His Majesty. Such an opportunity might not present itself again. The prospect of the withdrawal of this favor was painful for Bernstorff to contemplate.[42] Fortunately for von Haven, the letter was dated June 21st 1763, twenty-seven days after his death from malaria in Mocha.

The little Cairo ship, with its apprehensive group of Europeans, set sail on the 10th of October for the Muslim Holy Land.

---

41. According to Hansen, Niebuhr feared tampering with his diary by the Danish authorities and sent only a provisional report. See Hansen, *Arabia Felix*, 186.

42. Ibid., 189.

14

Afterward

> The merchants of Cairo had given us letters of recommendation to two captains; we saw their accommodations, and secured for ourselves the highest cabin on the larger of the two ships, so that we could retire from the Mohammedans when we wanted to be alone. We had already traveled with Mohammedans by sea and land, and so were tolerably accustomed to their company; but we were never so fearful as for the upcoming journey from Suez to Jidda, since we were still under the impression that Mohammedans considered Christians unworthy to undertake this passage, which they considered holy. (*Travels*, vol. I, 255)

THE DANISH EXPEDITION HAD now passed out of Egypt and, consequently, out of the purview of this book. But a brief sketch of their fate, and of the fortitude and dedication of the sole survivor, will complete their story and help place their achievements in a larger perspective. The little group of four Cairo ships set sail on the 9th of October on a kind of trial voyage, each master adjusting the loads—as in a caravan—before judging the vessel to be seaworthy. The cabin on the main deck afforded some privacy and allowed Niebuhr to take his sightings unobserved. A group of women occupied the cabin below and when Niebuhr descended to the outboard toilet to answer a call of nature he was disconcerted to hear the nearby sounds of women's voices. He had hardly seen a woman unveiled in Egypt, but confessed that several times he saw a group of them naked and bathing through a gap in the partition.

The ship coasted down the Red Sea, rarely passing out of sight of land, but still avoiding the reefs with which the Red Sea littoral was dotted. Niebuhr took daily sights and recorded the positions in his logbook. Consisting mainly of latitudes and plotted on a modern nautical chart of the Red Sea they are, not surprisingly, still very accurate today. Baurenfeind made a sketch of the little settlement of Tor, where they stopped for water, and it was

Niebuhr in Egypt

included as Plate LI of the *Travels*. In the left foreground of the plate is a ship with what appears to be more European lines and flying an unidentifiable ensign at the stern. There was a moment of drama when Forsskal, who alone had gone ashore, overstayed and, was missed. Several of the janissaries with whom they traveled landed and sought him out. They found him botanizing in the nearby date plantations, and returned him safely to the ship. Niebuhr was moved to remark that Europeans might not have been so solicitous of strangers traveling in their midst as were these Muslim merchants.

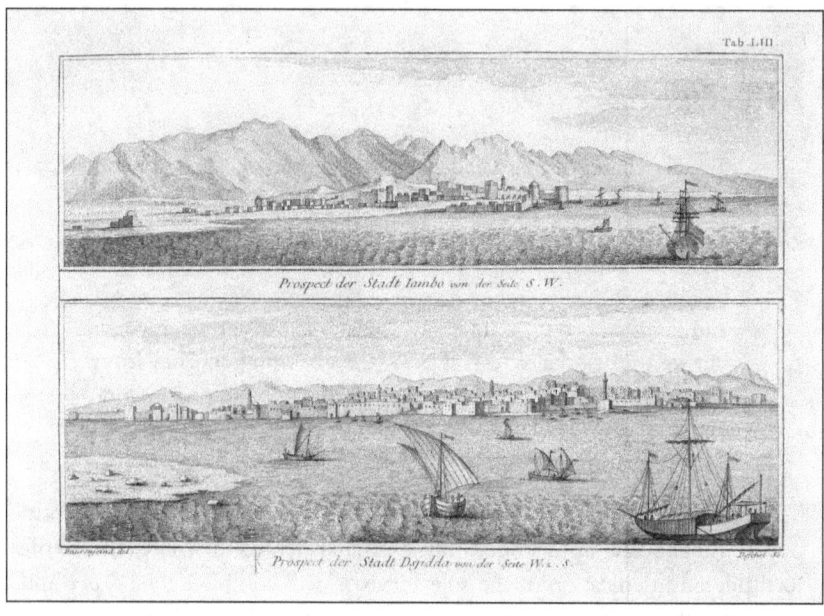

On the 22nd of October they stopped briefly in Yanbu, the port of Medina. There, in Plate LIII, Baurenfeind shows what appears to be the same ship we saw above, flying the same ensign, this time stern-to. It is interesting to speculate that it may have been the Cairo ship in which they traveled, or perhaps a sister ship. As Franks, the Europeans met with a none-too-friendly reception in Yanbu, which did not augur well for the future. They were now in the Hejaz, sensitive in Muslim eyes as the home to *al-Haramain al-Sharifain*, the two holiest places of Islam, Mecca and Medina. Egypt had been a frequent destination for European visitors for the past several centuries, and many passed through the country on their way into the Sinai or the Holy Land. But the western coast of the Arabian Peninsula was definitely off the beaten path, and Europeans were not generally welcome. They arrived at the breakwater off Jidda in the afternoon of October 29th, 1762. Customs formalities were lengthy and kept them aboard for two days. Finally, with much trepidation, they went ashore.

Afterward

Jidda

To their pleasant surprise, they were well received. Von Gahler, the Danish ambassador in Istanbul, knew the Pasha in Jidda and the connection proved invaluable. Letters provided by von Gahler were an introduction to such society as existed in this remote outpost of the Ottoman Empire. Moreover, the Pasha had a superficial knowledge of astronomy and invited Niebuhr to show him his instruments, although Niebuhr struggled to explain the technical terms in Arabic. Forsskal persuaded the Pasha to make a little garden in his courtyard and have sent from Medina cuttings of the shrub that produced the balm of Mecca. And the Europeans discovered that this part of the Hejaz, contrary to its reputation—and to the eternal scandal of the more religious Najd—was actually relatively cosmopolitan, and there was a small community of European consuls and traders living in Jidda. But what gave the city its particular character was the thriving community of former *hajjis* who had come from every corner of the Muslim world before settling in the holy land. There were *Jawis* from the East Indies, called after the island that produced the lettered class in the archipelago, and central Africans, called *Takruris*, some said from the Arabic verb *karra*, "to return, do repeatedly," since they poured across the Red Sea in a steady stream from an area that now comprises Mali, Niger, Chad, and the Sudan. They came as pilgrims and stayed on in Jidda, living in the outskirts of the city and serving as carriers of water and hewers of such wood as existed in the hills around Taif. There were also thousands of *Hadhramis* from the south of Yemen. They were mainly merchants, many involved in the slave trade with Zeyla and Tajura—now Djibouti—on the opposite coast of the Red Sea. There were also *Hindis* from the subcontinent, Indian Muslims who would increasingly play a prominent role in Islamic affairs as it reacted to the challenge of Europe. These were the most visible groups, but other Muslims had also come to the Holy Land from the four corners of the earth and many had stayed. As in Egypt, the ruling class in the Hejaz was Turkish.

The expedition stayed in Jidda until the middle of December. Niebuhr was free to wander the city, with the exception of the area around the Mecca Gate, and he spent his time absorbing the local color. He saw more evidence of geological change beyond the city walls—the same puzzling marine fossils he had seen near Suez—visited Eve's tomb, sketched a map of the city, and made an inventory of the goods passing in and out of the gates and of the coins used to pay for them. Baurenfeind was not idle and made sketches of the harbor, this time dominated by ships with lateen sails, of a Turkish pilgrim who had the look of a Greek of the classical period, and a local fisherman with the same look. But the members of the expedition were increasingly impatient to move on, now so near their ultimate goal, and rather than wait for a larger ship, on the 13th of December they embarked in an Omani vessel returning to Loheia on the coast of the Yemen. The ship in the right foreground of Plate LIII, with its high poop and three square-rigged masts, looks to be a predecessor of the Omani *baghlah*, and it may have been this or a similar vessel in which they traveled. The sailors were mostly Africans, probably from Zanzibar, who had a reputation for seamanship superior to that of the Yemenis, and the Europeans were advised by their friends to embark at once. They carried letters from the Kikhya and merchants of Jidda for the *Dolas*[1] of Hodeida and Loheia and the merchants of the two cities.

To the Yemen

The little ship set sail on the 14th and for the next two weeks Niebuhr busied himself alongside the pilot, taking sightings and recording the names and positions of the islands, anchorages, prominent headlands, banks of coral, and other geographical features on the southwestern seaboard of the Arabian peninsula. The wind was often contrary and the sightings were not always up to Niebuhr's standard. But he persevered, and the records of his position finding would ultimately fill 124 pages of vol. III of the *Travels*.[2] On the 20th they reached the vicinity of Ghunfude, at 19° 7' latitude, the first city of any size they had seen since they left Jidda. They dropped the anchor on the 23rd near Fej al-Jalbe, and dressed as simply as possible so as not to arouse the cupidity of the natives, Niebuhr and Forsskal went ashore. Their caution was well advised, for they encountered here people such as they had not seen before in the Orient: men with hair falling to their shoulders and naked with the exception of a piece of cloth covering their loins, and women unveiled and tattooed. They bought milk and bread. On the 27th they passed the city of Jizan (Niebuhr determined it to be at 16° 57') and reached Loheia two days later. And it was there, on the 29th of December 1762, that they finally stepped ashore in Happy Arabia, the ultimate goal of the expedition. It was now just six days short of two years since they left Copenhagen, the original duration of the expedition as envisaged by Professor Michaelis.

Yemen was the hub of the thriving coffee trade with the West, and Loheia did not suffer from the same reputation as the cities farther to the north. Their reception by the Amir Farhan was friendly enough. This

---

1. Used most frequently for "country" or "state," in this context it refers to a ruler who exercises authority in an administrative unit, in this case a city.

2. They would have a more immediate use when an English mariner in Bombay would later make a copy of the chart for use in his regular voyages through the Red Sea.

Tab. LIV.

Abbildung eines türkischen Pilgrims.

seemed to be a good portent. He was of the pure African type, very black, and was at first puzzled at the sight of these Europeans. He had seen European merchants before at Mocha, but none so strange as these men, with their full beards and Oriental dress, arriving from the Hejaz. They originally intended to continue to Mocha in the Omani ship. But, heartily tired of the life on board, they agreed to the Amir's suggestion that they disembark here at Loheia and proceed overland to the other city. One of the merchants, to whom they had letters from his correspondent in Jidda, offered them a house. So, after settling with the master of the little ship—they had not paid in advance—they settled down for a lengthy stay. The Amir facilitated the landing of their baggage and was afterwards treated to demonstrations of the telescope, microscope, and astrolabe. At the customhouse their chests were opened and the contents examined. Forsskal offered to show several officers a louse under the microscope, but they were offended when asked to produce such vermin themselves. However, one was forthcoming on promise of payment, and the next day a man offered to sell him an entire handful. Niebuhr showed them his astrolabe which, as we have seen, reversed the images, and the men were astonished when they saw a woman walking upside-down and her skirts didn't fall. Each novelty produced cries of *Allahu Akbar!* Everywhere, their accounts of Europe were met with interest, not with the hostility of Egypt.

The relations with other Yemenis were at first very friendly, and the services of Kramer, the doctor, were especially in demand. As we have seen, this was what Michaelis intended: not only would the physician minister to the travelers, but he would also use his skills to gain the trust of the Arabs. Kramer administered a violent emetic to the Bash Kateb (chief scribe), to the latter's complete satisfaction, and he was afterward asked to treat the favorite saddle horse of the Amir al-Bahr (admiral). Kramer demurred but Berggren, the Swedish servant, had spent several years in the hussars and was familiar with animals. His efforts were successful with the horse and he was afterwards asked to treat men.

In Jidda they had to remain close to the walls of the city. But in the Yemen they were permitted to range over the length and breadth of the country, and they spent nearly two months in the vicinity of Loheia, from Niebuhr's description a kind of idyllic time in comparison with their later travails. But they were anxious to be on their way. They had collected many "natural curiosities" in the vicinity of Loheia and arranged to have them sent by sea with the rest of their baggage to Mocha. The Amir, at first disappointed that they wanted to leave the city, was afterwards his usual helpful self, sending a letter to the Dola of Mocha asking that their baggage be kept unopened at the customhouse until they arrived to claim it. So after an exchange of

Afterward

presents with the good Farhan—a bit of liberality that was later to cost them dearly—they left Loheia on the 20th of February, bound for Beit al-Fakih. They were mounted on donkeys. This was not Egypt, where they had been forbidden to ride horses. But asses, especially the large and hardy breed of the Tehama, were here the preferred means of overland transport.

Travel in the Tehama, the coastal plain of the peninsula, was as safe as in Europe and they traveled alone, not waiting for a caravan to accompany them. With apologies to the reader and appeals to his patience, Niebuhr carefully (to some, tiresomely) lists the little villages and coffee huts along the way, along with the compass bearings and distances between them. But rural Yemen was not familiar territory. It was largely unknown to Europe, and he was responsible for a map of the country. So he persisted with his map making, listing the details in the text of his journal. They would later be transferred to the map that would appear as the first plate in vol. I of the *Travels*. Their good fortune continued and at the little *mansale*s,[3] or inns, along the way the proprietors refused money, although they were persuaded to accept small tokens of gratitude. The reception by the Amir Farhan and their treatment since they entered the Yemen seemed too good to be true. Unfortunately, as we will see below, it was.

They arrived in Beit al-Fakih on the morning of February 25th, 1763. The city was the commercial center of the coffee trade, the entrepôt to which the beans were brought from the mountains to be sorted and graded before being sent to Mocha for onward shipment to an increasingly worldwide market. Niebuhr reports that it contained merchants from Europe as well as the Hejaz, Egypt, Syria, Constantinople, Fez of Morocco in Barbary, Abyssinia, the east coast of Arabia, Persia, and India. Most importantly, it was near the mountains where the interests of the expedition lay, and the party made the city a kind of base of operations. They soon settled into the pattern they had established during the year in Egypt. Niebuhr and Forsskal were indefatigable in their researches, and they scoured the countryside, sometimes together, more frequently alone. Forsskal collected everything, whether flora or fauna, that seemed of interest. Niebuhr made a map of the city (Plate LXII) and the surrounding fields and villages. Kramer ministered to the injured foot of the Dola and Baurenfeind occupied himself with his drawings of Forsskal's collections. Von Haven seemed unable to bestir himself from his bed. Niebuhr devoted his time to his map of the Yemen and to the search for inscriptions, which had become something of an obsession with him. In Beit al-Fakih itself he made a copy of a Kufic inscription without attracting the kind of attention that had so plagued him in Cairo. Beginning with a short trip to Ghalefka and

---

3. A stopping place or way station, from the verb *nazala*, "he dismounted."

Niebuhr in Egypt

then Hodeida, he gradually expanded the scope of his travels and eventually reached Zebid, once the first commercial city in the Tehama. But not a trace of its ancient splendor, reported by Abulfeda, remained.

The countryside in the early spring was beautiful, this being the brief season when water in the Wadi Zebid irrigated the surrounding fields. Indigo was the most common crop. Niebuhr completed his second trip in early March and immediately thought about a third, although he was concerned that he would find it difficult during the fasting month of Ramadan, which began that year on the 16th of the month. But the Yemenis were not nearly as scrupulous as the Egyptians, and readily indulged the dispensation from fasting granted to travelers. So he set off once again for Kahme where there were reported inscriptions. As with his later work with the trilingual cuneiform inscriptions at Persepolis, Niebuhr's contribution would be significant. Here, it was not for the copies he made, but the mere report of the existence of inscriptions that stirred Europe's interest. His report of unknown inscriptions in the ruins of Zafar brought other seekers after him and led to the first copies for the scholars of Europe, made by Seetzen in 1810.[4] When Forsskal sent a pleasant report of the weather in the coffee plantations in the mountains, Baurenfeind and Kramer joined him there. Niebuhr set out to meet them and the four made a rendezvous in the mountains near Bulgose. Baurenfeind made a beautiful sketch of the terraced coffee trees that clung to the slopes around the village (Plate LXIII).

---

4. *First Encyclopedia of Islam*, vol. VII, 10.

Afterward

After a short stay they returned to Beit al-Fakih. The location was ideal, the inhabitants so used to seeing Europeans that the comings and goings of the members of the Danish expedition went virtually unnoticed. Hardly believing their good fortune, the two intrepid explorers, Niebuhr and Forsskal, now decided to extend their research to the southeast. The journey would be useful to both men, with Niebuhr continuing his map-making and search for inscriptions while Forsskal botanized in this unfamiliar and virgin territory. They set out on the 26th of March and early on the next day, near the village of Meshal, they saw running water for the first time in the Yemen. The stream was over twenty feet wide, but would soon be lost in the thirsty sands of the Tehama. As they began the climb into the mountains, the difficulties of map making increased: unlike the flat where determination of distance was a relatively easy function of elapsed time, Niebuhr now had to deal with an erratic pace, steep rises and falls, as well as frequent turns in the road. At Udden, fourteen German miles (or sixty-five English miles) to the southeast of Beit al-Fakih, they found the area that produced the best coffee in the Yemen. But there were other crops as well and on either side of the road the terraced fields were full of ripening rye, a beautiful sight in the early spring. The beards of these Europeans had also grown full and, in their Oriental garb, they were assumed by anyone to be Turks or Eastern Christians, perhaps Armenians. But they might just as well have been Europeans for all that the Yemenis seemed to care. On the 2nd of April, they arrived for the first time in Taizz, in itself a virtual botanical laboratory. They spent only a few hours in the outskirts of the town, knowing that they would return, and moved on.

On the 4th of April, Forsskal was rewarded with a discovery that seemed to make all his work to date worthwhile. Niebuhr tells the story:

> About a mile to the N.W. by W. 1/2 W. of Oude, Mr. Forsskal found a great balsam in full bloom. After some examination, he said he believed he had discovered the tree that produced the balm of Mekka; he was delighted with the discovery. The tree was in full flower and he could not have been more pleased. It allowed him to prepare a complete description of the tree in its very shade, and to take away number of blossoms as proof of his discovery.[5]

Also known as the balm of Gilead, it yielded a sap with purported medicinal properties and was mentioned in Jeremiah 8:22 and Ezekiel 27:17. Niebuhr adds that Forsskal wrote a letter to Linnaeus entitled detailing the discovery. Linnaeus would have been pleased, but so would Michaelis, since this was precisely the kind of "biblical botany" for which Forsskal was responsible. They left the mountains the next day but even now, in early April, they

---

5. *Travels*, vol. I, 351.

noticed the change: after the cool and green of the highlands, the heat in the Tehama had become oppressive. And it was on the next day, near Zebid, that Niebuhr felt for the first time the effects of a fever which afterwards recurred and "whose violence so enfeebled me that I could hardly move."[6] The change in weather, after the brief spring of their arrival in the Yemen, was also the signal of a change in their fortunes.

For "Happy Arabia" now proved to be malignant. After their first friendly reception the agents and authorities they dealt with were becoming more difficult. Everyone had heard of the gifts to the Amir Farhan and expected the same. It was a constant source of puzzlement to the Arabs that a group of Europeans would have come so far, at such obvious expense, without the prospect of gain. Much more understandable were the merchants, with whom the Arabs had a common commercial interest, and the expedition had entered the world of the petty trader *par excellence*, where every traveler or sailor carried goods on his own account, and the impulse to buy cheap and sell dear was as natural as drawing breath. The obvious disinterest in trade shown by Niebuhr, Forsskal, Baurenfeind, Kramer, and von Haven lends a certain naiveté, not to say poignancy, to their period in the Yemen.

To the local traders this same naiveté was only to be exploited, and they later became the prey of the unscrupulous wherever they went. Not only were they exploited, but the basest motives were assigned to their disinterestedness: if they weren't in the Yemen for profit, then there must be some other, devious motive. In this, they may have contributed to the problem. As we saw in Turkey, Europeans could hardly help themselves from seeing in military terms everything in this strange, and still vaguely threatening Orient. They made maps of the cities, paced off walls, and speculated on the capacity of Oriental fortifications to withstand the latest in European military science. To suspicious locals, the motives seemed obvious. The feeling was explicitly stated by a native of Hafshid u Bekil in the eastern Yemen: "Of what interest is my country?" he asked Niebuhr, "do you want to go there and conquer it?"[7]

The Amir of Loheia, who was interested in their scientific instruments, may have been friendly. But the relations with the other Yemenis quickly soured. Having exhausted their interest in the area around Beit al-Fakih and finding von Haven now in a state to travel, they left for Mocha on the 20th of April. On entering that city on the 25th they were forced to follow "the humiliating custom of Cairo, that is, the Europeans must dismount and proceed on foot."[8] It was an ill omen. A young merchant from Mocha whom

6. Ibid., 353.
7. Ibid., 301.
8. *Travels*, vol. 1, 359.

Afterward

they met in Jidda and had at first trusted, abused the trust and cost them dearly in time and ill-advised bribes to the authorities. Painting the inhabitants of Mocha in the vilest possible terms and assuring the Europeans that only he could keep them from being exploited, the merchant was all the while manipulating them to his own advantage. It was an object lesson in the dangers to which credulous foreigners were subject and it would be a month before Niebuhr and Forsskal freed themselves from his malign influence. After four months of travel in the country, they probably knew the Yemen better than he did. Disastrously, many of the specimens collected by Forsskal were destroyed as their baggage was roughly handled in an inspection at the Mocha customhouse. It was discovered that the specimens were preserved in alcohol, and this caused the liveliest disgust among the authorities. In spite of the letter from the Amir of Loheia, the Dola of Mocha became actively hostile, largely a result of the fracas in the customhouse, and the remaining days of the expedition in the Yemen were dogged by sullen lack of cooperation or outright hostility. Michaelis's admonition to behave with the utmost circumspection and to avoid giving offense, was being sorely tested. And to complicate matters, their health now began to fail.

The strain of living in an utterly foreign clime, manageable while they were healthy, now began to take its toll. Like many Europeans in the East, they attempted to live like Europeans, a practice which probably contributed to their fate. The Arabs who survived to adulthood had probably built immunities to many of the diseases to which the Europeans were subject. But there were still measures foreigners could take to guard against disease, and most of these measures the members of the Danish expedition failed to take. They carried their own supply of wine and brandy, going to great pains to find the most secure method of transporting liquids in the desert (well-waxed skins, as we saw above). The European preference for alcohol had a reason beyond mere taste or the pleasant effects of drink. They drank alcohol for the same reason that they preserved meat with spices: it was safer than the unadulterated item. But in not exposing themselves to the local water they were prey to bacteria and parasites when the wine ran out as it did, Niebuhr tells, us shortly after they reached the Yemen. It is difficult to say what effect the exhaustion of the alcohol had on their collective health. But Niebuhr, for one, seems to believe that they would have been better off without it all along and he was, in any case, extremely temperate in his use of wine. In fact, he became a convert to local habits during his travels, we suspect as much out of necessity as of conviction. However accidental, the conversion was heartfelt:

> I myself, while my companions were still alive, wished to live as they did, in the European fashion, and I suffered several grave

illnesses; but from the moment I was surrounded only by Orientals, and learned the precautions that must be taken in these countries I traveled in Persia, and from Basra overland to Copenhagen in good health.[9]

They did not guard themselves against "fevers" like the Arabs did, and slept outside, unprotected at night. It was a fatal lack of precaution, since malaria was—as it remains today—a great killer in the torrid regions of the globe. Transmission of malaria typically takes place between dusk and dawn when mosquitoes are most active. It was the practice of the members of the expedition to travel at night because of the great heat during the day. They also slept on the roof whenever possible. Niebuhr mentions that von Haven slept outside and "had not learned to sleep with his face covered as the Arabs do." This would have increased the chances of his being bitten. Niebuhr himself was "always sure to cover my face during the night, especially in the open air, and to guard against dew and pernicious winds."[10]

They remained for the next month in the Tehama, the low-lying, pestilential part of the Yemen that lies to the west of the mountains that climb to the highland plateau. As we have seen, the change in temperature was already apparent in early April. All felt the effects of the heat and humidity. But the discomfort was as nothing compared with the danger of the microscopic enemy that now attacked their weakened constitutions. The Tehama was then a fertile breeding ground for the mosquitoes that carry the malaria microbe. Both Beit al-Fakih and Mocha were near sea level and both fell within the band of temperature—sixty-eight to eighty-six degrees Fahrenheit—and relative humidity that bracket the optimal conditions for transmission. Niebuhr's careful records show temperatures in Beit el Fakih in March 1763 of between seventy-five and eighty-nine degrees, with readings taken at 7 a.m. and 10 p.m. It was certainly humid, with only a few months in the year in this area when it was not unbearably so. While water was present in the wadis of the Tehama in great volume only during the rainy seasons in April and September, Niebuhr noticed stagnant water in some places in March. In short, it was an ideal breeding for the mosquitoes that carry the disease and all the members of the expedition became intermittently sick with what from Niebuhr's account appears to have been malaria. The first to succumb was the philologist von Haven who died in Mocha on May 25th. He had fallen ill in Beit al-Fakih several months before, and now had difficulty withstanding the heat of the day. In the evenings he appeared

---

9. *Description of Arabia*, Preface, x.
10. *Travels*, vol. I, 341.

Afterward

refreshed, and slept outside. But there was no staying the deadly course of the disease. Niebuhr tells the story of his end:

> He risked passing the nights of the 24th and 25th on the terrace, and afterwards became so ill, probably because of severe cold, that he had to be assisted to his room by two servants. He became weaker and weaker and by 8 o'clock in the morning his pulse seemed to have completely ceased, although it started again after he was bled. An hour later he made his will, and we had not entirely lost hope of his recovery when, at 8 o'clock in the evening, he began to speak deliriously, sometimes in Arabic, then in French, Italian, German, and Danish. After this he fell into a deep sleep, or rather a swoon, and died about 10 o'clock.[11]

The "cold" with which von Haven was afflicted can only have been chills: Niebuhr's records show that at 9 p.m. on the two nights in question, the 24th and 25th, it was eighty-nine and ninety degrees respectively. The drop in blood pressure is consistent with malaria, although the bleeding of a patient already suffering from a drop in red cells probably hastened the end. Von Haven's weakness and delirium are also consistent with the disease. Malaria can mask many other diseases or combination of diseases, and the evidence is too sketchy to point to anything but a tentative diagnosis. But malaria, certainly, must be the prime suspect. Whatever the cause, it had to have been a bitter end for the man who entertained such a high opinion of his own abilities and in whom the Danish authorities had placed such high hopes. It is perhaps merciful that, as we have seen above, Bernstorff's stern reprimand did not reach him before his death. The English in Mocha sent six Catholic sailors to act as pallbearers and they buried von Haven the next day, the 26th of May, in the little cemetery for the Franks outside the city.

Unfortunately this was only the beginning of their travails. After the problems with the Dola and von Haven's death, some members of the party—we suspect Kramer and Baurenfeind—wanted to return to Europe at once. Niebuhr doesn't say who they were, but we can be certain it was not he and Forsskal, and the "perfect democracy" with which the expedition should conduct its affairs was being sorely tested. Instead, they decided to go up to Sana at once. If they found it to their liking, they could stay. If not, they could still return to Mocha in time to leave with the last English ship for Bombay. But the relative liberty with which they had previously traveled was now revoked, on the excuse that the Dola would have first to receive permission from the Imam in the capital before they could leave. So, claiming that the heat was unendurable and had already caused the death of one of their number, they asked

---

11. *Travels,* vol. I, 369.

for permission to go to Taizz. This, too, was refused. Finally, on the 9th of June they were permitted to leave for Taizz with the proviso that they go no farther without permission and take a servant of the Dola with them. This man, they felt, was little more than a spy. They left the bulk of their ready money with an Indian broker in Mocha against letters of credit with his countrymen in Taizz and Sana and departed after a most unpleasant stay in the city.

To Sana

Taizz was a relief after the stifling heat of the Tehama. It was a walled city of some antiquity and Niebuhr paced off the walls, visited the most prominent mosques, and listened to stories of the recent and violent political history of the city. There were only two gates and the Arabs considered its walls impregnable. However to this particular European it was clear that it could not be held for long against a modern army. The Eid al-Adha, or the Greater Bairam, fell on the 21st of June and they celebrated the feast in the city, buying a sheep and cakes for their Muslim servants. Maddeningly, Forsskal was repeatedly refused permission to visit the nearby Mount Saber, which the Arabs said, grew "every plant found in the world." Soon, even the delights of Taizz had begun to wear thin and the little band decided to leave for Sana. But bureaucratic delay after bureaucratic delay was employed to keep them in the city. First the Dola refused permission for them to leave, then refused to grant them an audience to explain the reason for his decision. They could not travel without camels, and the Dola himself headed the syndicate of camel brokers.

After a series of fruitless attempts at negotiation with the Dola, they enlisted the aid of the Kadi who, to their surprise, ruled in their favor and against the Dola. Niebuhr, with his usual scrupulous fairness, notes the incident for the record:

> The justness and readiness to serve of this Mohammedan judge amazed us . . . And I especially, who was alone fortunate enough to return to Europe, must cite this example to prove that not all Arab Kadis are as self-interested and unjust, as most Turkish Kadis, probably with reason, are believed to be.[12]

And so on the 28th of June, based on the ruling of this man of religion, they were abruptly given permission to leave and told that they must depart that same day. They had been packed for some time but, emboldened by the weeklong contest with the Dola, they had no intention of yielding to this peremptory order. They completed their preparations and left the next

---

12. *Travels*, vol. I, 393.

Tab. LXVIII.

Prospect eines Hauses zu Bîr el Assab bey Sanà.

Prospect des Castels und eines Theils der Stadt Ierîm.

morning, the 29th of June, mounted on camels. On the 30th there was a violent thunderstorm that, while it turned the ravines into torrents, left them relatively dry on the heights. On the way, Niebuhr looked for evidence of several cities mentioned by Abulfeda, but could find no trace.

Soon their numbers were, sadly, further depleted. Forsskal, who with Niebuhr had been untiring in the range and scope of his investigations, had fallen ill in Taizz but had not wanted his illness to be the cause for further delay. An arduous journey up the escarpment to Sana was not the best prescription for the sick man, although there is probably little that could have been done for him in any case. On the 2nd of July, on the road, he had become so weak that they sent him ahead to the next village slung from a camel. And now Niebuhr himself felt the onset of a fever. On the 5th of July he experienced violent nausea and vomiting, which he attributed to a cold he had caught from dressing too lightly in the mountains. He also suffered from a terrible thirst. However, all he could imagine was a recurrence of the "bloody flux," the dysentery that had nearly killed him at sea off Izmir. Forsskal arrived that evening in Jerim in a pitiable state, having been roughly handled by the Arabs, who objected to being required to carry a Christian. The illness of the two men kept them in Jerim for several days. Forsskal at first seemed to improve but then, like von Haven, his condition worsened and he suffered an onset so violent that they despaired of his recovery. On the evening of the 10th of July he fell into a deep sleep from which he never awakened. He died the next morning. After an unseemly dispute with the local authorities over the right to his personal effects, they buried him by the side of the road. They later learned that the corpse had been disinterred shortly after their departure and robbed of its shroud. Jews were made to re-bury it.

This latest loss was immeasurable. Forsskal had all the academic qualifications, if not pretensions, of von Haven but a great deal more besides. For the past two years he and Niebuhr had worked as a kind of team, and the latter was unstinting in his praise:

> His loss was a great blow to us; for in the process of gathering plants, he came into contact with the common people, and this made him the best in the company in understanding Arabic and its various dialects, and for this reason he often acted as our spokesman; but more, because of his great enthusiasm for the success of our journey. It was if he were born to undertake a journey to Arabia. He was not discouraged when conveniences were not available. He adjusted at once to the life of the inhabitants, and this is a necessity if one is to travel in Arabia with profit

and satisfaction. Without this quality, even the most knowledgeable scholar will not discover much in these lands.[13]

Hastily the survivors gathered themselves together and pressed on toward Sana. Near Damar they were met by a crowd of curious onlookers who had heard that Europeans were passing through the area. Foreigners may have been a familiar sight in the lowlands, but they were a novelty here. The crowd swelled and, even after they had ensconced themselves in a small rented house for the night, youths began to pelt the windows with stones to make the foreigners appear.

They were on the road again the next morning, leaving Berggren behind. He was too ill to travel and they feared repeating the experience of Forsskal. When he rejoined them he complained bitterly that he was denied lodgings out of fear that he might die during his stay—the expense of the funeral would be the responsibility of the landlord. They passed through the district of Belled Anes where Niebuhr had heard a report of unknown inscriptions. But they were too pressed for time to turn out of the way. A Jew he later met in Sana said he had seen the inscriptions and that they were not in the square Hebrew script. Niebuhr suspected they were Himyaritic and contented himself with noting their presence in his diary. They were now near the capital and hoped to reach the city on the next day, the 16th of July. So they began to make preparations for a suitable entry into this celebrated seat of the Imams.

They usually traveled according to the Arab custom, which meant to be poorly dressed, to avoid the attention of the curious as well as the avarice of the landlords and merchants with whom they dealt. But now they put on their Turkish clothes, which were in slightly better condition. Sending a servant with a letter ahead to announce their arrival, they found that they were already anticipated. The Imam had placed a house at their disposal in a suburb of the city, Bir al-Assab.[14] In spite of the fact that they were forced to dismount and walk a considerable distance to the house while the Yemenis remained mounted—a truly Cairene twist, observed Niebuhr—the house was as pleasant as anything they had seen in months. It was in a garden and here in the highlands the trees were heavy with fruit—grapes, hazelnuts, apricots, and pears. On the 17th the Imam sent them a present of a sheep, sweetmeats, and other edibles. To their inquiries as to leaving the house, however, they found that, as in Mocha and Taizz, protocol took precedence. They must have an audience with the Imam before venturing into the city.

---

13. *Travels*, vol. I, 404.
14. It remains today the neighborhood of foreign embassies in Sana.

Niebuhr in Egypt

Audience with the Imam

On the 19th it was announced that they must appear before the Imam. They were led to the palace and discovered, to their astonishment, that the most elaborate preparations had been made for the audience. The courtyard was so full of horses, officials and other retainers and hangers-on that a slave with a heavy staff had to make a way for the Europeans through the press of men and animals. Inside, on his throne reposed the imam, the very picture of an Oriental potentate:

> The audience room was a spacious, vaulted hall. In the middle was a fountain in a large basin, which threw water into the air to a height of about fourteen feet. Behind the basin was a platform about a foot and a-half above the level of the floor and four to five feet wide. Behind this was another small elevation just by the steps to the Imam's throne. The floor around the basin as well as on the throne was covered with beautiful Persian carpets. The throne itself was a four-sided platform covered in silk cloth, on the back as well as the sides of which were placed large, beautifully worked cushions. The Imam was seated on his throne among the cushions, with his legs crossed in the Oriental manner, wearing a bright green robe with long, wide sleeves in the Arab fashion. On the breast on each side was a heavy gold filigree of the kind sometimes worn by distinguished Turks on their traveling cloaks, and on his head he wore a large white turban.[15]

A profound silence reigned as the foreigners approached the throne. But as the first European kissed the hem of the Imam's robe and the palm, as well as the back, of his hand (a good sign), there was a collective gasp from the onlookers "God preserve the Imam." Niebuhr was at first disconcerted but, ever the fair-minded, reflected that this was not that different from the behavior of students at a German university. People were not so different in different parts of the world, after all. It is a signature observation of Niebuhr, to be repeated over and over again during his travels. They made presents of several watches and other small instruments but later learned that they were considered to be traveling as dervishes and nothing had been expected from them.

The interview was brief. They spoke through interpreters since the Imam spoke in classical Arabic and Niebuhr, now over two years in the Arab world and used to the colloquial language, understood only about one word in four. Even their servant from Mocha had trouble understanding him. Realizing that he would probably not comprehend the real reason for

---

15. *Travels*, vol. I, 413–14.

Afterward

their visit to his domains, they told the Imam that they were on their way to the Danish possessions in India and the Red Sea route was the most direct way from Europe. He replied that they were welcome to stay as long in his dominions as they wished and their affairs permitted. After the formal audience, they were asked to show him the instruments—the microscopes, thermometers, magnifying glasses, and printed Arabic books—that they had shown to the Amir Farhan in Loheia. Clearly, the Imam had been kept informed of their every move since they arrived in the Yemen. The demonstration was received with expressions of satisfaction. After a mutual exchange of gifts, they were permitted to leave. They could now enter the city and Niebuhr, in spite of the fact that he attracted a crowd everywhere he went, set about preparing his usual map and description of its environs.

And yet, in spite of the friendly reception and offer of cooperation, they were back in Mocha in just over two weeks. We should probably pause here and reflect on the change in their circumstances. They were now in the highlands, Sana itself being about 7,200 feet above sea level, and away from the discomfort of the Tehama. Niebuhr's records show a high of 85° Fahrenheit at noon and a low of 58° early on the morning of the 19th of July. It was ten to fifteen degrees hotter in Mocha, and far more humid. They had been received cordially by the Imam, provided with a house in a pleasant suburb, and invited to stay as long as they wished. And, most importantly, they had finally reached the part of the country that had been their goal since their departure from Copenhagen nearly three years before, the highlands of Happy Arabia. There was a large Jewish population—Niebuhr estimates them at 2,000 souls—in the village of "Kaa al-Ihud" near Sana, and the Europeans might have spent the time profitably in plumbing their knowledge of the Hebrew Bible.

But the strain of the past several months had clearly taken its toll, as much on the minds as the bodies of the surviving members of the expedition. They were all sick. And their spirit appears to have been broken as well as their bodies, in spite of the matter-of-fact tone of the Niebuhr account. The Imam seemed to be well disposed, but they didn't trust any of these autocrats after their experiences in Mocha and Taizz. The two professors were dead and any real hope of discoveries in the language and natural history of the country seemed to have died with them. Niebuhr had nearly completed his map of the inhabited areas of the Yemen. The other two members of the expedition, Kramer and Baurenfeind, had always played little more than a supporting role. As death took its inexorable toll, the prospect of leaving not only the fruit of their labors but also their bones in the Yemen began to prey on their minds. Their papers, not to mention Forsskal's specimens, filled many boxes and they had been very rudely handled by the customs authorities in Mocha. It was clear that they had seen Yemen, and the best they could do now was to get away with as much of their work as they could.

The departure of the last English ship that season for Bombay was scheduled for late August and that became the only goal to which they could now look forward. We suspect that the vote in their reduced democracy—now down to three persons—was unanimous. Such was their haste that they departed on the 26th of July on rented animals, in spite of the offer of the Imam to provide them himself. They were afraid that this might result in a delay and they were, in any event, now wary of gifts. The only concession to their responsibilities as an expedition was a descent by a different route so that Niebuhr could fill in a blank on his map. It was truly a descent, and the onset of the rainy season, with violent thunderstorms, made the going down the escarpment treacherous. Several times they had to make little bridges over portions of the road that had been washed away. In spite of the great volume of water that filled the wadis during the rainy season, it all seemed to be lost in the sands of the Tehama before reaching the sea. That information, at least, seemed to answer the question of European scholars about the possible existence of a perennial stream in this part of the peninsula.

After the relative cool of the heights, the heat of the Tehama came as a shock and they now traveled only at night. They reached Beit al-Fakih on the 2nd of August and were back in Mocha on the 5th. All took to their beds, under the tender ministrations of Mr. Francis Scott of the English Company. Niebuhr apparently arose long enough to make his usual review of the layout, history, and commerce of the port of Mocha. In an observation that would have made Fernand Braudel proud, Niebuhr observed one of the pillars of world trade in the eighteenth century, the recycling of precious metals from the New World, via Europe, to the Orient:

> When one thinks of the amount of silver that is sent annually from Europe to India and China, it is not astonishing to reflect that Europe would have long since been drained of gold and silver, were it not for the treasures of America?[16]

His review of the city ends, uncharacteristically for the well-disposed Niebuhr, with a parting shot at the Arab agents in Mocha and a warning to European merchants to resist their blandishments. The remnant of the Danish expedition, with the boxes of papers and specimens, took ship for Bombay on the 25th of August. Baurenfeind and Berggren died at sea shortly after their departure, and their bodies were committed to the deep. Kramer

---

16. *Travels*, vol. I, 446. In 1987 in Sana I myself bought a well-worn Spanish piece of eight probably minted in Peru.

Afterward

breathed his last in Bombay a few months after his and Niebuhr's arrival in that city on September 11, 1763.

Back to Copenhagen

The Danish expedition to Happy Arabia was now technically over. It was now nearly three years since it had left Copenhagen, only the last eight months of which had actually been spent in the Yemen. The six Europeans who set out so confidently in early 1761 were now reduced to a single man. The others were dead and Niebuhr would suffer for the remainder of his life from recurring fevers that would periodically confine him to his bed. Had he been a different man, he might have considered his responsibilities fulfilled and found the shortest way back to Europe. But, now alone, Niebuhr gathered together the fruit of the labors of his traveling companions and made provision for their dispatch to Copenhagen. He also summoned reserves of courage and dedication and determined to carry on the quest by himself. He was a cartographer and mathematician and had no illusion as to his fitness for the linguistic and scientific researches that were the subjects of the questions that belatedly reached him from Europe. But he had made it his habit to do whatever seemed necessary, and the painstaking copies of the hieroglyphs and the Nabataean inscriptions the expedition had produced thus far were entirely his work.

He would have been the first to grant that, had the other members of the expedition survived, the results might have been more substantial. But characteristically, he simply did his best. His best was, incidentally, good enough to set the standard for Arabian exploration for the next hundred years. The voluminous notes and specimens collected by Forsskal, the sketches of Baurenfeind and the manuscripts purchased by von Haven eventually reached Copenhagen, where the passage of time and the deaths of the professors seemed to reduce their value. They seemed a poor return on the extravagant expectations of the expedition that had been entertained in Europe. On his return, Niebuhr would supervise the publication of the Forsskal volumes with Baurenfeind's illustrations. He would also publish his own accounts. The first volume of his travel diary covered the period of the departure from Copenhagen to his arrival in Bombay. If he had done nothing else, Niebuhr had already produced a work that would long remain a classic of its kind. But he was only beginning to hit his stride.

He stayed in Bombay for fourteen months, making a short trip to Surat 150 miles up the coast. He also began the study of English and acquired the Anglophilia he was to retain for the rest of his life. Amid recurring

281

bouts of fever, he occupied himself with his, by now, familiar tasks. He made maps of the cities of Bombay and Surat, established latitudes, and sketched reliefs of the walls of the pagoda on Elephantine Island in the harbor of Bombay. He was not the first European to be interested in the rock-cut temples, and accounts had already been left by Ovington, Freyer, Hamilton, Prose, and Ives.[17] His sketches of the temple figures were not those of a trained artist, but the elevation of a typical column and entablature in Plate IV of vol. II of the *Travels* was laid out with the precision of the engineer that Niebuhr was. And the figures were surprisingly lifelike, even daring, with their depiction of the breasts and genitalia that Hindus accepted as natural but would raise the eyebrows of eighteenth century Europeans. An example appears in Plate X.

Bombay also stood out in his travels for another reason. In the city there were Parsis, Portuguese Catholics, Armenians, Greek Orthodox, Jews, Banyans, Hindus or pagans of various castes, and Sunnis and Shiites, all of them worshipping with complete freedom of conscience under the relatively benign rule of the English. The experience made a profound impression on Niebuhr. He remarked that while Europeans had not formed a very good opinion of the local pagans, he found them to be abstemious as well as "gentle, virtuous, and industrious," their pantheon able to accommodate other deities including the Father, Son and Holy Ghost of the Christians. It was a rare example of religious inclusiveness rather than exclusiveness, although their own belief system prescribed separation between castes.

Niebuhr had already seen the effects of religious intolerance and contention in Egypt, hatred of Jewish Rabbinites for their Karaite coreligionists, continual friction between Latin and Orthodox Christians, and would later witness the lethal hostility between Sunnis and Shiites in Turkish Iraq, not to mention the common hostility of these three "people of the book" for one another. He was not immune to these feelings himself, and his reservations about the activities of the Catholics are apparent throughout, not least when he is a witness of the poverty to which the clergy has reduced the peasantry in Poland. But he is never bitter. He stayed with members of Catholic religious orders throughout this travels, and always refers to them as the "good fathers." But we suspect he looked back to that placid atmosphere in Bombay as the epitome of religious tolerance.

17. *Travels*, vol. II, 24–25.

Abbildung der Figur bey 9 auf dem Grundris Tab. III.

II Th.

Tab. X.

The Persian Gulf

Leaving Bombay on December 8, 1764 in a ship of the English East India Company he made his way to Muscat where he found a few remnants of the Portuguese ascendancy in the East, including two churches, one now the residence of the governor, the other being used a warehouse. Then it was on to Bushire, Shiraz and Persepolis in Persia. He no longer cultivated the officialdom that had so plagued the party in the Yemen, confining his attention to the learned men who possessed the information he wanted. He traveled alone, without even the European servant he had taken in Bombay, and generally wore the local clothing and ate the local food out of conviction that conformity to local habits was easier and healthier. But there was no question that he had gone native. As for religious belief he retained the catholicity of spirit and willingness to see the beliefs of others—unless, perhaps, they were papists—in the best possible light. But he remained a Christian and a Lutheran to the end of his days.

From Bushire it was inland to Shiraz and he used the city as a kind of base from which to plumb the wonders of Persepolis, spending twenty-three days in the ruins drawing reliefs and copying cuneiform inscriptions, and incidentally damaging his eyesight in the refection of the sunlight off the brilliant white limestone. Just as he had done with the Egyptian hieroglyphs, he made a list of the individual letters and found them to be forty-two in number. Even without the knowledge of their underlying meaning his experience in copying scripts allowed him to conclude that the inscriptions were in three different languages. In fact, we now know they were in Persian, Elamite, and Babylonian. His extremely accurate copies were a kind of Rosetta stone of cuneiform and led directly to the deciphering of the texts themselves and the cuneiform script in general.[18] Again, it was an example of what he had hoped when he left Copenhagen: without claiming precedence for himself, he would contribute to the gradual assembling of information that would allow European scholars to unlock the secrets of the Orient.

On Easter the little flock of believers in Shiraz put him in mind of the first Christians, with the chief ecclesiastic doubling as a weaver and the others including a shoemaker, a cook, a servant, and a carpenter. It wasn't until the 14th of May that a sufficient number of travelers to make up a caravan had assembled and they left that day for Bushire and the coast. There he found the heat so oppressive after the elevation of Shiraz that he left almost at once in a small vessel for the island of Kharj, where he hoped to find passage an English ship bound for Basra. He departed Kharj on the last day of July and the wind

---

18. See Clark, *The Art of Early Writing*, 35.

was so favorable that they arrived at the mouth of the Shatt al-Arab on the evening of the following day, the 1st of August 1765.

Turkish Iraq

Upriver lay Basra, most of whose inhabitants appeared to be Sunni, although in fact many were Shiites who masqueraded as Sunnis, given the unrest in the area and the toxic state of relations between the two sects. Several years before the city had received an unexpected flood of Sunni refugees who had been driven out of Arabia by the adherents of a new dispensation founded by one Mohammed ibn Abdel-Wahhab. Niebuhr estimated that the old city, so famous under the early caliphs, was located a few miles to the southwest of the modern city. After making his customary map he proceeded up the Euphrates to Baghdad, along the way entering Mashed Ali (or Najaf)—being, he believed, the first European to do so—and Mashed Hussein (or Karbala) and speculating on the location of ancient Babylon. A German mile to the northwest were the ruins of the once-famous city of Kufa, which contained, among other things, the house supposedly built by Noah after the landing of the ark. There was a Jewish community living in the area and several places sacred to the Jews nearby, including the tomb of Ezekiel. But the Jews kept a low profile, a characteristic of the community he would witness throughout Iraq.

Niebuhr arrived in Baghdad on January 6th, 1766. It was now just over five years since the expedition had departed from Copenhagen. He stayed in city for almost two months making, among other things his usual map, although it lacked the detail of his map of Cairo. There, his curiosity had brought him too often to the attention of the authorities and he was not anxious to repeat the experience. Here, the walls were in a poor state of repair and the moat was dry. The local janissaries thought it stoutly defended but, in Niebuhr's opinion, the city could have easily been taken by a European army. He found that the Tigris rose and fell like the Nile and between the 19th and the 21st of the month it overflowed its banks as a result of snow melt in the mountains to the north, and inundated outlying areas of the city. The Baghdad of the *Thousand Nights and a Night*, built by the Caliph Mansur, lay to the west of the Tigris and there he visited the tombs of several famous men, including Abu Hanifa and Ahmed Ibn Hanbal, founders of two of the four orthodox schools of Sunni Islam. Of the city's once-celebrated reputation for learning, there seemed hardly a trace.

He left Baghdad at the end of February for Mosul in a small caravan consisting of thirty Jewish merchants. They reacted with their usual timidity to the taunts of the Muslim muleteers and Niebuhr was moved to a reflection that seems a rare departure from his usual perspicacity: "Such is the cowardice of the Jews under the Mohammedan yoke; they would consequently never take up arms against the Turks to conquer the Promised Land, however numerous they might become."[19] In Kirkuk lay the resting places of the prophets Daniel, Michael, Hananiah, and Azariah, although the Jews were not permitted to enter the tombs. There was also a small community of Nestorian Christians who, as Uniates, recognized the primacy of the pope. When they heard that a European was in town they assumed he was of their dispensation and paid Niebuhr a visit. He was his usual diplomatic self, not telling him that he was a Lutheran but suggesting that they make common cause with other Christians. In general he found these converts to be the most implacable enemies of those not in communion with Rome. The Beiram was celebrated in Altun Kupri and as part of the festivities Niebuhr took advantage of another Oriental specialty, a full-body massage that restored the circulation after an arduous day of travel.

To Aleppo

He left Mosul on the 11th of April in a large caravan bound for Aleppo. The 2,000 camels, horses, and donkeys were carrying gallnuts from Kurdistan, cloth from India, Persia, Baghdad, and Mosul, as well as coffee and miscellaneous wares. He had an odd traveling companion, a rabbi from Prague who had been visiting the tombs of Jonah, Ezekiel, and other prophets and supporting himself by selling Hebrew manuscripts to local communities. But he had been shunned by other Jews, some said because he put on airs, others because he was prodigiously avaricious. Niebuhr took the rabbi under his wing as far as Mardin and met him later in Aleppo, where he learned that he had been stripped bare by Turkmen, probably because he had refused to travel on the Sabbath and was proceeding alone. Niebuhr could only reflect than that if one presumes to travel in foreign lands he should make some concession to local practices.

Passing through Mardin, Diarbakr, and Urfa—the Edessa of the Crusades—Niebuhr crossed the Euphrates at al-Bir and reached Aleppo on the 6th of June 1766, to an astonishing reception by a European community that had long since given him up for dead. The second volume of his travel diary covers the period of his departure from Bombay to his arrival in Aleppo. It includes descriptions of governments, cities, commerce, religions, tribes, and historical sketches. The fifty-two plates, including maps, views and inscriptions, can only have been from his own hand.

19. *Travels*, vol. II, 336.

Afterward

Syria

Syria constituted a kind of interregnum on Niebuhr's return to Europe. On his arrival in Aleppo he was most cordially received by the Dutch consul, Herr von Masseyk, a native of Holstein. After years in the company of Orientals—and for all of his native ecumenism—he was happy to be again among his own kind. There were also Dr. Patrick Russell, nephew of Alexander Russell, the author of the highly-regarded *Description of Aleppo*, Mr. Dawes, chaplain to the English factory, the English consul Mr. Preston, and merchants of many nations. It would have been understandable if he had lingered in this cordial atmosphere. But Niebuhr was not a man to let the grass grow under his feet. With the exception of a visit to Cyprus to examine inscriptions thought by Copenhagen to be Phoenician (they were not), he spent the next five and a-half months exploring greater Syria, with a particular interest in its religious minorities. The last fifty-seven pages of vol. II of the *Travels* are devoted to his review of the *pashaliks* of Damascus, Aleppo, Tripoli, and Sidon. He was the expedition's cartographer but he had long since expanded his duties into a far-ranging interest in the Orient and he absorbed everything the area had to offer. In this latter part of his travels he could almost be said to have become an *amateur* in religious dispensations: from a specialist in physical geography he had become an expert in the geography of the sacred, but the territory he explored was often that existing between the ears of the believers.

Syria was a kaleidoscope of religious belief. Sunnis were the predominant sect, but there were also Shia offshoots—Metwalis, Nusayris, and Ismaelites—as well as Druse, Jews, Christians, and believers in smaller sects who kept their beliefs to themselves. The Metwalis had the same aversion to contact with those of other beliefs as their Persian coreligionists, and were distributed in several districts reporting variously to Sidon, Tripoli, and Homs. The Druse, originally followers of the mad Egyptian Caliph Hakim and whose tenets seemed to have been borrowed from all three "religions of the Book," were concentrated primarily in Mount Lebanon. They were said to be descendants of the Europeans of the crusader period, but Niebuhr found the suggestion unpersuasive. As with other heterodox sects in this cockpit of sectarian animus, stories circulated about their abhorrent practices, of their adoration of the figure of a calf and marriage with their own sisters and daughters. But the Druse themselves may have been responsible for some of the misunderstanding. They gathered in secret conclaves, which gave rise to suspicion of arcane practices and even led some Christians from Aleppo—familiar with similar practices among the English—to conclude that they were Freemasons. They seemed to be perfect religious chameleons, practicing male circumcision and worshipping as Sunnis when necessary, but also indulging wine and pork among their Christian neighbors and even permitting their sons to be baptized.

The Nusayris, or Alawites, were another heterodox Muslim sect that maintained their practices out of the public eye. Their sheikhs appeared to be Sunni and were relatively wealthy since they controlled the trade in tobacco, exported from Latakia. They were less isolated than the Druse and therefore more subject to abuse by the Turkish authorities. The Ismailites were yet another small Shia sect, also vulnerable to pressure from the Turks. They were concentrated in the north of the country, in the area lying between Latakia, Aleppo, and Antioch. Without access to their holy books Niebuhr could learn very little about their beliefs but he found that the greatest absurdities about all these little groups had been circulated, out of ignorance of their tenets and practices. All appeared to have taken portions of their eclectic faiths from their Jewish, Christian, and Muslim neighbors.

The Jews were primarily Rabbinites, or Talmudites, but there were also Karaites and Samaritans, the latter an ancient community concentrated in Sichem and Nablus. There were many Christians in Syria, the most concentrated single group being the Maronites in Mount Lebanon. In the early seventeenth century they had come under the protection of France and they recognized the primacy of the Pope in Rome, although their attachment to the papacy seemed purely nominal to Niebuhr. Their services were held primarily in Syriac although they spoke Arabic in their daily lives. Having embraced Rome they had regular contact with Europe, although that did not mean a greater familiarity with European art and science. There were places for thirty students at the Maronite college in Rome,[20] but only a few had matriculated, out of the fear of their parents that the students would be infected with European ideas. But there were contacts with Europe of a different sort in the steady stream of "Princes of Mount Lebanon" who circulated between European capitals, feted by the authorities and living on the public dole. Few were actual princes and most were simply taking advantage of a credulous public. They were not the last quasi–Europeans to be embraced by the West as simulacra of themselves in the little-understood East.

There were other Christians in communion with Rome, including some who were called Greeks since they once belonged to the Greek Church (although they understood as little Greek as the Maronites did Syriac) as well as Armenians, also in Mount Lebanon. The Maronites had their own patriarchs and they lived relatively peacefully—in a situation that could

---

20. As we have seen, von Haven studied briefly there, awaiting the arrival of the the rest of the expedition in Marseilles.

have prevailed only in Syria—under the protection of the Druse. There were also Greeks, Armenians, and Nestorians who recognized the patriarchs of Constantinople, Adsch Miazin, and Diarbekr and so were not protected by the Druse. Their lives were, therefore, more troubled.

Western religious orders were represented by Franciscans, Jesuits, Capuchins, and Carmelites. As elsewhere in the East they were zealously proselytizing among the Oriental Christians, to the dismay of the local congregations who saw their numbers dwindle. Apostasy was attractive, given the prospect of access to European learning, languages, and sponsorship, and probably had little to do with doctrine. It would be another century before American Protestants entered the lists, offering the same range of attractions, with educational activities that eventually resulted in such institutions as Roberts College in Istanbul and the American universities in Cairo and Beirut. As in Cairo, the Turkish authorities were indifferent to the conversions as long as the head tax was paid. They even benefitted financially from the contention between the various sects.

Jerusalem

Proceeding inland from Jaffa Niebuhr visited Jerusalem in the company of group of Franciscan monks from Calabria, whose intolerance was exceeded only by their ignorance. They had never heard of Denmark and, since Niebuhr was in his Turkish dress, concluded that it must lie somewhere in Anatolia. Oddly, his first act on entering the Holy City was to attend Mass: Protestants were not widely known in the city, and absenting himself from the service might have given offence. He was rewarded with a performance of excellent vocal and instrumental music, the organist and most of the soloists being Germans. After nearly six years of drought, the music came as a tonic for this sometime church organist. He made his usual map of the city within its sixteenth-century Ottoman walls, as detailed as time and circumstances permitted. It included a key to fifty-four places of interest including the seven gates, the Church of the Resurrection (or Church of the Holy Sepulcher), the Franciscan monastery where he stayed, and sundry holy sites including various Stations of the Cross, and *Elharam*, or the Dome of the Rock. Outside the walls lay the Vale of Jehoshaphat, the brook of Kidron, the garden of Gethsemane, and the Mount of Olives, which afforded the best view of the city. Even at its greatest extent, by comparison with Cairo, Constantinople, London, or Paris, Jerusalem was tiny.

In the process of map-making he was exposed to the unseemly state of relations between the Latin and Orthodox communities in the city: their disagreements led to frequent fisticuffs, even bloodshed. As usual, the Ottoman authorities mediated between the two and realized large sums for their services. In truth, Jerusalem was a religious customhouse: industry was almost non-existent in the city and the Turks lived off the expenditures of Jewish and Christian pilgrims, whose contributions dwarfed every other source of revenue. The Jews held Jerusalem in no less veneration than the Christians, although they were in possession only of a small remnant of the wall, a part of the onetime Temple of Solomon. Old Jews came to Jerusalem to die and to be buried in the Vale of Jehoshaphat. To Niebuhr there was an advantage to this very practical Ottoman approach to religion: if the Christians had been in charge other faiths would probably not be permitted to conduct their services here. He later paid a visit to Bethlehem and, ever considerate, bought rosaries for his Catholic friends at home.

Returning to Aleppo in early September Niebuhr spent a quiet interval polishing his maps and notes. Leaving in a caravan for Constantinople, he crossed Anatolia in three months, passing through Konya, Adana, Izmir, and Brusa. The party was made up largely of Greek merchants although Muslim *katerjis*, or muleteers, drove the animals and a *karawanbashi* handled formalities with the authorities they encountered along the way. The Greeks were cringing in public, although contemptuous of the Turks in private. In Konya there was the great dervish monastery of the Mevlevia order and Niebuhr made the acquaintance of two monks there. On a return visit the three consumed several bottles of wine. It was his first—and apparently only—exposure to this more mystical branch of Islam. The Pasha of Konya was of three tails but did not appear to be greatly feared, the dervishes daring to openly defy him. Niebuhr noted rather caustically the Turkish adage that a man courted abuse if he argued with a Sherif in Damascus, a janissary in Erzerum, a Jew in Salonica, or a dervish in Konya.

In Brusa the onset of Ramadan fell shortly after his arrival and at the khan where he stayed a Turkman sat outside during the day and smoked his pipe in full view of passers-by, a practice unthinkable in Arabia or Syria. On the evening of the 30th of January there was a sharp earthquake but the substantial stone buildings withstood it well. He arrived in the Ottoman capital for the second time on February 16th 1767, nearly six years since he had convalesced there on the outward voyage. He spent nearly four months in Constantinople, devoting his attention to areas he had not seen and refining his map of the city.

He left in early June with a caravan accompanying a janissary agha to a posting on the Danube. Among the first sights to greet them on the road to Edirne was the spectacle of two men impaled by the side of the road. They were accused of robbery and murder and had been executed that morning.

Afterward

It was grim introduction to travel in Rumelia. This was not Iraq or Kurdistan or Syria where single travelers may have been plundered but were rarely harmed, much less murdered. They were now on their guard. It was forty-four and a-half leagues or just over 135 English miles to Edirne, where they arrived on the 12th of the month after a journey through beautiful fields, most of which now lay fallow. The city had been taken by Sultan Murad in 1320 and alternated with Brusa as the imperial capital until Constantinople fell in 1453, and it was graced by many beautiful mosques.

They continued through mixed Turkish and Bulgarian villages, the Christians being forced to put up the agha and his retinue for only a nominal fee. Niebuhr stayed in public khans and paid for what he consumed. Increasingly, the populations were Christian but the baleful effects of rule by pashas continued, only now the tyrants were Greeks and Christians. On the 24th they reached the Danube, to Niebuhr's practiced surveyor's eye 800 to 1,000 feet wide and running in spate. The inhabitants were a mixture of Turks, Wallachians, and Bulgarians, the last two belonging to the Greek Church. On the next day he crossed the Danube to Ugru and Wallachia. The heavy weather broke in Ugru and he was able to make a sighting of the sun. Based on the reading he determined the latitude to be 43° 53'.

In Bucharest the plague was raging and he was in some haste to be clear of the city, although there were compensations for Niebuhr, including the friendly sound of church bells and the sight of beautifully dressed—and unveiled—women in the streets. On the 17th they reached Choczin on the Niester, the border with Poland. A fortress manned by janissaries was the last vestige of Ottoman dominion. The city was the typical polyglot border scene, full of desperados, Turks, Prussian officers serving as mercenaries, Gypsies, Kyrgiz, Jews, Lipker Tatars, and Christian Moldovans. Niebuhr was briefly felled by a bout of fever and vomiting, but he resolved to press on, now so close to home. On the 18th of July, he dismissed the drayman that had accompanied him since Edirne and crossed by ferry to the Polish city of Zwaniec. On this first evening in Christendom after nearly six years in the East, he set up his quadrant to take a sighting. He attracted a crown of over a hundred Christians and Jews who took him for an astrologer, wanting to know if their wives would be fertile, their daughters would marry, etc. etc. He was in Europe again but, for all of its pretense to scientific rigor, he might as well have been in Cairo or Suez.

In Poland he passed through Kamaniec, taken by the Turks in 1692, re-conquered seven years later, and full of Armenians and Russians in communion with Rome. There were also a few Lutherans, German manufacturers of quality wares, but the Polish authorities would not allow them to build a church, although Jews were allowed to build their synagogues and the Lipker Tatars their mosques. The country was also full of Jews, mainly merchants and tenant farmers, and most of the public houses seemed to be owned by Jews. The Polish clergy had prevailed on the Russian and Greek peasantry to reject their own clergy if not united with Rome, but most peasants seemed indifferent to the dispute. The constant pressure had even moved some Catholics to convert to Judaism.

From Kamaniec to Lemberg (or Lvov) he passed through an area of beautifully cultivated fields, it seemed as many as Niebuhr had seen between Edirne and the Polish border. He found travel in Poland less convenient than in Turkey with its regular system of caravans. Here one traveled alone and, without the regular pace of the loaded animals, this made the determination of distances more difficult. In the public houses Jews accommodated Christians even on the Sabbath, accepting payment in an apron that covered the hand, which apparently made the exchange lawful. In Lvov the peasants groaned under the yoke of the clergy. The wealth of the city seemed to be concentrated in magnificent churches and monasteries and the religious orders were busily increasing their holdings through forfeited debentures and estates left to the Church. Protestants were barely tolerated and attendance at Mass on feast days was as carefully monitored as mosque attendance in the Sunni lands.

On the 8th of August Niebuhr left for Lublin and then Warsaw. He arrived in the capital on the 18th and spent a very pleasant two and a-half weeks, finding the King to be unusually sagacious and welcoming to foreigners. If he was able to indulge his predilections for art and science Niebuhr was certain that Poland would soon take its place among the most enlightened nations in Europe. On the 6th of September he resumed his journey, reaching the German border four days later. From here to Copenhagen the road was well documented and only then does our *billigdenkender Reisender* (fair-minded traveler) lay down his pen. He reached the Danish capital on the 20th of November 1767, 6 years ad 313 days since his departure. A third volume, published posthumously in 1837, covered this last part of his journey. It includes the usual maps and descriptions and all the astronomical observations, listed to within a second of accuracy, made on this nearly seven-year odyssey.

15

The Results

> How many accounts of Palestine and Egypt do we have, all full of repetition and useless notions about the supposed holy places? . . . But if I may be permitted to observe the accounts of Happy Arabia are very small in number. Nature there has spread riches of which we are still entirely ignorant. Its history goes back to the highest Antiquity; and the idiom spoken there differs from that of Western Arabia, which we know; and as this idiom has been until now the surest light to guide us in the Hebrew language, how will new insights on that most important of books, I mean the Bible, be possible if we are not able to attain a knowledge of the dialect of Oriental Arabic to the same degree that we understand that to the West of this land? (*Fragen*, Introduction, v.)

The "Royal Danish expedition to Happy Arabia" merits, at best a footnote in the history of the European penetration of the peninsula. A number of anomalies characterize the title, including the fact that the majority of its members were not Danish and that they spent only a fraction of their time in Happy Arabia. Their work is remembered largely because of the contributions of a German, Carsten Niebuhr, most of which had to do with other parts of the Arab and Muslim Worlds. But it is useful to remember the original purpose of the expedition in any assessment of the results. This was nothing more or less than to equip European scholars with the knowledge to better understand the Hebrew Scriptures. There was plenty of what might be called profane knowledge of the Orient in Europe, and even a great deal of nonsense about the "supposed holy places" in Egypt and Palestine. But this was not what the expedition's prime mover, Professor Johann David Michaelis, had in mind. He was interested in that remote part of the Arabian peninsula where the dialect of the Eastern Arabs, the "surest idiom" to an understanding of Hebrew, would guide European scholars to an understanding of "that most important of books," the Bible.

Michaelis was unapologetic about this narrow frame of reference: to those who might object to the theological nature of his questions, he would say that all of them were important and would lead to a better understanding of the Scriptures. And an understanding of the Scriptures would itself contribute to more general scientific knowledge. The expedition would be the vicarious means by which scholars—men like Michaelis himself—would do their research. The travelers would be surrogates, prepared beforehand, schooled in the languages and disciplines necessary to gather the raw materials that would then be assembled and analyzed in the laboratories of Europe. So it is not out of place to assess, against its original purposes, the results of this extraordinary, not to say quixotic, undertaking, painstakingly planned by the foremost Oriental philologist in Europe, guided by a set of specific questions from scholars on the Continent, whose progress was periodically monitored in Copenhagen through the distant agency of the Danish ambassador in Constantinople.

Given this frame of reference, it seems clear from the outset that the expedition was bound to fail. Part of the reason lay in the lack of a practical, commonsense spirit that would animate the instructions drafted by Michaelis and issued by the King. It would also fail, first, because the expedition was made up of different and often conflicting personalities, with no clearly designated authority in the group. There was no guidance as to how they should regulate their internal affairs other than the utterly impractical admonition that

> There should be perfect equality among the scholars who undertake the journey; and no one among them shall arrogate to himself any kind of right or superiority over the others. The maintenance of peace and harmony shall be their constant and principal preoccupation . . .[1]

Indeed, harmony *was* their principal preoccupation, but not for the reasons the Danish authorities had in mind. As we have seen, the clashes of personality soon threatened to take what some members believed to be a murderous turn. The expedition was built around the two professors, Forsskal and von Haven—although some would argue that Niebuhr made it a triumvirate—with the others playing supporting roles. But the two were jealous of their prerogatives, obsessed with matters of precedence, and bitter enemies of one another. The others—the phlegmatic Kramer, the dutiful Baurenfeind,

---

1. *Fragen*, XXIII. They seemed as carelessly thrown together as consulting teams in the twenty-first century.

and the quietly industrious Niebuhr—tried, we suspect unsuccessfully, to stay out of the crossfire. The *Fragen* instructs the members of the team to decide everything democratically, by vote. But this, like so much else in that learned but impractical document, was unhelpful in regulating the everyday affairs of the expedition. The incident over the arsenic in Rhodes showed how raw were the nerves of the five men only a few months into the journey.

The year of their enforced stay in Egypt had been a productive interregnum, with the members living in separate quarters and carrying out their tasks independently. The fear of von Haven's allegedly malign intentions may have lessened by the time they reached the Yemen, their ultimate goal. But other hostile forces had arisen to take their place. The customs officials in Mocha were openly contemptuous of their scientific pretensions and scandalized at the preservation of their specimens in alcohol. As a result, their baggage was roughly handled when they finally arrived in Happy Arabia and many of the specimens they had so laboriously collected were destroyed. They were prey to superficially Europeanized merchants who used their knowledge of European ways to ingratiate themselves with the members of the expedition and lead them astray for their own purposes. The assistance of these alleged "friends" was no more helpful than the extortionate demands of the officials they would later deal with.

But their greatest enemies were the forces of nature. The expedition would also fail because the health of the members failed. By the time they arrived in the highlands of the Yemen, the two professors who were the keys to their quest were dead. The survivors themselves were ill, harried by officials on their journey up the escarpment to Sana and aware that the last English ship that year—their means of escape from the nightmare—would leave for Bombay in less than a month's time. They spent less than three weeks in the city in spite of the fact that their reception had been cordial and a house had been arranged to accommodate them for a stay of several years. Ultimately, their illnesses were not a failure of medicine, or of preparation, or even of a willingness to take proper precautions. Without the immunities possessed by Arabs, these Europeans were simply unable to withstand the diseases that were endemic in the areas through which they traveled. A physician, Kramer, traveled with them. But European medicine of the day was worse than useless under the circumstances, and seemed to consist primarily of cupping or bleeding the victims. For patients suffering from malaria, a disease that attacks and destroys red blood cells, bleeding was probably the worst prescription possible.

Of the five, only Niebuhr survived their eight-month stay in the Yemen. All the others died, probably of malaria, and Niebuhr suffered from recurring bouts of the disease for the rest of his life. Niebuhr's survival may have been purely a matter of chance, and he almost died of severe dysentery in the Greek Archipelago in the early days of the expedition. More than the others, he had accustomed himself to local conditions and so may have built up immunities to factors—dietary and otherwise—that contributed to the death of his colleagues. But, more probably, he was just lucky. Death seemed utterly impartial, carrying off Forsskal who was active in the countryside as well as von Haven who rarely seemed to leave his bed. The spectacle of a group of northern Europeans presuming to travel through the most inhospitable of climates, all the while resolutely refusing to make the most rudimentary concessions to local conditions, may seem astonishing today.

Niebuhr and Forsskal were the exceptions, but the rest continued to have their food prepared in the European fashion by their cook, drank only wine and brandy when available, and admitted the locals to their society only to the minimum extent necessary to what they conceived to be the success of their work. It is no wonder that they were constantly ill and misled. Niebuhr mentions that only by allowing a Bedouin to ride with him on his camel was he able to obtain reliable geographical information in the Sinai. For some of the others, such familiarity was beneath their dignity. It is tempting to say that their arrogance killed them. But arrogance probably had nothing to do with it: disease would have taken its relentless course, whatever the mindset of the victims. There seemed to be an utter capriciousness about their deaths. These Enlightenment scientists, armed with their science, their discipline, and their strange remorseless God were, in the end, defenseless against the most insignificant of insects. If they presumed to confront science in all of its complexity, let them feel its terrible effects. God and science were one in their formulation, and the expedition was an explicit attempt to reconcile that unity. However, only in Niebuhr do we sense the requisite suppleness, the resolve to push science to the limit and then, but only then, to abandon himself to the mercy of his God.

Finally, the expedition failed because it was based on a faulty premise: that there were inhabitants of the Yemen who spoke an idiom of "Eastern Arabic" that could provide enlightenment on the meaning of uncertain words in the Hebrew Bible. What better way to explicate difficult passages of the Hebrew Bible, reasoned Michaelis, than to ask native speakers of the language what particular words meant. In addition, since the flora of the Yemen did not differ greatly from that in other parts of the Near East, a botanist could profitably study the subject first hand and correlate the names he found in 1762 with those that appeared in the Bible. One of the happier events in Niebuhr's account is his description of Forsskal's delight at finding a balsam in full bloom, the plant that produced the Balm of

Gilead.[2] But its like was a rare occurrence and it is here, when measured against detailed questions of Michaelis, that the return on the investment in the expedition seems particularly slender.

The *Fragen*

The one hundred questions as fully developed in the *Fragen*, fill over 200 pages with minute detail. As an example, Michaelis's question XXXIII has to do with locusts and other insects that damage crops. It cites verses in Joel, Leviticus, Nahum, Jeremiah, and Deuteronomy where the depredations of these insects are mentioned. Michaelis refers to the Septuagint, the Greek translation of the Pentateuch made in Alexandria in the third century BC, as well as to Syriac and Chaldean translations of the same document, and to current Arabic words for insects. He is clearly interested in understanding the differences between locusts, caterpillars, maybugs, junebugs, click-beetles, worms, and lice as rendered in the several languages. He asks the travelers to note whether locusts appear in successive swarms in the same year, the time of the year in which locusts generally appear, whether caterpillars appear in the fields before the locusts and maybugs afterwards feed on what the locusts leave behind, and whether beetles eat roots in August and September, as in Europe, before turning their attention to wheat in the winter months. Finally, the Greek version of Joel and Deuteronomy contains a word that appears to refer to "corn blight" or "wheat rust" and they should ask modern Greeks if the word doesn't, after all, refer to an insect.

In response, Niebuhr devotes several pages in the *Description of Arabia* to locusts, from the great swarms he saw Cairo in December 1761 and in Jidda in November of 1762, to the swarm that arrived in Mocha in the summer of 1763 to feed on the ripening dates. They all came from the west, not from the east of Scripture. In Persia in 1765 he saw them in the spring. He describes their appearance, how they were prepared for the table by Arabs of the Yemen, and provides their Arabic names in al-Ahsa in the Arabian peninsula, and in Oman and Basra. He appears to answer some of the questions of Michaelis, although it is hard to see how the information he provides can help Michaelis unravel the question as to whether the successive scourges described in Joel 1:4 were four species of locust or (as it appears) four different insects. The prophet had said:

---

2. See above, p. 269.

Niebuhr in Egypt

> That which the palmerworm hath left hath the locust eaten; and that which the locust hath left hath the cankerworm eaten; and which the cankerworm hath left hath the caterpillar eaten.

We might ask today whether the passage expresses scientific fact, and can be related to the habits of palmerworms, cankerworms, locusts, and caterpillars, or is simply an imaginative way of expressing the devastation that will be visited upon the land of Zion. Michaelis clearly believes that it is the former and asks the travelers to look into the possible occurrence of several scourges visiting the fields in succession.

We probably have no choice but to accept Michaelis's contention that "to correctly understand Oriental languages, it is necessary to learn them all together ... Arabic, Ethiopian, Syriac, Chaldean, Hebrew, and Samaritan to properly understand the analogies that exist in these languages ..." or that a knowledge of "Eastern Arabic" had been the surest guide to a knowledge of the Hebrew of the Bible. However, this eighteenth-century attempt to relate current usages to a language spoken two thousand years previously was probably condemned to the same failure as the attempt to relate the physical characteristics of the Gulf of Suez in 1762 to the purported crossing of the children of Israel in the second millennium BC. If anything, the changes in language had been even greater than changes in the mean level of the tides. Niebuhr fills eleven pages of the preface to the *Description of Arabia* with his and Forsskal's answers to some of the specific question of Michaelis, and other answers are littered throughout the text, as we have seen above. But little of the information can have been of much use to the scholars who labored to understand the Hebrew of the Old Testament, if only because the answers seem perfunctory.

This is not to suggest that the study of the Bible as science or history was unimportant, or that it would not continue to interest scholars. The preface to a new edition of Edward Robinson's[3] translation of Gesenius's *A Hebrew and English Lexicon of the Old Testament* made it clear that the subject of the Old Testament as an historical document would remain an important one for many years to come:

> The need of a new Hebrew and English Lexicon of the Old Testament has been so long felt that no elaborate explanation of the appearance of the present work seems called for. Wilhelm Gesenius, the father of modern Hebrew Lexicography, died in 1842 ... the last revision of Robinson's Gesenius was made in 1854 and Robinson died in 1863 ... In the meantime Semitic studies

---

3. We saw him as that rare combination of a first-rate biblical scholar and traveler to the region. See chapter on Suez and Sinai.

have been pursued on all hands with energy and success. The language and text of the Old Testament have been subjected to a minute and searching inquiry before unknown. The languages cognate with Hebrew have claimed the attention of specialists in nearly all civilized countries. Wide fields of research have been opened, the very existence of which was a surprise, and have invited explorers. Arabic, ancient and modern, Ethiopic, with its allied dialects, Aramaic, in its various literatures and localities, have all yielded new treasures; while the discovery and decipherment of inscriptions from Babylonia and Assyria, Phoenecia, Northern Africa, Southern Arabia, and other old abodes of Semitic peoples, have contributed to a far more comprehensive and accurate knowledge of the Hebrew vocabulary in its sources and its usage than was possible forty or fifty years ago."[4]

The preface goes on to review recent developments in all these areas, with a special mention of the "brilliant and suggestive" work done in Germany. So, the work originally begun by Michaelis was carried forward by other laborers in the field and produced the substantial results described above. Michaelis himself is credited for his seminal efforts in the Gesenius work, and several travelers—including Charles Montague Doughty, J. L. Burckhardt, and a later traveler in Yemen, Joseph Halevy—would eventually be credited with contributions to the gradual increase in knowledge of the Semitic languages. Niebuhr is not mentioned, although we have already noted that Robinson was the only English-speaking scholar who appeared to be familiar with Niebuhr in the original German. But the tools with which Niebuhr and his companions were equipped were inadequate to a task that, on the face of it, was extremely difficult if not impossible. Even had their preparation been more complete—or had Michaelis himself traveled to the Yemen—there was simply not enough time to address linguistic questions as abstruse as those Michaelis poses.

But if the ostensible purpose of the Danish expedition to Happy Arabia was based on a flawed assumption, still by that measure it was not a failure. Even if not widely recognized, Niebuhr's work in transcribing other inscriptions—hieroglyphic, Himyaritic, and cuneiform—widened the field of the "cognates of Hebrew" and incorporated other Oriental languages into the corpus, an understanding of which would bear on the Hebrew Bible. And it is probably time that we set aside this narrow biblical frame of reference and examine the contributions of the expedition in a larger, more secular context. The manuscripts bought by von Haven and the collections of botanical and zoological specimens made by Forsskal—at least those that

4. Gesenius, v.

survived—eventually made their way to Copenhagen where they came to constitute the bulk of the royal collections. What do these collections tell us of the contributions of the other members of the expedition? What of Forsskal, who with Niebuhr had been tireless in his explorations? A student of Linnaeus[5] in Uppsula, Forsskal was not the first—nor would he be the last—student of the great man to perish on what ultimately constituted missionary work in the new secular religion of biology.

The Contributions of Forsskal

While he was in Egypt Forsskal himself sent packets of dried seeds to Copenhagen, as well as to botanical gardens in Uppsula, Paris, Leiden, and Montpelier.[6] The seeds sent to Uppsula and Copenhagen were germinated and examined by Linnaeus and it is on the basis of this examination that Linnaeus named a genus of plant, *Forskalea tenacissima* after the younger man. The adjective *tenacissima* was carefully chosen, Forsskal not being the kind of man to take no for an answer. In addition, an unknown number of chests containing plants were sent from Bombay by Niebuhr. These probably contained Forsskal's collections made in the latter part of the year in Egypt as well as material gathered in the Yemen. The chests arrived in Copenhagen and were delivered to the Royal Botanical Gardens where they were examined by the botanist Johann Zoega over the period 1768–70. The total number of plants collected on the journey is estimated at 1,500 to 1,800, the dry material probably preserved in paper, as collections would still be today. Some 1,300 samples are still preserved in the Botanical Museum in Copenhagen, as well as in other European herbariums in Lund, Berlin, St. Petersburg, and Paris. While eighteenth-century methods cannot compete with those of today, the Forsskal material remains as a well-preserved example of the passion for collections of the period. In fact, according to Danish botanical historian Carl Christensen, it represents the Botanical Museum's greatest treasure.

But it is for more than his contribution to the collection of a national museum that Forsskal should be remembered. Instead, he should be seen as a botanist of rare promise whose work was cut short by his premature death. Unfortunately, it was left to others, whose knowledge of the material was often incomplete, to preserve his memory. After he had seen his *Description of Arabia* through to completion, Niebuhr began in 1773 to assemble

---

5. See "Introduction" where we saw him as the most influential botanist and naturalist of the age and the developer of the system of binomial nomenclature still in use today.

6. See Friis "The Botanical Results of the Arabian Journey."

Forsskal's notes for publication. The first were his botanical papers. But Niebuhr was not a botanist and his knowledge of Latin was insufficient to an understanding of the elaborate classification system. So he engaged an unknown Swedish botanist to prepare the material for publication. This first work appeared in 1775 as *Flora aegyptiaco-arabica*. Unfortunately, many of the inconsistencies in the notes, matters which Forsskal himself probably would have seen and corrected, made their way through to publication. Even with the inconsistencies it is recognized as a classic of descriptive botany. It listed fifty-four new genera, of which twenty-four are still recognized as scientifically correct. Of 656 plant species listed, over 300 are still recognized. Martin Vahl in the late eighteenth century and P. Ascherson and G. Schweinfurth in the late nineteenth, devoted considerable labor to the review of Forsskal's material. Schweinfurth himself visited the Yemen and published his results over the period 1889–99 as *Results of an expedition to Yemen in memory of Forsskal* (translated from the Latin). In 1922, the Danish botanist Carl Christensen published the first complete review of Forsskal's herbarium. After the Second World War, the English botanists F. N. Hepper and J. R. I. Wood, with the collaboration of Danish botanists, undertook the task of identifying anew the plants of Forsskal's herbarium from both Egypt and Yemen.

According to the procedures for assigning botanical names, the name of the researcher who discovers a genus or species is added to the scientific name of the plant. The entry "*Catha edulis* Forsk.," for example, would indicate that this particular genus was discovered by Forsskal. A measure of the value of his work in Egypt in 1761–62 is the attribution to Forsskal of over 130 genera and species in Vivi Täckholm's *Students' Flora of Egypt*, more than to any other single researcher. The physical geography of Egypt—with its Mediterranean, Sudanese, Sinaitic, and Asiatic ecosystems—made Forsskal's work in that country particularly important, as having applicability to a wide band of desert and savannah stretching from sub-Saharan Africa to Central Asia. So Forsskal's work had wider application than to Egypt and Yemen alone. His collections from the mountains in the Yemen would have parallels even in the more mountainous and humid areas of Africa such as The Gambia and Cameroon. While Forsskal's botanical work broke no new theoretical ground, it was a practical achievement of no small importance in the development of the science. It can only be regretted that Forsskal himself, a man who combined energy with an inquisitive scientific mind, did not live to develop and publish what he had so painstakingly collected.

A second book, *Descriptiones animalium, avium, amphibion, piscium, insectorum, verminum quae in itinere orientali observavit P. Forskal*, described, as the title suggests, the animals, birds, amphibians, fish, insects, and vermin

Forsskal found in his travels in the Orient, at least those that survived the handling of the customs department in Mocha. It was published in 1775, also with the helping hand of Niebuhr. And finally, in 1776 Niebuhr followed with publication of the third of the Forsskal works, the *Icones rerum naturalium* based on Forsskal's notes on both plants and animals, complete with copper engravings prepared from Baurenfeind's drawings. As with the *Flora aegyptiaco-arabica*, portions of the material were reviewed by the botanist Johann Zoega.[7] Like Forsskal, Zoega was a student of Linnaeus and had been

---

7. Interestingly, as we have seen, it was another Zoega (1756–1809) who developed Bartélémy's correct conclusion that the cartouches in the Egyptian hieroglyphs contained royal names. The dates would suggest that these two Zoegas were not the same man, but the name is unusual enough that there was probably some connection.

appointed to the Royal Botanical Gardens at Amaliagade where he worked in the early 1770s identifying plants in Forsskal's collection. Unfortunately, in the turmoil surrounding the reorganization of the University at Streunsee, Zoega was dismissed and the review was left incomplete. However incomplete and unedited, the three publications stand as a lasting testament to the short, promising career of Petrus Forsskal.[8]

Von Haven's Legacy

What of the philological contributions of Frederik Christian von Haven, the Danish professor in whom such high hopes were entertained by the authorities of that nation. It is probably safe to say that Michaelis was not rewarded with any striking insights as a result of the contributions of the philologist. As we have seen, he was bitterly disappointed at the failure of von Haven to gain access to the library at St. Catherine's Monastery in the Sinai. From the Hansen book a picture of von Haven emerges as not only difficult in the extreme, manipulative, and self-important, but also uninterested in anything resembling the kind of hard work that animated Niebuhr and Forsskal. He was unable to endure life at sea on the outward journey and secured permission to proceed overland to Marseilles where he later met the ship on its arrival. In Cairo, he appeared to prefer the amenities of the French consulate to the working atmosphere in the quarters of the other men. This was in a city that probably offered more raw material for the student of Oriental languages—Arabic, Hebrew, Coptic, and ancient Egyptian, not to mention Turkish, Greek, Circassian, and the other languages of the elites—than any city in the world. After a year in the relatively cosmopolitan Cairo, the Yemen was an obvious exile and, as we have seen, von Haven rarely seemed able to leave his bed during the five months he spent in the Tehama.

But is this picture of von Haven fair? The broad outlines of this picture of his personality seem to be accurate, although there is nothing in Niebuhr's account that suggests particular difficulties with von Haven. But Niebuhr was too fair-minded to have left anything critical in print about another member of the expedition, although his silence may speak volumes. Privately, he appears to have been more forthcoming and his biographger suggests that he believed von Haven to be "indolent, haughty and conceited." But von Haven was not idle in the Orient. Even in matters of philology, the expedition was not without successes and some of the credit for these must go to him. If he was

---

8. A fuller account of Forsskal's contribution in the area of botanical research and nomenclature is provided in *The Plants of Pehr Forsskal's Flora Aegyptiaco-Arabica* by F. Nigel Hepper and I. Friis, published by the Royal Botanic Gardens, Kew in 1994.

not a tireless worker in the field, he at least was responsible for the purchase of manuscripts and here he left a solid record of performance. As he had with the botanical and zoological material collected and catalogued by Forsskal and drawn by Baurenfeind, so Niebuhr was able to send from Bombay many of the manuscripts the expedition had assembled over the two years since they had arrived in the East. In addition, von Haven's two-volume journal has survived as a record of the daily activities of the expedition. The first volume consists of copies of Arabic poems and manuscripts, historical extracts, miscellaneous notes on such things as the principal mosques in Cairo, a catalogue of the purchases of manuscripts, and lexicographic notes. The notes, filling 287 of the 462 pages, make up the bulk of the journal and include comments on Arabic-Latin, Arabic-Italian, Arabic-Danish, and Hebrew lexicographies. While they are a record of careful scholarship, it is probably safe to say that they contain no groundbreaking linguistic material.

The second volume is made up of maps, extracts in several languages—Arabic, Latin, French, German, and Greek—as well as miscellaneous notes on theology. But the bulk (404 pages out of 700) is made up of the "journal" itself, that is, von Haven's diary as the expedition traveled from Copenhagen to Cairo, Sinai, and the Yemen. The formal entries end on the 26th of March, 1762, but there is a separate, loose-leaf account of the journey into the Sinai and a large number of brief notes—obviously jotted down before being transferred to the journal—leading up to von Haven's death in Mocha in May 1763. In his own words as preserved in the journal, we have a record of von Haven as a man of some wit and perspicacity, although this was perhaps a slender return on the King's investment.[9]

Of the collections themselves, the von Haven material included seven valuable manuscripts of the Hebrew Bible purchased in Egypt. One in particular was to become one of the treasures of the Royal Library's collection. It was written on parchment in "precise Sephardic type"[10] and lavishly ornamented. The history of the document, contained in a note to the first volume, states that it was inherited by the son of Rabbi Joseph ben Rabbi Judah ibn Hanin, who emigrated from Spain to Tunisia in 1492. Presumably, both the man and the manuscript subsequently made their way to Cairo as part of the influx of Sephardic Jews that was to change the face of the Jewish community in Egypt in the last decade of the fifteenth century. The manuscript is dated to some time earlier in the same century. In an early example of scholarly cooperation, the seven manuscripts were lent by Bernstorff to Benjamin Kennicott of Oxford University shortly

9. As we saw above, the diary was published early in the twenty-first century.

10. This review of von Haven's work is based on Stig Rasmussen's *Den Arabiske Rejse*, 342.

after their arrival in Copenhagen. Kennicott, who had worked for many years to produce a standard edition of the Old Testament, added the seven to others he had examined in making a comparison of the various versions in which it appeared. The result was the publication of *Vetus Testamentum Hebraicum cum variis lectionibus* ("The Hebrew Old Testament with various readings").[11] So, there was a legacy for von Haven, after all, in the work of later men who would pursue the study of the Hebrew Bible. And lest the picture of von Haven be colored too greatly by his obvious distaste for the hard work of exploration, the conclusion of the reviewer in *Den Arabiske Rejse* is worth noting:

> A thorough study of F. C. von Haven's diary . . . leaves the impression of von Haven as a competent orientalist, well-versed in Arabic dialects and literature, and with a sound knowledge of Hebrew philology and Jewish practices. The Hebrew manuscripts he bought were chosen with great care and acumen, the evidence of a competent and knowledgeable man.[12]

Of memorials to Kramer and Baurenfeind, other than the engravings of the latter in the Forsskal and Niebuhr books, there is little record.

Niebuhr and the Orient

And what of Carsten Niebuhr, the painstaking Saxon? As we saw in the preface, in the spring of 1984 the Royal Library in Copenhagen, in cooperation with the Danish Foreign Ministry sponsored an exhibition "The Arabian Journey 1761–1767" in Riyadh, Saudi Arabia. The exhibition was expanded and shown in Kiel in Schleswig-Holstein and Meldorf in the Ditmarches where Niebuhr was posted after his return. It is perhaps fitting that the "Danish expedition to Happy Arabia" had metamorphosed into the "Arabian Journey," and that Saudi Arabia should have been the venue for the exhibition. This metamorphosis would seem to have forgotten the biblical overtones of the expedition and the predilections of its sponsor, Johann David Michaelis. It would, instead, remember primarily the contributions of Carsten Niebuhr, whose "Arabian" travels the expedition most particularly was. Forsskal, von Haven, Baurenfeind, and Kramer are, at best, footnotes to the story. There is perhaps justice in this. Niebuhr's *Description of Arabia* is, in some respects, the most important of his works and it is a truly a description of Arabia writ large—of the peninsula in general,

11. See Rasmussen, *Den Arabiske Rejse*, 345.
12. Ibid., 346.

the Yemen, the Hadramaut, Oman, the Trucial States, Najd, the Hejaz, and Sinai. His contributions were not in the area of biblical scholarship, the ostensible purpose of the expedition, although his work in the inscription of other "Oriental" scripts cannot but have helped to advance the process of understanding that Oriental document. His careful records of latitudes and occasional longitudes were used for years afterwards to correct maps of the eastern Mediterranean and the Red Sea, and his maps of Cairo, the Egyptian Delta, and the Yemen were unmatched for detail in his day. But he is not remembered primarily as a cartographer.

Instead, he put into liberal effect the admonition from Michaelis to interest himself in everything he saw in the East—the history, language, geography, commerce, form of government, religion—of the societies in which he traveled, and it is primarily as an "Orientalist," for want of a better term, that he is remembered. The Orient he saw included not just the Arabian peninsula, but also Anatolia, Egypt, the Levant, parts of India, Persia, Iraq, Kurdistan, Greater Syria, Anatolia, and Rumelia or European Turkey. Indeed, there are several Niebuhrs for us to see, from the Niebuhr of the Yemen, the subject of Hogarth's chapter in *The Penetration of Arabia*, to the Niebuhr of Egypt, the subject of the present work. But there was also Niebuhr the painstaking cartographer, Niebuhr the, at first diffident, then increasingly confident copyist of scripts and, finally, Niebuhr the interested observer of the religious kaleidoscope of the Orient. What he brought to all these efforts was a careful scholarship, a scientific temper, a native shrewdness, and a willingness to see other peoples not as threats to his own way of thinking but as manifestations of the diversity of human experience.

In the early exuberance of its deliverance from the shackles of ignorance, Europe of the Enlightenment approached the world with boundless self-confidence. It thought that it had all the answers and, in the absence of a competing view, perhaps the attitude was understandable. But, to those who would read between the lines, the work of travelers like Niebuhr was a regular reminder that Europe did not. He traveled as the surrogate of no country, not of Denmark his nominal sponsor or of his native Friesland in Lower Saxony. Rather, he brought his natural commonsense to bear on what he saw. Any tendentious use to which the information he gathered was put, was not his business. In the end, he resisted Michaelis's attempt to dictate what he should see and hear. We can be grateful for this, not because of Michaelis's shortcomings but because it gave the necessary scope to Niebuhr's native shrewdness and judgment.

Deprived of the society of his friends after Bombay, Niebuhr decided that the most effective way to travel was as a native and he did so for the remaining four years of his journey. While a willingness to see value in the ways of others was a characteristic of the man, and shines like a beacon throughout his accounts, his decision to accommodate himself to local ways of doing was not a statement of principle but a matter of survival. Not only did he learn to eat the local food and live in the local fashion, but his lower profile avoided exciting the attention of the powerful that had so plagued the expedition in the Yemen. His perspicacity from this lowered perspective remained intact. Niebuhr's careful records, collected over the seven years of his travels, are a testament to his powers of observation and his scientific temper. His endurance during the travels and, even more so, in the lonely years after his return to a remarkably disinterested scientific community in Europe, were testimony to his perseverance in the face of difficulty. In the end, this epitome of the believing, Enlightenment scientist might have conceded that God had a plan for him. If so, we are its beneficiaries.

Niebuhr's Work of Publication

Shortly after his return to Copenhagen in 1767, Niebuhr set about ensuring that theresults of the expedition would not be lost to the world of scholarship. Whether because so many of the members of the expedition had died, or perhaps because the hopes of the scientific community had died with them, the return of the *lieutenant des ingénieurs*[13] to Copenhagen caused hardly a ripple in scholarly circles. Professor Michaelis seemed curiously uninterested in the results of a journey he had been so instrumental in setting in motion. Mayer was dead. But circumstances had also changed. Frederick IV had died and Bernstorff's days were numbered under the new regime. He remained long enough to secure for Niebuhr a small grant to prepare his accounts for publication, but it apparently covered only the *Description of Arabia*.

Niebuhr paid for the Forsskal volumes and his *Travels in Arabia* from his own pocket until the money ran out. The financial scrupulousness that so impressed Bernstorff at the outset of the expedition was no less in evidence on his return. Niebuhr's modest final accounts—minus disbursements made for expenses he considered personal—were evidence that in Niebuhr the Danish government had made a very sound investment indeed. This did not, however, keep them from presenting him with a bill for the cost of the copper engraving of the plates that appeared in the several volumes. Niebuhr would thereby own the plates, but that investment was later lost in 1795 in a fire that consumed the Royal Palace in Copenhagen before spreading more widely in the city. The loss included not only those plates that had already

---

13. He was later made a *capitaine*.

been published, but also those for the planned third volume. It was a heavy blow. Volume III of the *Travels* was not published in Niebuhr's lifetime.

There were several reasons for the gap in publication. In the first place, in spite of the issuance of his own and the Forsskal works, he was still not confident of his skill as a writer. Niebuhr had also made a speculative investment in the Danish East India Company during the American war and when the bubble burst the investment was lost. He was not about to sacrifice the patrimony of his children to further publicize his travels. In addition, his contributions to the journal *Das deutsches Museum* early in the new century, consisted of portions of what eventually became vol. III and he saw no reason to publish them separately. Finally, his own concerns about the astronomical observations that supported the lunar distances method of longitude determination were not laid to rest until they were later checked and proved to be sound. They may not have been particularly useful as a method for navigators at sea, but they were accurate and the 124 pages of *Niebuhrs astronomische Beobachtungen* would eventually constitute Appendix I of the posthumous third volume. More than one offer to publish in English was received and entertained before Niebuhr decided that his first loyalties should be to the country that had financed the expedition and to the language he himself spoke. His second thoughts about the offers came too late and the third volume appeared only in 1837.

For the eleven years after his return Niebuhr occupied himself with the publication of his own and Forsskal's works. There was a tremendous amount of work to be done, not only in drafting the text from his and Forsskal's notes and putting the manuscripts into suitable form for printing, but also in engaging artists, printers, and engravers. As we have seen, the three Forsskal volumes were published in 1775 and 1776. They were a disappointment, largely due to the lack of a technical editor well versed in both Latin and the specialized language of botany and biology. The lack of collaborators was a constant problem. When Niebuhr asked Michaelis to carefully examine the draft of the *Description of Arabia* and correct any errors, he was disappointed to receive the document back virtually unchanged. With the rebuff, Niebuhr turned instead to Johann Reiske, a fellow Saxon and a schoolmate of Michaelis in Halle. A difficult and irascible man, whose Latin was supposedly barbarous and his religion suspect, Reiske nonetheless had the best knowledge of the Arabic language and literature of any European of his time, and his thirty-five major works included translations of Abulfeda and Marai and contributions to d'Herbelot's *Bibliotheque Orientale*. Reiske provided translations from the Arabic of two dedicatory texts on bridges over the Nile at Giza, as well as the Kufic inscriptions copied in the Yemen.

The Results

The absence of the hand of Michaelis in the publication of the results of an expedition that he had virtually created remains, on the face of it, a puzzle. But seen from his viewpoint, his attitude perhaps seems more understandable. The results cannot have been satisfactory to him. The two professors, Forsskal and von Haven had died in the Yemen and, if Forsskal had been tireless in his exertions, von Haven, his protégé in philology, had been viewed as a failure almost from the beginning. Von Haven did purchase the documents that, as we have seen, became the foundation of the collection of Oriental manuscripts in the Royal Library. But in terms of understanding that idiom of Oriental Arabic that he believed was the surest guide to an understanding of Hebrew, von Haven had made little contribution. Niebuhr, the mathematician whose lack of enthusuasm for the study of Arabic had already sorely tried Michaelis's patience, provided perfunctory answers to some of his carefully crafted questions in the Preface to the *Description of Arabia*. But the answers suggested that the questions either could not be satisfactorily answered or were not particularly pertinent.

Engineer-lieutenant Niebuhr had, in fact, already displayed alarming signs of an independence of mind. He had already shown the temerity to disagree in print with Michaelis on a subject dear to the older man's heart: the place of crossing of the Red Sea by the children of Israel in their exodus from Egypt. In fact, the whole subject of the crossing seemed to Niebuhr to be matter of faith, not science and, as we have seen, he disagreed fundamentally with the attempt to reconcile the two. This incident probably was an epitome of the more basic rupture between the two men. There were matters to which Niebuhr, the practitioner in the field, could profitably devote his time and—although he doesn't say so explicitly—there were those that were a waste of his time.

Most of the questions of Michaelis seemed to fall in the latter category. Seen in this light, Michaelis's disinterest in Niebuhr's work is understandable. And we should not forget the relationship between the scholars and the travelers: the latter were only the lens through which the former would view the Orient, effectively non-sentient instruments in the hands of Michaelis and his colleagues. Trained scholars—experts in their fields—would have been the first choice for the work of the expedition. But they could not be expected to withstand the rigors of a journey of this kind. The constitutions of younger men were needed and even they, as events proved, were unequal to the task. But this placed a premium at the outset on preparing the younger men: they did not know what to look for, and had to be equipped, as it were, with the eyes of the best scholars in Europe so as to see through the wholly unfamiliar phenomena of the Orient and register only what was truly worth recording.

But, in practice, this premise was also flawed. By the time the travelers had experienced the difficulties of travel in an unknown and suspicious world; of gathering information of any kind, much less that of a reliable kind; of the barriers that language placed in their way; of the lingering hostilities between Christianity and Islam; of the countless little—and not so little—misunderstandings that were the product of fundamental differences in culture; of the suspicions of the authorities and misunderstanding of their mission, by this time it was the travelers who were the experts, not the academics who attempted to pull the strings from their remote laboratories. The best that could be hoped for, under the circumstances, was that the travelers would develop the discernment that would allow them to filter the permanent and truly important from the transitory in what they would see in the Orient. The Danish authorities were fortunate in the choice of Carsten Niebuhr.[14] But this independence did not endear him to Michaelis, the man who had first set the expedition in motion but who, we suspect, had long-since given up any interest in its results by the time the sole survivor returned to Europe.

Many years later[15] Michaelis himself would provide a clue to the puzzle. In Part IV of his own *Bibliotheque Orientale* he included an extract of Niebuhr's *Description of Arabia*. With an apology for its cursory nature—after all, only so much could be said in an extract—Michaelis proceeded to devote the first eleven pages of a thirty-seven page document to a long, Jesuitical defense of his decision *not* to comment earlier on Niebuhr's effort. His reasoning is curious: in a legal proceeding would a judge advise a witness on his testimony? And, besides, he was busy. Throughout the extract he refers to Niebuhr as "Mr. N." and although he found much of value in the *Description*, it is difficult to read the document without the sense that Michaelis is damning with faint praise. Niebuhr admittedly made an effort to be of help, but the questions in the *Fragen* were really intended for the professors von Haven and Forsskal who would have understood them better; Niebuhr certainly mistook the thrust of the questions when he asked Jews about various words in the Hebrew Bible—Michaelis was interested not in Hebrew, but in an early, "Eastern" form of Arabic; he certainly believed the crossing of the Red sea was a miracle; Niebuhr was "unsure of himself" and of his writing style, which was understandable since he was so many years away from Germany but his "simplicity and candor" admittedly leant credibility to his accounts; etc. etc.

Comments elsewhere in the same document may shed some light on the atmosphere in which Michaelis was operating: he mentions literary

14. And also Forsskal. Two out of five is not bad.

15. He mentions a figure of seventeen years, although it is not clear whether this should be counted from the dispatch of the expedition, the publication of Niebuhr's *Description of Arabia*, or some other unspecified date.

gazettes whose authors were unable to conceal their dislike of him and whose only function seemed to be to "contradict and misinterpret" everything he writes. The fact that the expedition was sent with such high hopes, that four of the five members died along the way and that the results appeared to be so slender probably led to the suggestion that the King's money was poorly spent. Whatever the reason, we see in the extract a chastened Michaelis. Gone are the confident tone and the careful scholarship of the *Fragen* and in its place we find a weariness and querulousness that is painful to read.

Niebuhr's own works were his first order of business. The preparation of his journals for publication was not a matter of simple transcription of his handwritten notes, but of correlating the material with the work of others who had dealt with the same subjects. These sources, as listed in the *Description* and the *Travels* were well over a hundred, ranging from ancients like Herodotus, Strabo, Pliny, and the anonymous author of the *Periplus of the Erythrian Sea*, through Arab historians like Abulfeda and Sherif Idrisi, to recent travelers through the same area such as the Dane Norden and the Englishman Pococke. Niebuhr made himself the beneficiary of the best historical information of the day to complement his own experience. His familiarity with and use of these, as well as sources in other languages such as Greek, Latin, French, and English as well as his own native German, is testimony to the scholarly bent of this simple engineer.

In fact, it is probably long-since time that we stopped describing Niebuhr as a rustic who simply did his best. He may not have been a giant in the field of German Oriental scholarship, such as Richard Lepsius, who made seminal contributions to the knowledge of the area. But his contributions were nonetheless enough to merit the opinion of many that his works are classics in the study of the Orient. They fall somewhere in that intermediate category called travel literature, neither original research but more than *Eothen*.[16] As a traveler, the breadth of his interests is of a different and higher order and sets him apart, in the company of the likes of Pococke, from others in the genre. He may not have had the gift for languages or the swagger of Burton, or the academic preparation—not to mention the benefit of seventy years of scholarship—of Lepsius, but his contributions were nonetheless significant. We would also do well to recall the testimony of another man who followed in his footsteps. As we have seen, William Gifford Palgrave dedicated his *Central and Eastern Arabia* to Niebuhr, "in honor of that intelligence and courage which first opened Arabia to Europe."[17]

---

16. A clever, if superficial, nineteenth-century work on the Levant by William Kinglake.

17. Palgrave, *Central and Eastern Arabia*. See Preface.

His first book, the *Beschreibung von Arabien aus eigen beobachtung und in lande selbst gesammleten nachrichten*, appeared at Michaelmas (September 29th) 1772 in German, Niebuhr's native tongue. This was the *Description of Arabia*, hereafter called the *Description*. It is probably the best known of his works. It included a total of 431 pages and was a compendium of the Arab world. The first part covered Arabia in general and such diverse subject as its boundaries, the quality of the soil, social practices, its scripts, agriculture, and science. The second part was a description of individual provinces in the peninsula, including the Yemen, the Hejaz, the Hadramaut, Najd, Oman, and Sinai. Some areas—the Yemen, the Hejaz, Oman, and Sinai—he visited himself. For the others, he was careful to note, he based his remarks on material collected from Arabs of the area. Niebuhr intended that the *Description* be read as his introductory remarks to all of the volumes. It was illustrated with twenty-five copper-engraved plates, many prepared from the drawings of Baurenfeind and more still from Niebuhr's own sketches. Complete French translations, entitled *Description de l'Arabie*, soon appeared, in Copenhagen in 1773 and in Amsterdam in 1774. A translation into Dutch also appeared in Amsterdam also appeared in 1774 entitled *Beschryving van Arabie*.

A second work *Reisebeschreibung nach Arabien und andern umleigenden Länderen*, or *Travels in Arabia and Surrounding Lands*, called here the *Travels*, appeared in Copenhagen in two volumes over the years 1774–78. It was essentially Niebuhr's *Tagebuch*, or diary.[18] Each member of the expedition had been required to keep a diary, and Niebuhr alone lived to transform the diary into a finished work. Volume I, as mentioned above, covered the period from the expedition's departure from Copenhagen in 1761 to the survivors' arrival in Bombay and was published in 1774. Volume II, Niebuhr's account of his travels through Muscat, Bushire, Shiraz, Persepolis, Basra, Baghdad, Mosul, and Diarbakr to Aleppo, was published in 1778. As with the *Description*, translations soon followed. Complete French versions appeared in Amsterdam and Copenhagen in 1776–80 and in Switzerland in 1780, entitled *Voyage en Arabie & en d'autres Pays circonvoisins*. A complete Dutch translation also appeared in the same period entitled *Reise naar Arabia en andere omliggende landen*.[19] A third volume in German with the subtitle *Reisen durch Syrien und Palestina* or *Travels through Syria and Palestine* was published posthumously in 1837. It includes Niebuhr's travels through Jerusalem, Konya, Antalya, Karahissar, Rutabya, Brussa, Constantinople, Bulgaria, Wallachia, Molda-

---

18. The original is kept in the University of Kiel.
19. A translation of the portions dealing with Persia appeared in Farsi in 1975.

via, and Germany as he made his way back to Denmark. To my knowledge this volume has not been translated into another language.

A partial English translation of the *Description* and vol. I of the *Travels* appeared in Edinburgh in 1792 as *Travels through Arabia and other Countries in the East, performed by M. Niebuhr*. They were bound together in two octavo volumes, with notes by the translator, Robert Heron. The portion devoted to the *Description* includes approximately three-fifths of the original material, and omits the preface and most of the plates. The portion of the *Travels* includes about half of the original length, is also without the plates. Moreover, it is hardly a translation in the true sense of the word and, while it is more than Dr. Johnson's epitome, it is still more of a summary than a translation. The text has been considerably rearranged, where it is not very roughly translated into English. Only eight of the original ninety-seven plates in the *Description* and *Travels* are included. A second partial translation appeared in John Pinkerton's *A General Collection of the Best and Most Interesting Voyages and Travels*. The collection appeared in London over the period 1808-14 and also included portions of the *Description* and the *Travels*. While the Pinkerton extract is a more faithful translation of the original, it includes even less of the original material than the Heron version. The translator apparently shares Heron's concerns about the attention span of the audience. None of the ninety-seven plates is reproduced, and none of the original Arabic is included in either version.

Collections of which Pinkerton's represented one of the better sort were issued, frequently in competition with one another, for the attention of the reading public. The versions of the Niebuhr works available in English represented a product of this demand. But it is primarily the things that make Niebuhr important, and a product of his scientific temperament, which are left out, since they would be of little interest to the general reader: nearly all of the mapmaking including latitudes and longitudes; all of Niebuhr's work on the hieroglyphs and the trilingual cuneiform inscriptions; and all of the foreign words, including the Arabic, which are often alone the infallible guide to place names. The detailed position finding, directions and distances *did* make tedious reading and words written in unfamiliar scripts would have been utterly lost on the general public. The serious student could presumably consult the original German or the French translation for the missing detail. But the bowdlerized versions that appeared in English gave readers in that language only a scant appreciation of Niebuhr's contribution to the study of the Orient.

As might be expected, given the translations that rapidly followed, the books created immediate interest in Europe. The journeys of exploration and discovery beginning in the late fifteenth and early sixteenth centuries, had led to an immense popular appetite for accounts of this kind. European compendia, beginning in the sixteenth century Italian work of Ramusio, appeared with regularity in most languages. Niebuhr himself used a German collection, *Samlung aller Reisebeschreibung*, translated from English and French sources. Translators and writers were employed in a virtual cottage industry to cater to this appetite. As we have seen, Samuel Johnson's first published work was a translation of Joachim le Grand's translation of Jerome Lobo's travels in Abyssinia. It appeared in 1734 as *A Voyage to Abyssinia*, and Johnson lent his considerable weight to the importance of the genre. Moreover, he did not merely translate le Grand but also weighed into the lists, supporting the controversial notion that Lobo had actually seen the sources of the Nile in springs in the highlands of the country. His writings are full of references to travel literature. The Johnson work is also interesting for his opinions on translation (he provided an epitome—a short statement of the main points of the work—not a translation), on travelers (they should be discerning and not credulous), and on translators (who must undertake to correct misrepresentations in the text). Johnson himself goes to some length to correct the editorializing of le Grand, taking a particularly Protestant point of view in a work that, at least as it passed through its first translator, was full of Catholic polemic. We have seen that Niebuhr was not immune to considerations of religion, and to a religion of narrowly defined sort. He may have aspired to be the supreme rationalist, that rare man for whom narrow sectarian contention is so foreign that we suspect it never enters his thoughts. In fact, he was only too human and manifested that common tendency to magnify differences the nearer the religion to his own: he was supremely rational about Islam, anxious to see its manifestations in the best light and attributing its shortcomings to ignorance or human frailty. He was not nearly so understanding of the religion of the Papists.

Niebuhr's Later Life

By 1778 Carsten Niebuhr had established himself in Copenhagen and had published six thick volumes of the results of the expedition, three of his own and three of Forsskal's. But his funds had been exhausted before the third volume of his own travels could be brought to the public. Contact with Abdurrahman Aga, the ambassador of Tripoli in Copenhagen, rekindled his interest in geographical explorations, this time into the interior of Africa.[20] However, personal interests intervened. In 1773 he had fallen

20. At this time a virtual terra incognita. Early in the new century Burckhardt would travel to the Near East under the sponsorship of "The Association for Promoting

in love and married Christine Sophie Blomenberg, and she soon produced two children. Declining an offer to lead a geographical survey of Norway, he instead applied for a post as tax collector in the district of Ditmarsches. He was accepted and, leaving Denmark in the summer of 1778, he returned to Meldorf, a locale not thirty miles from the place of his birth. A modern map shows it to be in present-day Schlesswig-Holstein on Heligoland Bay, about 20 miles north of where the Elbe exits into the North Sea. There, he built himself a house with a small observatory. The area had been subject to the vicissitudes of war and plunder for centuries, and it was poor, quiet, and deserted. After the excitement and turmoil of the previous seventeen years, this seemed to appeal to Niebuhr's contemplative nature. His official duties included the management of wetlands and valuation of land holdings, both of which he carried out with his customary scrupulousness and diligence.

The overriding concern became the education of his children, particularly his son, Barthold Georg, born in 1776. And it was through this remarkable son that Niebuhr may have contributed an equal part with his own accomplishments in a kind of joint legacy. The two were, in many ways, opposites and the father's intense practicality often clashed with the predilections of the more imaginative and intuitive son. "The boy gets on wonderfully,"[21] he wrote to an acaquaintance, "but he requires to be managed in a particular way . . . Oh if he could but learn to control the warmth of his temper." In spite of this temperamental difference, Barthold Georg's childhood appears to have been a happy one in remote Meldorf where, almost always alone with of his father, he imbibed a regimen of learning more characteristic of an adult than a child:

> 'I well remember,' says Niebuhr in the Life of his father, 'how he used to tell stories in my childhood about the East and the structure of the universe; particularly in the evening, just before bedtime he would take me upon his knee, and feed my imagination with these instead of fairy tales. The history of Mahomet, of the early Caliphs—especially Omar and Ali, for whom he had the deepest reverence—of the conquests and spread of Islamism, and the virtues of the heroes of the new faith, with the history of the Turks, were early imprinted on my memory in the most

---

the Discovery of the Interior Parts of Africa." As we have seen, his actual assignment was to proceed to Cairo and there await the departure of the caravan west through the Fezzan to Timbuctu. He grew tired of waiting and never made it to the west. Instead, he went south then east and became the first European, at least since the Crusades, to see Abu Simbel and Petra.

21. Winkworth, *The Life and Letters of Barthold Georg Niebuhr*, 9

lively colors; nay, works on those subjects were among the first books put into my hands.[22]

He adds that his own first juvenile scribblings were made in the margins of extra sheets of the Forsskal volumes, left over from the print runs.[23]

His father's first intention was to prepare Barthold Georg for travel, as he had prepared himself. But the boy's delicate constitution—and the influence of his mother—soon put an end to that hope. But he was a prodigy of learning and, unlike his father, soaked up languages like a sponge, although Niebuhr failed in an attempt to teach him Arabic. In a letter in 1807 his father, while apologizing for his enthusiasm—"I do not mean to boast of him"—enumerates the twenty languages that Barthold Georg, exposed to the wider world, now speaks: German, Latin, Greek, Hebrew, Danish, English, French, Italian, Portuguese, Spanish, Persian, Arabic, Dutch, Swedish, Icelandic, Russian, Slavonic, Polish, Bohemian, Illyrian, and Low German.

However, he later combined these linguistic accomplishments with a heavy dose of his father's practicality and in a career that spanned twenty-seven years from 1804 to his premature death in 1831, he was succesively director of the Danish National Bank, a senior member of the Prussian State Service, and the Prussian ambassador to Rome, before ending his days as a professor at the University of Bonn. His real interest was history and, after joining the Berlin Academy of Sciences, he began a series of lectures that eventually resulted in his *Roman History*, the first edition appearing in 1811–12. It was a seminal work in historiography, with pioneering philological contributions to history and introducing social history as an entirely new discipline. Michaelis, dead since 1791, might have been more accepting of this second-generation Niebuhr. No less a personage than Goethe was extravagant in his praise. Later critics shared the enthusiasm and Barthold Georg Niebuhr, by common consent, is regarded as the seminal figure in the development of historiography as we know it today and a model for the practitioners who followed in his footsteps.[24]

In the English-speaking world some of the same reservations were expressed about this latest Niebuhr as had been, earlier, expressed about his father: the English translator of the *History* doubts that there is much English appetite for German literature of the "graver cast," before he plunges

---

22. *Lives of Eminent Persons*, 24.

23. A virgin copy of Forsskal's 1775 *Flora Aegyptiaco-Arabica* I purchased in 2004 was unbound, the roughly-sewn end papers consisting of just such extras. Apparently it was common in the late eighteenth century for purchasers themselves to pay for the binding, printers not hazarding the expense before the book was sold.

24. See Gooch, *History and Historians in the Nineteenth Century*.

ahead with the translation. The texts *are* dense, evidence of Niebuhr's wide learning and fecund imagination, and as a lecturer it was said that the ideas came so quickly that he often had trouble connecting them. However, after the disaster of the Napoleonic wars he saw the Prussian contribution to the world as primarily intellectual, and he was anxious to lead the charge. Other publications followed, including a translation of the Armenian Eusebius, a volume of Byzantine historians, and reflections on the geography of Herodotus and the history of the Scythians. But, more than the history, it is the letters of Barthold Georg Niebuhr that reveal him to be an intellectual of the first water, a profound humanist and a man who combined the sturdy independence and intellectual honesty of his father with a nearly photographic memory and a learning as deep as it was wide. His father would undoubtedly have been proud if he knew that in the *Allgemeine Deutsch Biographie* his own entry would be dwarfed by that of his son. There, half a page is devoted to the *berühmter Reisender*—famous traveler—while fifteen dense pages are needed to cover the accomplishments of the *ausgezeichneter Staatsmann, Geschichtschreiber und Altertumsforscher*—outstanding statesman, historian and antiquarian.

In his self-imposed exile in Meldorf, Niebuhr had only occasional contact with his intellectual equals, although the appointment in 1781 of Heinrich Christian Boie as *Landvogt*, or governor, of the province with headquarters at Meldorf led to a welcome return to learned society and aceesss to a considerable library. Boie was founder of the journal *Das deutsche Museum* to which Niebuhr became a regular contributor. He also corresponded with many of the most prominent explorers and scientists of the day and Silvestre de Sacy, the French Orientalist, used Niebuhr's material extensively in his history of the conquest of the Yemen. Niebuhr also maintained contact with many friends made on his eastern journey. They included the shipowner Francis Scott who had befriended him in Mocha (when Barthold Georg was in Scotland in 1799 it was the Scotts with whom he stayed); in Bombay there was Captain Howe of the Royal Navy, brother of Admiral Lord Howe and General Sir William Howe, commanders, respectively, of the British sea and land forces during the American Revolution; and in Aleppo, Dr. Patrick Russell, author of an important work on the plague and editor of his uncle Alexander Russell's *Description of Aleppo*. When Barthold Georg was in London it was his father's influence, through contacts with Russell and Joseph Banks that led to his introduction at the Royal Geographic Society. It is probably no accident that these men were all British, and Niebuhr retained to the end of his days the Anglophilia that he acquired in Bombay. In spite of a deep respect for their mathematicians and Orientalists, he mistrusted the French and saw little good either in the

Niebuhr in Egypt

French Revolution or the prospect of an improvement in Egypt's lot with Napoleon's invasion in 1798.[25]

In 1799 Niebuhr bought a piece of marshland and set to work to drain it. But physical infirmities increasingly plagued him in the cold, harsh climate. He was lamed by an arthritic hip and his eyesight, damaged at Persepolis years before, increasingly troubled him. Boie died in 1806. Niebuhr's wife, of a delicate constitution, was unable to withstand the rigors of this rural life and she died a year later in 1807. In his last years there was some measure of official recognition for his contributions. Niebuhr's work won belated acceptance, particularly the maps and the determinations of latitude and longitude, and a revised map of the Eastern Mediterranean was based on his readings of forty years before. He was made a corresponding member of the French *Institut national* in 1802 and was appointed a Royal Counselor in 1808. In 1809 he was made a Knight of Danebrog.

When his eyesight finally failed completely Niebuhr, characteristically, tendered his resignation. However, the government refused to accept it and instead appointed an assistant with whose aid he was able to discharge his duties to the end. His daughter, particularly after the death of her mother, was a great help and she acted in his last years as his eyes and contacts with the wider literary world. But when he died on April 26th, 1815, at the ripe old age of 82, Carsten Niebuhr was nearly alone and completely blind, able to picture only in his mind's eye the fantastic sights and shapes and scenes of his more than Sindbad's odyssey.

---

25. One reader has suggested that these anti-French sentiments have been overplayed, and his real antipathy was to revolution itself, which was "surely not his cup of tea."

Appendix A

Questions[1]

Contained in this Work

    I   What is the סוף which has given its name to the Red Sea, and what is the *Suph*[2] that lives in the Nile?
    II On the ebb-tide which occurs at the northern extremity of the Red Sea. On the timing and level of this ebb-tide. On the depth and the bottom of the Sea in the place crossed by the Israelites. On corals.
    III On the flood of Egypt near el-Arish.
    IV On the fish living in the Red Sea.
    V The Sumana,[3] a bird of Happy Arabia.

1. A complete set of the list of questions prepared by Michaelis for the travelers were published in Frankfurt in 1762. They did not reach the expedition until Niebuhr was in Bombay, long after the two professors von Haven and Forsskal, for whom they were primarily intended, were dead. But Niebuhr and Forsskal had seen a portion of them and responses to twenty-five by Niebuhr and thirty-five by Forsskal (with some overlap) were included by Niebuhr in the Preface to his *Description of Arabia*. A review of the questions confirms their biblical reference point, about which Michaelis was unapologetic. His biographer suggests that the questions

> seemed to offer details of an erudition as arid and minute as it was profound ... they are in truth full of interest, sagacity, and of such precision that they do not leave a traveler for a moment in doubt as to the point of a difficulty or the essential object of his researches.

Of their erudition, there can be no doubt. The individual questions were elaborated into a disquisition that eventually filled 208 pages of text. Based, as they were, on Biblical passages, however, the likelihood of their being satisfactorily answered by travelers in the eighteenth, or any other, century was another matter. As mentioned above, the answers provided by Niebuhr and Forsskal seem perfunctory. I am not a scholar of Hebrew and the below references in Gesenius are merely tentative suggestions to the reader of the general tenor of the questions.

2. In Gesenius "reeds or rushes" (693).

3. A kind of quail.

Appendix A

VI. Who were the Seleucides?[4]
VII. On swarms of flies, and on the Myiagre,[5] their enemy.
VIII. What is תחש?[6]
IX. Arab huts.
X. On relations with women during their time of menstruation, and the ill effects that result from this in the Orient.
XI. On leprosy.
XII. On leprosy of houses and clothing.
XIII. Extraordinary multiplication of wheat in Asia and Africa.
XIV. On the sorting of seed.
XV. On the ivraie or zizania.[7]
XVI. Of barley bread.
XVII. The medicinal use of oil.
XVIII. On a wood that makes saltwater sweet.
XIX. On waters alternately sweet and bitter.
XX. On the springs of Sinai, and the stream in the Valley of Rephidim.
XXI. On the stones of the twelve mouths:[8] supposed monuments of the miracles of Moses.
XXII. On the sciniphes[9], or a kind of small wasp of Egypt.
XXIII. On the species of vine called sorek.[10]
XXIV. On the samum,[11] a pestilent wind.
XXV. On the two trees, march and aphar.
XXVI. On various sorts of Arab manna.
XXVII. On wild honey.
XXVIII. Additional questions on leprosy and illnesses related to it.
XXIX. On incense.
XXX. On flying quadrupeds.
XXXI. On the feet of locusts.

4. The dynasty of successors to Alexander the Great who ruled in the Near East from 312 to 65 BC.

5. Unclear.

6. In Gesenius "a kind of leather or skin and perhaps the animal yielding it" (1065).

7. Both are varieties of grass that grow in wheat fields, the first a kind of rye and the second a kind of wild rice.

8. Still to be seen today with its twelve horizontal apertures in the *Wadi al-Arba'in* behind the monastery of St. Catherine's. Niebuhr did not see it.

9. Unclear. Perhaps *sinf*, "kind, sort, specimen." Hans Wehr, p. 527.

10. Unclear.

11. The hot wind in several Arabic speaking countries. Occurs in three passages in the Koran.

Appendix A

XXXII. On the nature of locusts, and their different species.
XXXIII. On other insects ordinarily taken for locusts.
XXXIV. Economic almanac.
XXXV. On the scabies of the face called λειχν.
XXXVI. On elephantiasis.
XXXVII. On the manatee, on תחש,[12] and the sirens.
XXXVIII. On the thoës, بنات الاوي and on the hyena.
XXXIX. On gold and other metals of Arabia
XL. On the valleys which in the rainy season amass water, because there is no place of outflow.
XLI. On the machines used by Arabs and Egyptians to irrigate their fields.
XLII. On the plant which the Arabs call *murar*.[13]
XLIII. On agallochum אהלים.[14]
XLIV. On אלה or אילון terebinthe.[15]
XLV. On כפר, cyprus, or alhinna.[16]
XLVI. On the ראם, ריס, or ريم,[17] a wild animal of the genus of oxen.
XLVII. On the רוש and on hemlock.
XLVIII. Why are *bitter* and *venomous* synonyms in the Orient?
XLIX. On absinthe and לענה.[18]
L. On the vena medinensis.[19]
LI. On the insect the worshipper of God.
LII. On the physical utility of circumcision of boys and girls.
LIII. Additional questions on manna, and the manner of preparing it for consumption.
LIV. On the methods of castration.
LV. On שעטנז.[20]

12. See n. 6 above. Among the animals speculated as the owner of the skin was the dugong, a class of seaweed-eating sirenian (or sea nymph) mammals living along the shores of the Indian Ocean (Gesenius, 1065).

13. Probably *murr*, "myrrh."

14. A kind of plant. Unclear in Gesenius.

15. A tree of the cashew family whose bark yields a kind of turpentine.

16. In Gesenius, "Al Henna," a shrub with fragrant "whitish" flowers. In Arabic, the word indicates a reddish-orange cosmetic from the shrub used to dye the hands and hair. It is still widely used today (499).

17. An addax, a large white antelope of the Sahara with twisted horns (Gesenius, 910).

18. In Gesenius, "wormwood" (542).

19. Literally, "the Medina vein," but referring to a worm thought to affect the nerves.

20. Unclear, at least to me, in Gesenius.

Appendix A

LVI. On proof of virginity retained after marriage.
LVII. On a certain manner of blessing the people which the priest must avoid.
LVIII. On the act of spitting.
LIX. On the habit of removing shoes during the assignment of rights.
LX. On the marriage of a woman with the brother of her first husband.
LXI. On the phrase נדד עקב or נגד עקב.[21]
LXII. On the hemorrhous, or cerastes, and also that kind of serpent which is called שׁפיפון in Hebrew, and in Arabic سف.
LXIII. On the serpent called *charmon* in the Syriac language.
LXIV. On the vines of the countryside, and the vine of Sodom.
LXV. Views from the highest mountains, especially Mount Sinai.
LXVI. On the Arabic names of *dipsas*.[22]
LXVII. On various maladies with which the Israelites were threatened, Levit. XXVI and Deut. XXVIII, and called שׁחפת.[23]
LXVIII. לקתד Lev. XXVI. 16 Deut. XXVIII. 22.[24]
LXIX. דלקת Deut. XXVIII. 22.[25]
LXX. חרחר Deut. XXVIII. 22.[26]
LXXI. עפלים Deut. XXVIII. I Sam. V. 5, 6, 9, 12. VI. 5, 11.[27]
LXXII. חרס Deut. XXVIII. 27.[28]
LXXIII. If לחם لحم means the earth[29]
LXXIV. Arrack.

21. Unclear.

22. A serpent with a bite that was said to produce intense thirst; subject of tales by Classical Greek authors.

23. In Gensnius, "a wasting disease."

24. In Gesenius, "a fever," (869).

25. In Gesenius, an inflammation (196).

26. In Gesenius, a violent heat or fever, (359).

27. I could not find this word in Gesenius.

28. Deuteronomy 28:15–68 is a catalogue of curses that shall be visited on the children of Israel if they do not "hearken diligently" to the voice of God. The particular word sought in this verse comes from the following:

    27 The LORD will smite thee with the botch of
    Egypt, and with the emerods, and with the
    scab, and with the itch, whereof thou canst not be healed.

None of the entries in Gesenius appears to carry with it the physiological meaning here intended.

29. In neither Hebrew nor Arabic do these words mean the earth, but, respectively, food and meat (Gesenius, 536).

Appendix A

LXXV. Are there cases where the venom of certain serpents, inflicted through a bite, can produce beneficial effects?

LXXVI. On the illnesses which protect against the plague.

LXXVII. On the diseases of wheat שִׁדָּפוֹן יֵרָקוֹן.

LXXVIII. On the antholops or jachmur.

LXXIX. First and last rains.

LXXX. Is Syriac still a living language?

LXXXI. On אקו, יעלה,[30] and the tragelaphus.

LXXXII. On the basilisk صل.[31]

LXXXIII. On flying serpents.

LXXXIV. On the saltpeter of Egypt and of borith.

LXXXV. On דישׁן, תאו and זמר Deut. XIV. 5. and of certain names of animals used by the ancient interpreters, particularly the زرفة or camelopard,[32] *camelopardalis*.

LXXXVI. The names of the stars in the Arabic language.

LXXXVII. The קיקיון, (Jon. IV), Kiki, Alkeroa, الخروع,[33] with other plants that grow and wither in a short period of time.

LXXXVIII. On the aurora borealis and the atmosphere of Arabia.

LXXXIX. On dew.

XC. On אר, is the fir or the cedar?

XCI. On the wood almuggim, or algummin.[34]

XCII. On the mountain mouse with two legs, called by the Arabs يربوع.[35]

XCIII. On several terms in Oriental languages by which the Hebrew word שׁפן is translated.[36]

XCIV. The inundation known by the name سيل العرم

XCV. Clean and unclean animals; the means by which one discerns them; and the stain contracted by contact with carrion.

XCVI. Pavement of glass in rooms where a throne is erected.

XCVII. On sapphire and lapis lazuli.

XCVIII. On precious stones in general.

---

30. A kind of wild goat, listed among the clean animals, Gesenius (70).

31. A mythical fire-breathing, lizard-like monster. In Arabic, a variety of viper.

32. The giraffe, so called because it had a neck like a camel and spots like a leopard.

33. In Arabic, the caster-oil plant. Also Gesenius, "a plant" (884).

34. Probably the Arabic *kashab al-mugna*, or mahogany. The inversion, or reversal of the order of consonants, is fairly common in spoken Arabic. Fluency in colloquial Egyptian is often described as *liblib*, a reversal of *bulbul*, as in "he speaks like a *bulbul*", or "like a nightingale."

35. The Arabic *yarbu'*, or jerboa.

36. In Gesenius, a *hyrax syriacus*, or unclean animal (1050). The Arabic is *wabara*.

Appendix A

XCIX. On certain precious stones named in the Bible.
C. On certain unclean birds that are mentioned in Levit. XI. and Deut. XIV.

Appendix B

Keys to the Map of Cairo

On the Map of City of Cairo or Masr[1]

A. Residence of the reigning *Pascha*

B. The quarter of the *Janissaries*, or in a narrow sense, the *Citadel*

C. The quarter of the *Assabs*. A, B, C are located on a rocky promontory commonly called the castle or citadel.

D. The place قراميدان *Karameidan*

E. The place رميلة *Romele*

F. قلعة الكبش *Kalla el Kabsch*, a ruined castle near the *Tulun* Mosque

G. سلطان حسن *Sultan Hassan*, an imposing Mosque. Not far from here is the Suk Selahh where the merchants gather.

H. جامع الازهر *Dsjamea al ashar*, a celebrated Mosque and Academy

I. The Patriarchal Church of the Copts. The Greek Patriarch of Egypt lives in this quarter.

J. [no listing]

K. St. Nicholas, a Greek church. Not far from here is the *Oqal²hamsaui*.

L. A Coptic church under which there is an Armenian church

M. The residence and church of the Greek Bishop of Mount Sinai

---

1. *Travels*, vol. I, 109–12 Niebuhr's transliteration of the Arabic, similar in both the original German and the French translation, have been preserved as they appeared in these documents. Thus, for example, "Dsjise" rather than "Giza," or "Bab el futuch" rather than "Bab el-Futuh." N.B.: there were no listings in Niebuhr's original for J, U, and W.

2. Clearly from the Arabic root *wakala*, and so like the more common pronunciation, a caravanserai or inn where merchants were able to stay, stable their animals, and store their wares.

Appendix B

- N. The residence of the *Kadi*
- O. *Chan chalil* خان الخليل
- P. *El muristan*, or the hospital
- Q. Residence of the Consul of France, and of the merchants of that nation.
- R. Residence of the Consul of Venice
- S. Quarter of the Jews
- T. *Kubbet el assab.*, in a former day the residence of the Assab Corps, with a small fort and a large Mosque. Today, it is an utter ruin. It is here that the inhabitants of Cairo receive the *Pashas* who arrive overland.
- U. [no listing]
- V. Venetian and Coptic cemeteries
- W. [no listing]
- X. Place where a great number of animals are slaughtered
- Y. An oven where eggs are hatched. There is also a powder factory in this area.
- Z. A lime kiln

Names of the Bridges over the Canal that passes through Cairo[3]

- a. قنطرة فم الخليج *Kantaret fum el chalidsg*
- b. قنطرة الجنينة *Kantaret ed sjeneine*
- c. قنطرة السباع *Kantaret es sabba*
- d. قنطرة امرشي *Kantaret amerschi*
- e. قنطرة الجماميز *Kantaret ed sjamemis*
- f. قنطرة سنقر *Kantaret Sunqur*
- g. قنطرة عبد الرحمان كخيا *Kantaret Abdrachman Kichja*
- h. قنطرة باب الخرق *Kantaret bab el chark*
- i. [no listing]
- j. [no listing]
- k. قنطرة الامير حسين *Kantaret el Emir Hossein*
- l. قنطرة المسكي *Kantaret el muski*

3. N.B.: there were no listings in Niebuhr's original for i, j, and v.

Appendix B

- m. قنطرة الجديدة *Kantaret ed sjedide*
- n. قنطرة باب الشعرية *Kantaret bab es scharie*
- o. قنطرة الخروبي *Kantaret el charube*
- p. قنطرة الضاهر به برس *Kantaret ed dahher Bebers*

Name of Lakes or Ponds

- q. بركة الشيخ قمر *Birket es schech kammer*
- r. بركة الرطلي *Birket er roteli*
- s. بركة اليزبكية *Birket el jusbekie*
- t. بركة الفوالة *Birket el fawale*
- u. بركة ابو شوارب *Birket abu schauarib*
- v. [no listing]
- w. بركة النصرية *Birket en nassarie*
- x. بركة القصارين *Birket el kassarin*
- y. بركة ايوب بة *Birket Aijubbeh*
- z. بركة الفيل *Birket el fil*

Names of the gates of Cairo

1. باب النصر *Bab el nasr*, a large and beautiful gate
2. باب الفتوح *Bab el fituch*, a still more magnificent gate; see Plate XIII
3. باب المدبخ *Bab el medbach*
4. معمل النشى *Bab en nascha*
5. باب الشعرية *Bab es scharie*, an old gate, strong and low
6. باب البكري *Bab el bakri*, today outside the city in the gardens
7. باب الشيخ شايب *Bab es schech schaiib*
8. باب الحديد *Bab el hadid*
9. باب ولاد عنان *Bab aulad anan*, in the gardens
10. باب الهوى *Bab el haua*
11. باب الفوالة *Bab el fawale*

Appendix B

12. باب سوق البكري *Bab suk el bakri*
13. باب المدابغ *Bab el medabegh*
14. باب الشيخ ريحان *Bab es schech rihan*
15. باب الناصرية *Bab en nasrie*
16. باب غيط الباشا *Bab gheit el bascha*
17. باب ايوب ة *Bab aijubbeh*
18. باب ستي زينب *Bab setti seinab*
19. باب طياون *Bab teilun*
20. باب الخليفي *Bab el chalifa*
21. باب القرافة *Bab el karafe*
22. باب الجبل *Bab ed sjabbel*
23. عرب ليسار *Arab lissar*
24. باب قراميدان *Bab kara meidan*
25. باب العزب *Bab el assab*
26. باب الانشارية *Bab el inkscharie*
27. باب الوزير *Bab el Wisir*
28. باب الحطابة *Bab el hattab*
29. باب المحروك *Bab el machruk*
30. باب الغريب *Bab el ghreiib*
31. باب السويلي *Bab es sueli*, a very beautiful gate near the center of the city

In Bulak, Masr el atik, and Dsjise

32. *El helle*, a ruined building where the inhabitants of Cairo receive the Pashas who arrive via the Nile
33. A warehouse for wood
34. An old arsenal
35. The salt storehouse
36. The customhouse
37. A large covered market *Kissarie*
38. A brickyard

Appendix B

39. Places where stone from Memphis and other ancient cities is unloaded, and where the people of Cairo go to fetch Nile water with camels
40. Country house of a Bey in which, when we were there, the deposed Pasha lived
41. *Mastabe*, a large space where the leading men of Cairo go with their servants and slaves to practice shooting the arquebus and the bow
42. *Kasr el ain*, a large building with a cupola, inhabited at the present time by *Darwisches*
43. A simple house that the Pasha occupies when the barrage in the canal has been opened. Between the barrage and the Nile stands the so called the Bride when the channel is dredged
44. Churches and cemeteries of the Copts; here it is believed that the bones of the dead move about on a certain day of the year
45. A very high structure with five wheels, by means of which water from the Nile is raised above the walls and sent to a cistern near the citadel
46. A large Mosque
47. The Mosque of *Abu saki* who built the aqueduct
48. A Coptic church
49. The *Amru* Mosque
50. A spacious area, surrounded by a wall like an old castle, occupied today only by Christians
51. The so called corn storehouse of Joseph
52. The *Basar*, or street of the merchants
53. The customhouse
54. A sal ammoniac oven in *Dsjise*
55. Several pottery factories. Only the *Birkets* or ponds that still contained water in February and March have been shown. It was impossible to show on the map the size of the enclosed gardens. There are gardens wherever trees are shown. The letters LLLLL represent cemeteries. The streets of Cairo generally take their names from nearby gates and bridges.

Bibliography

Abulfeda. *Takwim al-buldan*. Frankfurt am Main: Goethe University, 1992.
Africanus, Leo. *Della descrittione dell'Africa*. French. Frankfurt: Goethe University, 1993.
Albright, William Foxwell. *The Proto-Sinaitic Inscriptions and Their Decipherment*. Cambridge: Harvard University Press, 1966.
Allgemeine Deutsch Biographie. Vol. 23. Leipzig: Dunder & Humblot, 1886.
Ali, Abdullah Yousef. *The Holy Qur'an: Text, Translation & Commentary*. Beirut: Dar al-Qur'an al-Karim, 1403 A.H. Reprinted, Lahore: Amana, 1983.
D'Anville, Jean-Baptiste Bourguignon. *Memoires sur l'Egypte, Ancienne et Moderne, suivi d'une description du golfe arabe*. Paris: Imprimerie Royale, 1766.
Armstrong, Karen. *A History of God*. New York: Knopf, 1993.
Barnes, John and Malek, Jaromir. *Atlas of Ancient Egypt*. Oxford: Phaidon, 1984.
Behrens Abu Seif, Doris. *Azbakiyya and its Environs from Azbak to Isma'il, (1476–1879)*. Cairo: Institut Français d'Archéologie Orientale, 1985.
Biographie universelle ancienne et moderne. Vol. 30. Paris: Desplaces, 1854–1865.
Blunt, Lady Anne. *A Pilgrimage to Nejd: The Cradle of the Arab Race*. London: Murray, 1881.
Boeckh, August. *On Interpretation and Criticism*. Translated by John Paul Pritchard. Norman: University of Oklahoma Press, 1968.
The Book of a Thousand Nights and a Night. Translated by Sir Richard Burton. London: Burton Club, 1885.
Bowman, Alan K. *Egypt after the Pharaohs: 332 BC—AD 642*. London. British Museum Publications, 1986.
Braudel, Fernand. *The Mediterranean and the Mediterranean World in the Age of Philip II*. 2 vols. Translated by Sian Reynolds. New York: Harper & Row, 1982.
Bryant, Jacob. *Observations and Inquiries Relating to the Various Parts of Ancient History, Containing Dissertations on the Western Euroclydon and on the Island Melite, Together with an Account of Egypt in its most Early State and of the Shepherd Kings*. Cambridge: Archdeacon, 1767.
———. *New System, or, an Analysis of Ancient Mythology wherein an Attempt is made to divest Tradition of Fables; and to reduce the Truth to its Original Purity In this Work is given an History of Babylonians, Cannanites, Ledeges, Chaldeans, Helladious, Dorians, Egyptians, Ionians, Pelasgi also of the Scythae, Ethiopians, Indo-Scythae, Phenecians. The whole contains an account of the principal Events in the first ages, from the Deluge to the Dispersion; Also of the various Migrations which ensued, and the Settlements made afterwards in different Parts: Circumstances of*

Bibliography

great Consequence, which were subsequently the Gentile History of Moses. London: Payne, 1774-1776.

Budge, E. A. Wallis. *An Egyptian Hieroglyphic Dictionary*. New York: Dover, 1978.

Burckhardt, John Lewis. *Travels in Syria and the Holy Land*. London: Murray, 1822.

Burton, Richard. *The Land of Midian [Revisited]*. London: Paul, 1879.

———. *Personal Narrative of a Pilgrimage to Al-Madinah & Meccah*. Memorial Edition. London: Tylston & Edwards, 1893.

Butler, Alfred J. *The Arab Conquest of Egypt and the Last Thirty Years of the Roman Dominion*. Oxford: Clarendon, 1902.

Campbell, Stafford. The *Yachting Book of Celestial Navigation*. New York: Dodd, Mead, 1984.

Cezzar, Ahmed Pasha. *Ottoman Egypt in the Eighteenth Century: The Nizhamname-i Misr*. Edited and translated from the original Turkish Stanford J. Shaw. Cambridge: Harvard University Press, 1962.

Clark, Cumberland. *The Art of Early Writing*. London: Mitre, 1938.

Clayton, Richard, Lord Bishop of Clogher. *A Journal from Grand Cairo to Mount Sinai and Back Again: Translated from the Manuscript written by the Prefetto of Egypt*. London: Bowyer, 1810.

Cromer, The Earl of. *Modern Egypt*. London: Macmillan, 1908.

Davies, W. W. *Egyptian Hieroglyphs*. London: British Museum Publications, 1987.

Drower, Margaret S. *Flinders Petrie: A Life in Archaeology*. London: Gollancz, 1985.

E. J. Brill's First Encyclopedia of Islam 1913-1936. Edited by M. Th. Houtsma, T. W. Arnold, R. Basset, and R. Hartmann. Leiden: Brill, 1987.

The Encyclopedia Britannica, Eleventh Edition. New York: Encyclopedia Britannica, 1910.

Fargeon, Maurice. *Les Juifs en Egypte: Depuis des origins jusqu'à ce jour*. Cairo: Barbey, 1938.

Flaherty, James C. *The Quarrel of Reason with Itself: Essays on Hamann, Michaelis, Lessing, and Nietzsche*. Columbia, SC: Camden, 1988.

Flaubert, Gustave. *Flaubert in Egypt*. Edited and translated from the French by Francis Steegmuller. London: Haag, 1972.

Forbes, Eric G. *Tobias Mayer (1723-1762): A Pioneer of Enlightenment Science in Germany*. Göttingen: Vandenhoek & Ruprecht, 1980.

Forsskal, Petrus. *Descriptiones animalium, aium, amphibian, piscium, insectorum, verminum quae in itinere orientali observavit P. Forskal*. Copenhagen: ex officina Mölleri, 1775.

———. *Flora Aegyptiaco-Arabica*. Edited by Carsten Niebuhr. Copenhagen: ex officinal Mölleri, 1775.

———. *Icones rerum naturalium*. Copenhagen. Editied by Carsten Niebuhr. Copenhagen: ex officinal Mölleri, 1776.

———. *Resa till Lycklige Arabien. Petrus Forsskal's Dagbuk 1761-1763*. Edited by Arvid Uggla, Uppsala: 1950.

Frangsmyer, Tore. *Linnaeus, The Man and His Work*. Berkeley: University of California Press, 1983.

Friis, Ib. "The Botannical Results of the Arabian Journey." In *Den Arabiske Rejse*. Copenhagen: 1990.

Furber, Holden. *Rival Empires of Trade in the Orient: 1600-1800*. Minneapolis: University of Minnesota Press, 1976.

Bibliography

Gardiner, Alan. *Egypt of the Pharaohs.* Oxford: Clarendon, 1961.

Gesenius, William. *A Hebrew and English Lexicon of the Old Testament as Translated by Edward Robinson.* Edited by Francis Brown. Oxford: Clarendon, 1906.

Gibbon, Edward. *The Decline and Fall of the Roman Empire.* An abridged version, edited and with an Introduction by Dero A. Saunders. London: Penguin Classics, 1988.

Gill, John. *A Dissertation Concerning the Antiquity of the Hebrew Language.* London: Keith, 1767.

Gooch, G. P. *History and Historians in the Nineteenth Century.* London: Longmans, 1967.

Habachi, Labib. *The Obelisks of Egypt.* Edited by Charles C. Van Siclen III. New York: Scribners, 1977.

Hall, H. R. "Egyptian Chronology." In *The Cambridge Ancient History.* Vol. 1. London: Cambridge University Press, 1927.

Hansen, Thorkild. *Arabia Felix: The Danish Expedition of 1761–1767.* Translated by James and Kathleen McFarlane. New York: Harper & Row, 1964.

Haven, Frederick von. *Frederick Christian von Havens Rejsejournal fra den Arabiske Rejse 1760–1763.* Edited by Stig T. Rasmussen and Ann Husland Hansen. Copenhagen: Vandkunsten, 2005.

Hepper, F. Nigel, and I. Friis. *The Plants of Pehr Forsskal's Flora Aegyptiaco-Arabica.* Kew, UK: Royal Botanic Gardens, 1994.

Herodotus. *The Histories.* Translated by Aubrey de Selincourt. Revised with an introduction and notes by John Marincola. London: Penguin, 1996.

Heron, Robert. *A Collection of Late Voyages and Travels...* London: Watson, 1797.

Hobbs, Joseph. J. *Mount Sinai.* Cairo: American University in Cairo Press, 1996.

Hoffman, Michael. *Egypt before the Pharaohs.* New York: Knopf, 1979.

Hogarth, David George. *The Penetration of Arabia: A Record of the Development of Western Knowledge concerning the Arabian Peninsula.* New York: Stokes, 1904.

Holt, P. M. *Studies in the History of the Near East.* London: Cass, 1973.

Jabarti, Sheikh abd el-Rahman. *Merveilles Biographiques et Historiques ou Chroniques.* Traduite de l'Arabe par Chefik Mansour Bey, Abdulaziz Kalil Bey, Gabriel Nicolas Kalil Bey et Iskander Ammoun Effendi. Cairo: Imprimie Nationale, 1888.

Kaempfer, Englebert. *Amoenitatatum exoticiarum...* Lemgo: Meyer, 1712.

Katchen, Aaron. *Christian Hebraists and Dutch Rabbis.* Cambridge: Harvard University Press, 1984.

King, Lester S. *The Medical World of the Eighteenth Century.* Chicago: University of Chicago Press, 1958.

Lane, Edward William. *Manners and Customs of the Modern Egyptians.* Cairo: Livres de France, 1989.

La Roque, Jean de. *A Voyage to Arabia the Happy, by Way of the Eastern Ocean and the Streights of the Red Sea, Performed by the French for the First Time, AD 1708, 1709, 1710.* London: Strahan and Williamson, 1726.

Lehner, Mark. *The Complete Pyramids.* Cairo: American University in Cairo Press, 1997.

Lepsius, Richard. *Letters from Egypt, Ethiopia, and the Peninsula of Sinai,* with extracts from the *Chronology of the Egyptians.* Translated by Leonora and Johanna D. Horner. London: Bohn, 1853.

Lewis, Bernard. "The Islamic Guilds." *Economic History Review* 8 (1937) 20–37.

Lobo, Jerome. *Relation Historique de'Abyssinie.* Paris: Coustelier & Guerin, 1728.

Bibliography

Luther's Works. American Edition. Edited by Helmut T. Lehmann and Walter L. Brandt. Philadelphia: Muhlenberg, 1962.

Lyell, Charles. *Elements of Geology*. London: Murray, 1838.

Lyster, William. *The Citadel of Cairo: A History and Guide*. Cairo: Palm Press, 1993.

Maillet, Benoit de. Description de l'Egypte. Paris: Genneau & Rollin, 1735.

Manetho. *Manetho with an English Translation by W. G. Waddell*. Cambridge: Harvard University Press, 1940.

Marsot, Afaf Lutfi al-Sayyid. *Egypt in the Reign of Mohammed Ali*. Cambridge Middle East Library. Cambridge: Cambridge University Press, 1984.

Meinardus, Otto. *The Historic Coptic Churches of Cairo*. Cairo: Philopatron, 1994.

Michaelis, Johann David. *Fragen an eine Gesellschaft Gelehrter Männer, die auf Befehl ihro Majestät des Königes von Dannemark nach Arabien resisen*. Frankfurt: Garbe, 1762.

Michaelis, Johann David. *Literarischer Briefwechsel von Johann David Michaelis*. Edited by Gottlieb Buhle. Leipzig: Weidmann, 1794–1796.

Muir, William. *The Mameluke or Slave Dynasty of Egypt 1260–1517 A.D.* Karachi: South Asian Publishers, 1983.

The New Standard Jewish Encyclopedia. New rev. ed. edited by Geoffrey Wigoder. New York: Doubleday, 1977.

Niebuhr, Carsten. *Beschreibung von Arabien aus eigenen beobachtungen und im Lande selbst gesammleten Nachrichten abgefasset*. Copenhagen: Moeller, 1772.

———. *Reisebeschreibung nach Arabin und andern umliegenden Ländern*. Graz, Austria: Akadamische Druck- u. Verlagsanstalt, 1968.

Norden, Frederick Lewis. *Voyage de l'Egypte et de Nubie, par Mr. Frederic Louis Norden, Capitaine des Vaisseaux du Roin, Ouvrage enrichi de Cartes et de Figures sur les lieux, par l'Auteur meme*. Copenhagen: la Maison Royale, 1755.

Palgrave, William Gifford. *Narrative of a Year's Journey through Central and Eastern Arabia (1862–63)*. London: Macmillan, 1865.

Palin, M. le Comte de. "Essai sur le moyen de parvenir a la lecture et a l'intelligence des Hieroglyphes Egyptiens." *Memoires de 'Academie* 29 (1764). Quoted in Budge.

Petrie, William Flinders. *Researches in Sinai*. London: Murray, 1906.

———. *Seventy Years in Archaeology*. London: Sampson, Low, Marston, 1931.

Philby, H. St.J. B. *The Heart of Arabia*. London: Constable, 1922.

Pinkerton, John. *A General Collection of the Best and Most Interesting Voyages and Travels . . .* London: Longman, Hurst, Rees, and Orme, 1808–1814.

Pococke, Richard. *A Description of the East*. London: Bower, 1743–1745.

Rasmussen, Stig. *Carsten Niebuhr und die Arabische Reise 1761–1767*. Holstein: Boyens, 1986.

———, editor. *Den Arabiske Rejse 1761–1767: en Dansk ekspedition set I idenskabshistorisk perspektiv*. Copenhagen: Munksgard, 1990.

Raymond, André. "Architechture and Urban Development: Cairo during the Ottoman Period, 1517–1798." In *Problems of the Middle East in Historical Perspective: Essays in Honour of Albert Hourani*. Oxford: Ithaca, 1992.

———. *Cairo: City of History*. Cairo: American University in Cairo Press, 2001.

Redford, Donald B. *Egypt, Canaan and Israel in Ancient Times*. Princeton: Princeton University Press, 1992.

Redhouse, Sir James W. *A Turkish and English Lexicon*. Beirut: Librairie du Liban, 1974.

Bibliography

Riley, James C. *The Eighteenth-century Campaign to Avoid Disease.* New York: St. Martin's, 1987.

Robinson, Edward. *Biblical Researches in Palestine, Mount Sinai and Arabia Petrea.* London: Murray, 1867.

Saad El Din, Mursi et al. *Alexandria: The Site & the History.* Milan: Ricci, 1992.

Sale, Koran. *Commonly Called the Alcoran of Mohammed.* "Preliminary Discourse" to his translation. London: Ackers, 1734.

Sharpe, Gregory. *Two Dissertations: I. Upon the Origin, Construction, and Relation of Languages. II. Upon the Original Powers of Letters, wherein is proved from the Analogy of Alphabets, and the Proportion of Letters, that the Hebrew ought to be without Points.* London: Millan, 1751.

Smyth, Charles Piazzi. *Life and Work at the Great Pyramid.* 3 vols. Edinburgh: Edmonston & Douglas, 1867.

Society for the Diffusion of Useful Knowledge. *Lives of Eminent Persons.* London: Baldwin, 1833.

Täckholm, Vivi. *Students' Flora of Egypt.* 2nd ed. Beirut: Cairo University, 1974.

Uggla, Arvid Hj. *Resa till Lycklige Arabien. Petrus Forsskal's Dagbuk 1761-1763.* Uppsala: Almqvist & Wiksell, 1950.

Valle, Pietro della. *The Pilgrim.* Translated, abridged and introduced by George Bull. London: Hutchinson, 1989.

Villoteau, Guillaume André. *Musique des Egyptiens et des Orienteaux.* Paris, 182?.

Wehr, Hans, *A Dictionary of Modern Written Arabic.* Edited by J. Milton Cowen. London: Macdonald & Evans, 1960.

Weil, Raymond. *Recueil des Inscriptions Egyptiennes du Sinai.* Paris: Société nouvelle de librairie et d'édition, 1894.

Whitehurst, John. *An Attempt towards Obtaining Invariable Measures of Length, Capacity & Weight.* London: Bent, 1787.

Williams Caroline. *Islamic Monuments in Cairo: A Practical Guide.* Cairo: American University in Cairo Press, 1985.

Wilkinson, John Gardner. *The Manners and Customs of the Ancient Egyptians, including their private life, government, laws, arts, manufactures, religion, agriculture and early history, derived from a comparison of the paintings, sculptures, and monuments still existing, with the accounts of the ancient authors.* London: Bent, 1837.

Winkworth, Susanna. *The Life and Letters of Barthold Georg Niebuhr and Selections from His Minor Writings.* London: Chapman & Hall, 1852.

Winter, Michael. *Egyptian Society under Ottoman Rule, 1517-1798.* London: Routledge, 1992.

Woolf, Harry. *The Transits of Venus: A Study of Eighteenth-century Science.* Princeton: Princeton University Press, 1959.

Wortham, John David. *British Egyptology, 1549-1906.* Norman: University of Oklahoma Press, 1971.

Subject Index

Abdel Rahman Katkhuda, 109, 131, 137–38
Abulfeda, 83
Abydos, 209
Abyssinia, 106, 126
Aegyptiaca, see under Manetho
Ajerud, 229, 230–31
al-Azhar, mosque of, 109, 142
Aleppo,
   Niebuhr's arrival in, 286
Alexander (the Great), 56
Alexandria,
   description of, 56–61
   doctrinal disputes in, 68–69
Ali Bey, 136–38
Alidade, Niebuhr's, 181
Amr Ibn al-As, 97
Anatolia, 17, 290
Antiochus, 74
d'Anville, Jean-Baptiste Bourguignon, 45n17
Arabia,
   Peninsula of, 19, 24
   South Arabia, 24
Arabic,
   Oriental, 24–25
Aramaic, 13
Arctic Circle, 40
Arctic Ocean, 40
Arnold, Gottfried, 12
Arrian, 56, 80
astronomical unit, 23
Augsburg, confession of, 7
Azbakiyya,
   birket, 2, 97, 100, 101, 114, 194
   development of, 112–13
   gardens, 141

Babylon, Roman fortress, 60, 96, 115
Bach, Johann Sebastian, 4
Baghdad, 285
Balm of Mecca (Balm of Gilead), 269, 296
Basra, 285
Baurenfeind, Georg Wilhelm,
   selection for expedition, 26–27, 35
   sketches, 40, 79, 173, 191, 199, 261, 267, 268
   death of, 280
Bedouins, depredations of, 65–68
Beit el-Fakih, 267–70
Ben Ezra Synagogue, 153
Benjamin of Tudela, 75, 84, 108, 208
Bernstorff, Count Johann, 19, 23, 37, 260, 304
Berossus, 70
Beschreibung von Arabien.. see *Description of Arabia*
bey(s), beylicate, 126–28
   listing of, 130–34
Bible
   and the expedition, 7–10, 24, 30–32
   as history, 206–11
   account of Egypt, 201–2, 224, 225–27
   and Sinai, 231,233, 236, 241, 242, 248, 254
birket(s) in Cairo, *see* map of Cairo
Blomenberg, Christine Sophie,
   Niebuhr marriage to, 315
   death of, 318
Boeckh, August, 6n4
Boie, Heinrich Christian, 317–18
Bombay, 281, 282
Breitenbach, Bernhard von, 61, 85

Subject Index

bridges in Cairo, *see* map of Cairo
Bruggen, Lars, 22, 266, 277
   death of, 280
Bryant, Jacob, 90–91
Budge, E. A. Wallis, 212n20, 219–20,
Bulaq, 97, 107, 117
Burckhardt, Johann Ludwig, 144n6, 244n18, 314n20
Byzantine Greeks, 15

Caesarion, church of, 60
Cairo,
   description of, 93–120
   birkets (ponds), 113–15
   bridges over *khalig*, 106–7
   churches, 110, 116
   gates, 104–6
   major landmarks, 107–11
   map of, 93–120
   outlying areas, 115–19
   Old Cairo, 115, 99, 100, 140
Canopus, 79, 178
Capuchin(s), 62, 98, 141, 191
Catholic(s), 18, 42, 54
Chaldean, 5, 10, 13, 70, 225
Champollion, Jean-Francois, 213, 219
Cheops, pyramid of, 81, 88, 222–23
Children of Israel
   in Sinai, 230–31, 233–35, 241, 253–54, 258–59
Christian(s),
   focus on the Orient, 7
   scholars of the Bible, 8–10
   Scriptures, 10, 13
   -Muslim divide, 15–16
   chroniclers, 72–74
   churches in Cairo, 110
   Orthodox/Catholic divide, 62
   desecration of pagan Alexandria, 69–70
   trading class in Cairo, 137, 166
Christianity,
   persecution of in Egypt, 69
   triumph of in Egypt, 69
Circassian(s), 54n32, 123–24, 128, 129
citadel, of Cairo, 1
   description of, 107–9
Clayton, Robert, 235

cloth, manufacture of, 161–62
coffee, 25, 196
   as a commodity, 157–59
commerce of Egypt, 156–69
   transit trade, 157–60
   raw materials, 160–61
   industry, 160–63
   and world economy, 163–64
   relationships, 165–66
   technology, 166–69
Constantinople, 45
   description of, 46–52
   Niebuhr's map of, 48
   second visit, 290
Copenhagen,
   Royal Library in, 3, 20, 33, 239, 300, 305, 307, 314
   Niebuhr's return to, 282
Copt(s), 2, 13
   calendar, 69, 140
   in Cairo, 144–48
   language, 7, 215
   quarter in Cairo, 97, 113 *see also* Haret en Nassarah
corsairs, in Mediterranean, 44, 54
Crusades, 15, 16
Cuxhaven (West Ludingworth), 3, 258
Cyprus, 307

Dahshur, pyramid of, 87, 89
Damietta, 2, 171, 173
dance, Oriental, 198–99
Danish East Indies Company, 35, 308
Dardanelles, 45, 52
Das deutsche Museum, 317
della Valle, Pietro, 93
Delta, 3
   Niebuhr's map of, 175, 178
   to Damietta, 171–77
   to Rashid 177–78
   history of, 178–80
Demotic, 213, 225
Denmark
   King of, 28, 34, 304, 307
Description de l'Egypte, cited, 236n9
Description of Arabia, 20n5, 231, 237, 249, 298, 300, 307, 310
   publication of, 308
diminution of waters, theory of, 224

Subject Index

dimmi(s), (see also Copts and Jews) 13, 62, 14
Diocletian, Emperor,
 column of, 64
 persecutions of, 69, 147
Diodorus, Siculus, 80, 211
distance, units of, 184–86
diwan of Cairo, 134
Doughty, Charles Montague, 19n4, 298
dress, Oriental, 189–94
Dulcigno, 52

egg hatching, industry of, 162
Egypt
 biblical accounts of, 207–8
 European knowledge of, 85–92
Egyptology, development of, 202–5
Enlightenment, 11–13, 20–22
Eutychius, Patriarch, 82
Exodus, book of, 72, 201–2, 204, 207–9, 231, 233, 241, 258

Fathers of the Propagation of the Faith, 41
Fatimid, city, 1, 2, 96, 97, 101–9, 111, 144
 dynasty, 123
Fayoum, 68, 89
fellaheen, 64–65, 123, 126
Feran, Wadi, 249–50
Flaubert, Gustave, in Egypt, 190
Forsskal, Petrus
 selection for expedition, 26, 28, 34
 feud with von Haven, 40, 45, 53, 97
 work of, 62, 98, 176, 221, 262, 267, 269
 on the Copts, 147
 on the Jews of Egypt, 152–53
 death of, 276
 assessment of contributions, 300–303
  publications of,
Flora Aegyptiaco-Arabica, 301
Descriptiones Animalium, Avium . . ., 301
Icones Rerum Naturalium, 318

Fragen an eine Geselleschaft . . ., 28–35,184, 199, 257, 259, 297–98
Franciscans, 62, 141, 191
Friederik V, 19, 23
Fustat, 61, 100, 115

Gallus, Aelius, 24, 57n
Gebel el Mokatteb (also *Mocatab*), 32, 235, 236, 240, 242, 254–55, 259
Genesis, book of, 117, 207
geology, development of, 223–25
Gesenius, William, *Hebrew and English Lexicon of the Old Testament*, 314
Gezira, island of, 115
Ghunfude, 283
Gibbon, Edward, 14, 89n19
Giza, 100, 105
 pyramids of, 221
Golius, Jacob, 9
Goshen, land of, 151, 178, 207, 210
Göttingen, university of, 5, 6, 18–19, 26, 30
Gozo, island, 42
Grand Masters, Knights of St. John, 43
Greaves, John, 86, 208
Greek(s), 13, 57
 in Alexandria, 56–57, 68–70
 in Cairo, 140
 of Suez, 237
Greek Archipelago, 44–45, 53–54
Greenwich, 39n5
Grönland, HMDS, 35, 37, 41
Grotius, Hugo, 9
gum Arabic, trade in, 159

Halle, university of, 18
Handel, George Frederic, 195
Hannover, 5
Hansen, Thorkild, *Arabia Felix*, 25n12, 36n1, 40n8, 53n31, 97n6, 229, 239, 260n41
Happy Arabia, 3, 16–17, 19, 24, 31, 32, 58 (see also Yemen)
hara, residential quarter in Cairo, 140
 Harat al-Ifrang, 2, 96
 Haret en-Nassarah, 141
 Harat ar-Rum, 110, 140

339

Subject Index

Harat al-Yahud, 111, 141
Hebraists, Christian, 9
Hebrew(s), (see also Israelites)
   language, 5, 8–10, 13, 16, 24, 225, 226, 235, 248, 298
   Scriptures, 5–10, 13, 19–20, 30–32, 68, 70, 154, 201–2, 293–94, 297–99, 304
Hejaz, 262–63, 312
Heliopolis, 179
d'Herebelot, Barthélémy, Abbe J. J., 14, 213
   Geographia Nubiensis, 83, 174, 182
Herodotus, 71, 80–82, 203, 222n42
hieroglyphs, Egyptian,
   decipherment of, 211–13
   Niebuhr's copies of, 214–20
Hippodrome (*at Meidan*) in Constantinople, 51
Holy Family, sojourn in Egypt, 117
Hyde, Thomas, 86, 195

Ibn Tulun, mosque of, 1, 100, 107, 109, 111, 141, 146
Idrisi, Sherif, *Geographia Nubiensis*, 83, 115, 174, 311
Imam, of Yemen
   audience with, 278–79
inhabitants, of Cairo, 140–55
   Muslims, 142–44
   Christians, 144–48
   Jews, 148–54
inscriptions:
   in Sinai, 235, 247, 254–57
   Himyaritic, 277
   cuneiform, 284
Islam
   Christian attitudes toward, 14–16
   conflict with, 42, 54, 172, 200
   role in Egypt, 142–43
   in the Hejaz, 263
   in the Yemen, 267
Israelites
   purported sojourn in Egypt, 151–53, 201–2, 206–11
   crossing of Red Sea, 234–35
   in Sinai, 236, 241, 242, 252–54
Izmir, 45

Jabarti, Sheikh Abdel Rahman,
   cited, 67–68n, 114n35, 118, 137, 138n40–42, 143
Janissary(ies)
   defined, 46n20, 109, 125n1
   quarter in Citadel, 108
   -merchant alliance, 129, 164, 262
Jerusalem, 289–90
Jesuits, 18, 141
Jews,
   Christians and the Bible, 9
   of Medina, 19, 193
   of Constantinople, 49
   on Rhodes, 53
   of Alexandria, 57, 68, 70–73,
   of Cairo, 110, 121, 140, 148–50
   in Egypt, 152–54
   in the Hejaz, 263
   in Yemen, 267
Jidda, description of, 262–63
Joseph, corn storehouses of, 206, 208
Josephus, Flavius, 72–73, 204

Kaaba, 43
Kant, Immanuel, 11
Karaite(s), 151n23–24
Karnak, 88, 209
Kasr el-'Ain, 116
Katkhuda, defined, 131 (see also *kikhya*)
Kennicott, Benjamin (Kennicott Bible), 154, 304
Kepler, Johannes, 3n1, 23
khalig, 2, 3, 96, 100, 105, 113
Khan al-Khalili, 107, 141
Kharj, island, 284
Khephren, pyramid of, 81, 88, 222, 223
kikhya, defined, 131,
   Jewish, 150
Kirchner, Anastasius, 212
Koran, 1, 10, 15, 25n4
Kramer, Christian Carl,
   selection for expedition, 26
   in Cairo, 98
   in the Yemen, 266–67
   death, 280
Kulzum, 232

Subject Index

Laroque, Jean de, 25n11
Lane, Edward William, 98, 105, 188, 216
latitude, determination of, 38
Laud, Archbishop William, 87
Leiden, 18
Leo Africanus, 84
Lepanto, battle of, 42
Lepsius, Richard, 71n23
  cited 122, 208, 209n15, 328
Levant
  trade, 41, 42
  Company, 87
Linnaeus, Carl, 8, 21–22, 269, 300, 301
Livorno, 159–60
Lobo, Jerome, 106n19
Loheia, 264–66
longitude, determination of, 38–39
Luther, Martin, 8n6
Lyell, Sir Charles, *Elements of Geology*, 224–25

Maimonides (Moses ben Maimon), 149
malaria, 23, 273
Malta, 42–44
Mamluk(s), 1, 2, 59
  building activities, 97, 111–12
  history of, 123
  rule in Egypt, 123–25, 127, 129–35
Manetho, *Aegyptiaca*, 71–75, 203–5, 209
manna, 236, 259
Mansura, 172
mapmaking, Niebuhr's, 180–87
Maronite(s)
  contributions to Oriental studies, 18, 83
Marraños, 9, 15, 160
Marseilles, 40
  Egyptian trade with, 159–60
Masoretes, 10
Matariya, 117, 179
Mayer, Johann Friedmann, 12
Mayer, Johann Tobias, 27, 39–40, 180
Medina
  Jews of, 19
  Prophet's tomb, 59

Mediterranean World, 24
  warfare in, 43, 44
Melchites, in Egypt, 137
Memphis,
  controversy over location, 80, 89
Meshed Ali (Najaf), 285
Meshed Hussein (Karbala), 285
Meskelyne, Nevil, 39
Michaelis, Christian Benedict, 5
Michaelis, Johann David
  early training, 5–8, 11–12, 14
  and the expedition, 18–19
  instructions to expedition, 29–31, 293–94 (*see also Fragen*)
  and the Red Sea, 231, 233, 258–59
  and the hieroglyphs, 225–26,
  and Sinai, 248, 251
  disinterest in results, 309–11
Milton, John, 11
Mingrelia, 128n17
Mocha, 270–74
Mohammed, the Prophet, 3n1
  Christian attitudes toward, 14
Mohammed Ali Pasha, 58, 64n14, 121n1, 129, 138
Montague, Lady Wortley, 193
Moriscos, 15
Mosheim, Johann Lorenz von, 12
Mouseion, 69–70
multazim (tax farmer), 130, 164
Muqattam, 1, 100, 140, 223, 230
Muscat, 284
music, Oriental, 196–98
Mycerinus, pyramid of, 81, 88, 203

Nabataean(s), 24
  script, 249, 255
Nasiri canal, 112
New Testament, 10
Newton, Sir Isaac, 3n1, 11, 23
Niebuhr, Barthold Georg, 3n1, 315–17
Niebuhr, Carsten
  early life, 3–5
  selection for expedition, 27–34
  work of publication, 307–14
  later life, 314–18
  death, 318

Subject Index

Niebuhrs astronomisch Beobachtungen, 308
Nile, 93, 96, 115–16
  annual rise of, 105–6
  Niebuhr's travels on, 171–78
  branches, 177n11
Norden, Frederick Ludwig, 87–88
North Sea, 37, 38, 258

obelisk(s)
  in Alexandria, 62–65
  of Tutmosis III, 49–51, 178
  of Senwosret, 179
  in Rome, 83
Ockley, Simon, 14
octant, Hadley's, 180
Old Testament (*see* Hebrew Scriptures)
Omar, Caliph
  "conditions" of, 2, 121–22
Onias, temple of, 152, 179
Orient
  definition of 13–14, 16, 18
Oriental languages, 6, 13, 298, 303
Ottoman(s)
  rule in Egypt, 125–28

Palermo Stone, 209
Palestine, 16, 19, 84, 89, 207, 293
Pascal, Blaise, 11
Pasha(s), rule in Egypt, 125–28
Persepolis
  trilingual inscriptions, 284
Petrie, W. Flinders, 75, 205
  biblical predilections, 207, 209–11
  wrestling with Exodus, 74, 209–11
  excavation of Serabit el-Khadem, 247
pharaoh, meaning of, 75–76
  of the Exodus (see Lepsius), 208
Pharisees (Talmudites), 148, 151
philology, sacred, 5–10
philosophy, defined, 6, 11
Pietism, and Michaelis, 18
Pinkerton, John, *A General Collection of Travels . . .*, 313
Pitts, Joseph, 200
plague, in Cairo, 139–40

Pliny, 80, 202, 228
Pococke, Edward, 14, 82, 87
Pococke, Richard, 87, 88–90, 215n30, 235
Poland, 291–92
Protestantism, and the Bible, 8
proto-Sinaitic script, 247–48
Ptolemies, 57
Ptolemy, Claudius, 58, 80
Ptolemy, Philometer, 68, 70
pyramid(s), of Giza, 88, 94, 96, 220–23
  Niebuhr's measurements of, 222

Qalaat al-Kabsh (Fortress of the Ram), 1, 109
qarafa (cemetery) in Cairo, 99n11, 117, 121n1, 138n42
Qaytbey, Sultan, 1
  fort in Alexandria, 58
  building program, 58–59
  exactions of, 59, 150
  madrasa, 109, 215
  mausoleum, 118
quinine (Peruvian bark), 23

Radziwil, Count, 85
Ragusa, 52
Ramusio, 84
Rashid (Rosetta), 61, 78, 79, 177
Red Sea, 25n11, 38, 261
  trade, 137, 156–58
  Israelites' crossing of, 233, 258
  tidal movements in, 258
Reimarus, Hermann, 12
Reisebeschreibung nach Arabien . . ., see *Travels in Arabia*
Reiske, Johann, 18, 308
Rhoda, island, 101
  nilometer on, 115
Rhodes, 52–53
rice, trade in Egypt, 160
Robinson, Edward, 245, 254
  translation of Gesenius, 298
Roman(s),
  elite in Alexandria, 57
Rosetta Stone, 217
Rumelia, Ottoman province of, 290

Subject Index

Russell, Alexander, 286, 317
Russell, Dr. Patrick, 286, 317

Sabaean(s), 35
sabil, 49, 131n22, 143
Sacy, Silvestre de, 213, 317
Salah ad-Din,
  building program in Cairo, 85, 103–4, 123
sal ammoniac, manufacture of, 160–61
Sale, George, 7, 14, 15, 25n11, 87
Samanud, 84, 179
  birthplace of Manetho, 203
Sana, sojourn in, 277
Saqqara, 80n5, 89, 203, 209
Saxony, 1, 3, 306
Schultens, Albert, 18, 85
Scott, Francis, 280, 317
Seldon, John, 9
Septuagint, 70, 240, 297
Serabit el-Khadem, 242–49
Serapaeum, 60, 69
Shafaei, Imam, sepulchre of, 118, 138n42, 143
Sharpe, Gregory, 8–9
Shaw, Thomas, 89–90
Sheba, Queen of, 19, 24
Sinai, peninsula
  Children of Israel in, 253
  Mount, 239, 252–53
Sionita, Gabriel, 83
slave(ry)
  female, 54
  galley slaves, 44
  rule in Egypt, *see* Mamluk
Smith, Adam, 3n
Smyth, Charles Piazzi, 75, 206, 209n16
Sneferu, Pharaoh, 87, 247
Spinoza, Baruch, 11
St. Catherine's, monastery in Sinai, 236, 251
  library of, 236, 251–52
St. Elmo's fire, 40
St. John, Knights of, 53
  cathedral of Malta, 43
St. Mark, Coptic church, 60, 62
St. Vincent, Cape, 40

Strabo, 57, 80, 202, 223
Suez, 126, 232–33
  trade of, 25
  Greeks of, 237
  tidal movements near, 233, 257–59
Sufism, in Egypt, 143
Suleiman the Magnificent, 53
Sultan Hassan, mosque, 107, 109
Syria, 287–89
Syriac, language, 5, 10, 13

Taizz, 274
Tanis, 173
Tanta, tomb of Ahmad al-Badawi, 59, 138
technology, in Egypt, 166–68
Tehama
  travel in, 267, 269, 272, 280
Theodosius, Emperor, 51, 69
Tor
  problems with Beduins of, 228
  monastery, 237
Tranquebar, 24
Travels in Arabia
  publication of, 307–12
Turab al-Yahud, 153
Turan, 13
Turin, Papyrus of, 209
Turk(s)
  in Cairo, 140, 141, 143
  popular attitudes toward, 128, 144
  (*see also* Ottomans)

Ulama, influence in Egypt, 142, 164
Ulugh Beg, 86n14
Upper Egypt,
  Holy Family sojourn in, 117
Ussher, Bishop James, 205, 224

Venetians,
  sack of Constantinople, 13
  in Levant trade, 42
Venus, transit of, 23–24, 41–42
Voltaire, Francois Marie de, 12
Von Gahler, Sigismund Wilhelm, 45, 53, 263
Von Haven, Friedrich Christian
  selection for expedition, 25–26, 28, 35

343

Subject Index

    episode in Rhodes, 53
    in Cairo, 97, 214, 221n40
    in Sinai, 239–57
    diary, 33n26, 304
    death, 272–73
    assessment of contributions, 303–5
Von Moltke, Adam Gottlob, 33, 34

Wadi Feran, 249–50
Wadi Mukatteb, 249
War of Austrian Succession, 23
War of Polish Succession, 23
War of Spanish Succession, 23
War, Seven Years, 23, 40–41
Weill, Raymond, and Serabit, 244n18
Wilkinson, John Gardner, 214
Wren, Sir Christopher, 3n1

Yanbu, 262
Yemen (Happy Arabia)
    expedition in, 264–80
Young, Dr. Thomas, 213

Zoan, 108, 173
Zoega, Georg, 213, 302
Zoega, Johann, 302–3

www.ingramcontent.com/pod-product-compliance
Lightning Source LLC
Chambersburg PA
CBHW071150300426
44113CB00009B/1156